Praise for *Food for Thought*

"Camila brings all the arguments for not eating animals and animal products together in one interconnected story, much like the issues are themselves. Her words call on our compassion but also our ability to connect hard facts to action. She makes the reader question their ideas about oppression and how it shows up in our lives, both among human and non-human animals. She challenges you to question the status quo in our beliefs, as well as the often misplaced trust we have for industries to do the right thing. You'll walk away educated with real life examples and empowered to share and utilize that information in your own life." —**Tessa Graham**, Director of Compassion Innovation at The Pollination Project

"*Food for Thought* could not have come at a more critical time. Humankind has reached its point of greatest damage to the planet and our kindred beings, and a sustainable future will not be possible without a major transformation of our global food systems. With impeccable detail, Dr. Camila Perussello presents comprehensive evidence of the impacts of animal exploitation on health and the environment, and effectively links it to a broad range of issues, including cultural identity, politics and governance, economics, supremacy and racism, human rights, and social justice. The graphic photographs, from undercover investigations by Animal Equality, drive home the point that commodifying animals is above all an act of unspeakable cruelty. It's difficult to imagine how any reader would come away from this book and not be compelled to go vegan. To that end, the last chapter of the book contains practical advice and tips for making the transition, and offers hope that the possibilities for a more compassionate world do indeed begin with the food we choose to put on our plates."—**Dr. Joanne Kong**, editor of *Vegan Voices: Essays by Inspiring Changemakers*

"*Food for Thought* is a well-researched wake up call to the issues related to the food we eat. Its message is as eloquent as it is empowering: in the fight for your planet and its denizens, you too can—and should—play a role!"—**Tobias Leenaert**, Author of *How to Create a Vegan World* and Co-Founder at Proveg International

"At a time when animal-based diets are contributing very significantly to a potential climate catastrophe and other environmental threats, an epidemic of life-threatening diseases, risks of future pandemics, the very wasteful use of increasingly scarce land, water, energy, and other resources, and the massive mistreatment of farmed animals, Dr. Camila Perussello's wonderful book, *Food for Thought,* is a very welcome addition to the literature on vegetarian and vegan diets. Her passion, compassion, and concern for a decent world for future generations are evident throughout the book. It is extremely comprehensive and well-documented. I have been reading books and articles on plant-based diets for over 40 years, but still learned many new things. I heartily recommend the book. It has the potential to help many people switch to healthier diets and to help shift our very imperiled planet onto a sustainable path."—**Richard H. Schwartz**, Professor Emeritus of Mathematics, College of Staten Island

"A meticulously researched tome that illuminates production methods used for animal foods and their associated moral, ethical, and environmental concerns. Using passionate and lucid prose, Perussello offers compelling arguments to persuade readers to consider a vegan diet as an alternative."—**Paul Singh,** Distinguished Professor of Food Engineering, University of California, Davis

FOOD FOR THOUGHT

Planetary Healing Begins on Our Plate

Camila Perussello, PhD

Lantern Publishing & Media • Brooklyn, NY

2022
Lantern Publishing & Media
128 Second Place
Brooklyn, NY 11231
www.lanternpm.org

Cover design: Rebecca Moore
Copyediting: Pauline Lafosse
Proofreading: Liza Barkova

Printed in the United States of America

Library of Congress Cataloging-in-Publication Data

Name: Perussello, Camila, author.
Title: Food for thought : planetary healing begins on our plate / Camila Perussello, PhD.
Description: Brooklyn, NY : Lantern Publishing & Media, 2022. | Includes bibliographical references.
Identifiers: LCCN 2021053837 (print) | LCCN 2021053838 (ebook) | ISBN 9781590566626 (paperback) | ISBN 9781590566633 (epub)
Subjects: LCSH: Animal rights. | Food industry and trade—Moral and ethical aspects. | Food industry and trade—Health aspects. | Food of animal origin—Moral and ethical aspects. | Food of animal origin—Health aspects. | BISAC: HEALTH & FITNESS / Diet & Nutrition / General | NATURE / Environmental Conservation & Protection
Classification: LCC HV4708 .P476 2022 (print) | LCC HV4708 (ebook) | DDC 613.2/622—dc23/eng/20211228
LC record available at https://lccn.loc.gov/2021053837
LC ebook record available at https://lccn.loc.gov/2021053838

For Brenda
whose 19 years of unconditional love
made me a better human being
When we met at that cat shelter
I was the one who was rescued
I love you and you are greatly missed.

Parental guidance is advisable for children viewing these images.

TABLE OF CONTENTS

ACKNOWLEDGEMENTS

I am deeply grateful to friends, family, and colleagues who encouraged me and thoughtfully commented on earlier drafts, contributing their opinions and insights to the completion of this book.

I want to express my deepest gratitude to Lantern Publishing & Media, especially to Brian Normoyle, for believing in my work. I am grateful to veterinary and animal rights lawyer Dr. Renato Pultz, from VegVets Brasil, for his support throughout the writing of the manuscript. My sincere appreciation goes to Animal Equality's team, who provided me with the pictures of undercover investigations used in this book. In specific, a special thanks to Sean Thomas and John Hopkinson—you make this world a better place.

My sincere thanks to Tessa Graham, from The Pollination Project, for her great generosity in proofreading the first chapters. I am profoundly grateful to Dr. Richard Schwartz for being so kind and warm, reading the manuscript, making suggestions, and recommending my work. Thank you also to Tobias Leenaert; I appreciate your being so friendly and honest in sharing your experience as a book author with me.

A special thank you to my beloved husband, Fernando Grando, who supported me on many levels to write this book. Thanks for our insightful discussions over the years and your endless patience in listening to "just one more paragraph" as I researched and wrote this book. My heartfelt thanks to my mom and dad, Regina Perussello and Décio Perussello, for raising me with the core values of forgiveness, honesty, and kindness.

Thanks to my soul sister Juliana Ganci for making my life better and always being on my side, no matter how far away we are. I also thank my dear friend Magda Dzieciuchowicz for encouraging me and celebrating every victory along the way, like a true kindred spirit. My wholehearted thank you to Brenda, Felipito, Isabella, Beethoven, Dominique, and all the other nonhuman friends who came before them for their many teachings and loving presence in my life.

I am grateful to the many people—writers, academics, philosophers, activists, teachers, spiritual masters, and visionaries—who have inspired me, despite the physical and temporal distance. Thank you for paving the way to a grander reality.

My profound gratitude to our fellow animals, who awakened me to my true essence and helped me become a higher expression of human nature. Thank you to our beautiful planet Earth, home to my mistakes and successes. I truly hope I am giving you something back.

PREFACE

Throughout history, we have behaved as if we were superior to others, based on a plethora of unconvincing arguments—skin colour, social status, nationality, gender, species, and so forth. We have constantly objectified individuals to legitimise oppression and exploitation.

During the first wave of European colonialism, Europeans supposedly doubted that Amerindians and Black people had souls. This claim was employed to validate slavery. Dalits, members of the lowest caste in India and Hindu regions of neighbouring countries, are long-time victims of violence and discrimination, including difficulties accessing education and employment. Also called *untouchables*, they are considered "unclean." The alleged impurity ironically disappears in the ordinary events of sexual abuse, including gang rape, by upper-caste men. To this day, nations engage in violent conflicts over political and religious issues—not unlike the Crusades in the Middle Ages. The opponents' lives are often considered less worthy. Before the eighteenth century, women's suffrage was not even contemplated, as their intelligence and judgment were commonly considered inferior to men's—especially regarding politics. The men's vote was enough to represent their interests. As a matter of fact, the fight for gender equality is still ongoing, even in wealthy, industrialised nations. In dozens of countries, it is disputed whether gay people have the right to marry or adopt children. Across the globe, people with physical disabilities or mental illnesses are commonly denied their civil and political rights. Nowadays, nonhuman animals are subjected to a legal status of things in most nations. Even in those countries where legislation acknowledges the scientific consensus that animals do think and feel, animals remain treated as things at our service. Male chicks are macerated alive in the egg industry because they do not lay eggs and grow too slowly to become meat. Cows spend days screaming in agony looking for their calves, taken away a few hours after birth so we can drink their milk. Sows spend most of their lives in tight crates, gestating, giving birth, and nursing piglets, who will soon be stolen to become bacon.

The underlying mindset in all of the above examples is that of *supremacy*.

As we shut ourselves to compassion and cooperation, we must create mechanisms to refrain from making obvious connections. And by choosing to limit our perception and intelligence, we have faced many crises of our own making. Deprived of our natural discernment of not harming others and blinded by our appetite for wealth, power, and animal flesh and secretions, we have prompted the onset of the sixth mass extinction.

But how can something so trivial as eating be associated with a problem of such proportions?—you might be asking. Although the consequences of our dietary habits remain hidden from the general population, the role of animal agriculture and fisheries in the climate emergency and biodiversity loss is well known by scientists, industry, and governments alike. *Food for Thought* will help readers understand the implications of their food choices. The book unveils how meat, dairy, fish, and eggs are produced and how they impact the environment, the economy, and human health beyond animals' inherent suffering. It also elaborates on the relationship between microbial pathogens, antibiotic resistance, pandemics (including COVID-19), and the consumption of animal products, rendering this a timely and urgent reading.

The first step towards a mature society is to move past our outdated illusion of separateness. As I have been inspired by other authors and visionaries that came before me, I hope this work is a beacon of light to readers interested in taking part in our world's moral, social, and ecological regeneration.

INTRODUCTION

Our illusion of separateness has led us to where we are now—hunger, war, diseases, pandemics, deforestation, social inequality, animal abuse, climate crisis, and the list goes on. No, I am not pessimistic; this is pure and simple observation. Humanity insists on living in denial, sweeping crises under the rug. But if we were wise, we would instead scan problems in the search for solutions. We can only find a way out if we dare to look the truth in the eye.

For example, the climate and ecological calamities facing the world have been treated with a blend of science denial and apathetic acknowledgement. Some naively say that climate change is a conspiracy, while others cross their arms to something they think is too big for them to fight against (or that they do not care enough about). A third group, such as legislators, recognises the problem but is implementing only minor measures and at a slow pace. However, to the people losing their lives or their loved ones due to the environmental collapse, there is a real and *urgent* crisis. What does our denial or defeated stance say about our empathy?

To reconnect with our sense of unity, it can be helpful to educate ourselves about the profound connections between the climate, the natural world, and us. *We are part of Nature*—all species are intertwined and dependent on healthy ecosystems. Changes to global temperatures and habitats can affect populations and even entire species, disrupting the whole web of life. And more importantly, the value of a species goes beyond its ecological role. Each species is made of *individuals*, and these matter on their own. Luckily, even when we fail to see this, information can trigger compassion towards other living beings by revealing that we all want the same thing: to live, be free, and feel safe.

To the awakened person, it is crystal clear that the supremacist paradigm (some are deserving of moral consideration, others are not) is only that: a paradigm. But for most of us, who are still in the awakening process, learning about the suffering (and destruction) embedded in a piece of meat or cheese can accelerate our liberation from a cynical system that objectifies everything—animals and their consumers. As stated by Oscar Wilde, *"A cynic is a man who knows the price of everything and the value of nothing."* Let us redeem our value by recognising others' intrinsic rights.

Climate change affects everyone

On the 20th of May 2020, a super-cyclone named Amphan hit the coast of India and Bangladesh (Ellis-Petersen & Ratcliffe, 2020). Winds at a speed of 190 km/h caused storm surges of up to 5 metres, killing 5 people. Around 2.5 million people were evacuated to shelters either because their homes were destroyed or the location was too risky for them to stay. Weather authorities warned the cyclone could push seawater 25 kilometres inland, flooding cities including Kolkata, one of the largest in India. Residents and volunteers faced a challenging situation given the risks of coronavirus spread during rescuing and sheltering. Specialists also warned of potential damage to the Sundarbans Forest Reserve, a UNESCO world heritage site located on the border between India and Bangladesh. The forest is home to more than a hundred endangered animal species.

The safety of nearly a million Rohingya—refugees from Myanmar in Bangladesh, 80% of which are women and children—was worsened because of Amphan, since most of them live in makeshift shacks or shelters. Rohingya, a stateless Muslim minority, have endured military persecution, horrific violence (including rape, arson, and murder), discrimination, and extreme poverty for decades (UNHCR, 2018). During the pandemic, they faced a terrible dilemma between the risk of catching COVID-19 in crowded shelters or being massacred by the super-storm.

Experts state that tropical cyclones have become more destructive over the past four decades due to human actions increasing ocean temperatures, which give storms more energy (Readfearn, 2020). Between March and early July 2020, more than 200 people died in Bihar, the poorest state in India, due to lightning strikes driven by rising global temperatures (India Today, 2020).

Extreme weather events are increasing in number and severity as climate change escalates. Between 2020 and 2021, while this book was being written, unprecedented heatwaves were recorded in India, Pakistan, the US, Canada, and Western Europe; severe floods happened in Central Europe and the delta regions of West Bengal and Bangladesh; wildfires hit Siberia, Australia, and North America; locust infestations wiped out crops in East Africa, India, Pakistan, the Arabian Peninsula, Paraguay, and Argentina; water scarcity and floods were reported in several parts of Africa; atypical droughts affected Southern Brazil, Argentina, and North America; and snowstorms occurred in China during summer, just to name some of the climate change–driven disasters.

Researchers claim that climate change–led catastrophes are likely to intersect with the COVID-19 crisis throughout the pandemic (Phillips et al., 2020). How do we maintain social distancing while rescuing and sheltering people from extreme weather events?

A series of massive bushfires ravaged Australia in 2019 and 2020 due to record-breaking temperatures. Some people argue that natural droughts, unrelated to climate change, caused the bushfires. But they fail to consider that the country recorded one of the driest summers in history because of climate change. The 2020 fires killed over 500 million animals and left koalas in a state of near-extinction (The University of Sydney, 2020). Glaciers melted or turned brown in New Zealand due to the Australian bushfires.

Rapid changes in the Himalayas' climate are melting the glaciers, resulting in abnormal shrub and grass growth in the mountains. This vegetation speeds up melt rates, increasing the risk of flooding. Research shows that a global warming of 1.5°C above pre-industrial temperatures can cause 36% of the high mountains of Asia to disappear by 2100. At higher emission scenarios, this figure rises to up to 65% (Kraaijenbrink et al., 2017). Such projections pose potential consequences for more than 1.8 billion people living in the Hindu Kush Himalaya or the river basins downstream (Wester et al., 2019). Worthy of note, the world has already warmed by more than 1°C.

In January 2020, Jakarta, Indonesia's capital, saw the worst flood in a decade. Tens of thousands of people were evacuated from their homes, and 66 lives were taken (Berlinger & Yee, 2020). Extreme rain and climate change are linked: a warming atmosphere increases evaporation, rendering more water available for precipitation.

Record-breaking heavy rains, landslides, and flooding in 3 South-Eastern Brazilian states killed around 150 people in early 2020 (Phillips, 2020). Even though tropical rains kill Brazilians every summer in favelas built on steep hillsides, rising global temperatures are leading to more severe storms. Some months after the floods, unusually low rainfall levels affected Southern Brazil and Argentina. As a result, the flow of Iguazu Falls, the largest waterfall in the world, dropped from the typical 1,600 m^3/s to only 288 m^3/s (Bateman, 2020).

On the 29th of June 2020, a snowstorm in the middle of the summer killed nearly 500 livestock animals in Xinjiang, China. And that is not the only extreme weather event striking the country in a *seven-day* period. A colossal sandstorm hit the city of Hotan on the 30th of June; Yichang recorded its heaviest rainfall in twenty years on the very same day; and over one million people were affected by severe flooding in Hubei in the first days of July (Celestial, 2020).

In February 2021, Texas was hit by extreme winter weather, killing 36 people and leaving millions without power or water. Temperatures were around 20°C colder than the average for this time of the year. As the state is not prepared for harsh winters, householders, businesses, and hospitals faced broken water mains and burst pipes. Heavy rain, thunderstorms, sleet, graupel, and even snow were seen across the state of Texas (NOAA & NWS, 2021).

In June and July 2021, unprecedented heatwaves hit the west of the US and Canada, with temperatures over 50°C recorded in some places. The high temperatures caused fires, droughts, heat-related illnesses, and hundreds of deaths (Pilkington & Canon, 2021; Bekiempis, 2021). Until the 22nd of July, 79 large fires had burned over 5,860 km^2 in the US (NIFC, 2021).

Also in July 2021, massive rains affected Germany, Belgium, and France, flooding cities and killing about 140 people (The Local, 2021).

Events like these will become the norm, unless we wake up before it is too late.

We cannot deny climate change is happening. We are now heading to what scientists call the Anthropocene extinction or the sixth mass extinction (Wagler, 2018). Except for the one that killed the dinosaurs, all other mass extinctions were caused by climate change. However, the planet is now facing the first climate crisis caused by *human* action. Experts say we have less than ten years to reduce carbon emissions drastically. If the average global temperature increases 2°C from pre-Industrial Revolution levels, the environmental damage will be *irreversible*. For this reason, the 2015 Paris Agreement, promoted by the United Nations (UN), advises that we limit warming to 1.5°C (UN, 2015; IPCC, 2015).

Impoverished communities are more affected by climate change, at least directly, as their livelihood commonly depends on the natural world (e.g., agriculture). Furthermore, economically vulnerable people frequently live in less robust homes, sometimes on hillsides or near the sea. While many blindly deny, from the comfort of their bubbles, that climate change is occurring, its effects are part of the daily routine of millions of people worldwide. Communities that live along Central America's coast, for example, are facing rising sea levels and depleted oceans. South Asia is being struck by stronger winds and tides every year, leaving people homeless, while prolonged droughts and locust infestations across the African continent are destroying farmland and intensifying hunger (Magome, 2019; National Geographic, 2020).

Climate disasters, be they driven by anthropogenic emissions or naturally occurring, affect food production (Raiten & Aimone, 2017). Agriculture is affected by climate and, at the same time, impacts the environment, increasing the globe's temperature. As global temperatures rise, ecosystems are affected, leading to a cascade effect. Higher temperatures alter the water and carbon cycles, melt glaciers, increase sea levels, induce stronger winds, excessive or insufficient rainfall, cause floods, bushfires, and droughts, endangering species and reducing biodiversity. The extinction of bird and insect species, which play a vital role in the pollination of food crops, exacerbates the problem. Food crops are hugely affected, aggravating the social and economic crises.

The good news is that we can revert this vicious scenario by reconsidering our consumption patterns—but we must act fast. Will we support the transition to the more peaceful and just world in which we say we want to live? Or waste the opportunity?

Agri-food production and public health

Danish anthropology researchers Jonas Ø. Nielsen and Henrik Vigh published a paper in the *Global Environmental Change* about human adaptation to climate variability (Nielsen & Vigh, 2012). The authors use the example of a small Sahelian village in Burkina Faso, called Biidi, to discuss how humans can adapt to climate change in terms of food production and acquisition, but still face problems with market prices.

The climate crisis is ultimately inviting us, collectively and individually, to re-examine our food and energy sources. We should not try to *adapt* to unfavourable climate conditions, but instead *stop fuelling* the crisis.

In a globalised economy, agricultural production fluctuations go easily unnoticed by privileged consumers. Most of us do not ponder where food comes from or how it is produced. Nonetheless, climate alterations and the misuse of natural resources are resulting in diminished crops and fishless oceans, leaving subsistence farmers and fishers in a state of food vulnerability. Fishes, as sentient beings, should not be exploited as food in a civilised world in the first place, but this is a topic for the following chapters. The point is: everything is interrelated in Nature. We cannot evade the consequences of our impact on the planet's balance.

Our current agri-food systems, heavily based on animal products, are draining the planet and warming the atmosphere at unprecedented rates. Although mainstream media focuses on fossil fuel burning, animal agriculture is also a leading cause of climate change.

Experts state that animal agriculture is responsible for 14.5% to 87% of anthropogenic greenhouse gas (GHG) emissions, depending on what the estimations include, as well as the timeframe considered (Rao, 2021; Reisinger & Clark, 2018; FAO, 2013; Goodland & Anhang, 2009; FAO, 2006). I will address this issue in detail in the chapters that follow. Even if we take the most conservative estimate, it would be imprudent to disregard the sector's role in the climate emergency. A "mere" 14.5% can be the difference between survival and mass extinction.

The world's five major producers of meat and dairy emit more annual GHGs than big oil corporations, such as Shell, ExxonMobil, and British Petroleum (IATP & GRAIN, 2018). And unlike fossil fuels, animal farming is not only associated with carbon emissions and ocean pollution, but also with deforestation, soil degradation, species extinction, the onset of pathogenic bacteria and viruses, antibiotic-resistant diseases and, of course, animal abuse. In fact, animal farming and capture fishing are the *leading causes* of biodiversity loss due to habitat destruction, trophic-level disruption, and genetic manipulation of the fauna (Machovina et al., 2015; Williams et al., 2015; McCauley et al., 2015; Bonhommeau et al., 2013).

Until the beginning of the Neolithic period, humans relied primarily on hunting and gathering of wild plants to obtain food. Archaeology studies indicate that crop domestication and animal herding may have started concurrently, around 11,500 years ago (Zeder, 2011). Although, back then, animal farming helped provide food throughout the year, despite changing seasons, humanity has relied on cereals, pulses, and tubers as staples for millennia. Even some decades ago, meat, dairy, and other animal products were luxury items reserved for special occasions. But things have drastically changed. Since 1961, global meat consumption has increased more than fourfold (Ritchie & Roser, 2019).

The average US American consumes nearly twice the protein needed, mostly in the form of animal protein. A similar pattern is observed in Europe, Argentina, and Brazil (Ranganathan et al., 2016). At the same time, three-quarters of North Americans fail to meet the minimum daily intake of fruit and vegetables (HHS & USDA, 2015). *Per capita* consumption of meat has shifted from 24 kg/year in 1964–66 to 36 kg/year in 1997–99 and is expected to reach 45 kg/year by 2030. Global milk consumption is also on the rise due to increasing *per capita* consumption in developing countries (WHO, 2020a). And we still ask ourselves why food-related diseases, such as obesity, heart disease, diabetes, and cancer, are growing rampant.

The consumption of animal products makes us sick but also impedes our natural empathy and intellectual capacity. To be able to consume animal parts and products from the reproductive cycle of females (such as eggs and milk), we need to convince ourselves that the violence against other feeling beings is justifiable. We must then repress any reference to the source of those "foods" or cling to the fantasy that exploited, enslaved, and tortured creatures are treated humanely. And of course, we must also deny the mounting evidence that animal agriculture is destroying our planet.

By suppressing our mental faculty, we become an easy target to manipulation. Furthermore, our lack of compassion for non-human animals is inevitably projected towards other humans, spiralling into a world of selfishness and apathy.

As I write this page, there are more than 1 billion cattle and 23 billion chickens in the world, mostly exploited for food purposes (Cook, 2021; Foer, 2019). Approximately 2.6 billion cattle, 56.8 billion birds, and trillions of fish are killed every year to become food (Mood, 2010). In fact, some 90% of global fish populations are overexploited (FAO, 2012). Our unnecessary dependence on animal products brought about a series of environmental and social problems. Currently, animal farming is responsible for nearly 80% of worldwide deforestation and around 90% in the Amazon rainforest, taking into consideration both livestock grazing lands and cropland areas for animal feed production (Foer, 2019). Grazing lands account for up to three times the area used to grow crops (Theurl et al., 2020). We have plenty of room to feed the global population in a healthy and varied way, but we are choosing the wrong path.

The production of meat, fish, milk, and eggs consumes much more resources—water, energy, land, nutrients, fossil fuels, machinery, medicines, supplements, etc.—than that of plant foods. To sustain our demand for animal products on a *finite* planet, we have increasingly recurred to extreme confinement of animals in so-called factory farming. The unnatural conditions in which they live, the chronic stress, and the overuse of antibiotics have all driven the development of dangerous pathogens and antibiotic resistance in humans (Greger, 2020; Saenz et al., 2006).

The vast majority of pigs, cows, fishes, and chickens worldwide are raised in factory farms: for example, 99% in the US (Anthis, 2019). A staggering 73% of all antimicrobials sold globally is used in animals raised for food, with sales expected to rise by 11.5% between 2017 and 2030, totalling 94,418 tonnes (Tiseo et al., 2020; Van Boeckel et al., 2017). Selective breeding, traditionally used in animal agriculture to increase product yield, is an additional factor leading to disease spread, as the natural genetic diversity is reduced within the herd.

Zoonotic diseases are increasingly spreading to humans through direct (e.g., farmers and veterinarians) and indirect contact (foodborne, waterborne, and vector-borne). Tuberculosis, brucellosis, typhoid fever, salmonellosis, listeriosis, and infections by *Escherichia coli, Bacillus cereus, Yersinia enterocolitica,* and *Campylobacter jejuni* are typically associated with the consumption of animal products. The use of wild and farmed animals as food is responsible for viral and bacterial epidemics among humans, such as AIDS, Ebola, H1N1 Swine Influenza, Avian Influenza, MERS-CoV, SARS-CoV-1, and SARS-CoV-2 (which caused COVID-19) (Johnson, 2021; WHO, 2020b; Sun et al., 2020; Reperant & Osterhaus, 2017; Greening & Canon, 2016; Loh et al., 2009).

In other words, animal farming is a major public threat.

COVID-19, animal products, and labour abuse

In a period marked by the coronavirus pandemic, slaughterhouses and meatpacking facilities are hotspots for virus spreading. Not surprising, given that these places are overcrowded, cold, and filthy. Yes, filthy—because blood, guts, muscle, heads, and legs are still body parts and secretions regardless of the hygiene standards adopted. Nonetheless, abattoirs and meat processing plants keep functioning worldwide by government order, although a massive part of their staff has contracted COVID-19.

The workers, many of them migrants, have not had the privilege to stay home and safe as many of us have. The meat industry traditionally relies on low wages and poor occupational health and safety to secure its profit margins. Migrants and underprivileged groups are attracted to jobs no one wants to do. In the US, 52% of meatpacking workers are migrants, 3 times the general workforce rate (17%). In addition, 44% are Hispanic and 25% are Black (Fremstad et al., 2020).

At US Tyson Foods, for instance, several hundred workers at the Wilkesboro chicken plant tested positive for COVID-19 in 2020. As many of them were asymptomatic, they were spreading the virus to other people outside work. Facing heavy criticism, the company felt the need to address the issue (although only a few months after the onset of the pandemic). Workers who tested positive will now receive paid leave and will not return to work until further notice (Hollan, 2020). At Camilla's Tyson Foods, in southwest Georgia, workers were being paid USD 500 bonuses to report to work amid the pandemic. The company is suppressing information on COVID-19 cases, as stated by the meatpacking employees and the vice-president of the Retail, Wholesale, and Department Store Union, Edgar Fields. According to Tara Williams, who has been working at the Camilla plant for five years, *"for us employees that work in production, we are treated like modern-day slaves"* (Laughland & Holpuch, 2020). Between March and September 2020, coronavirus outbreaks affected nearly 500 meat plants in the US, killing at least 203 meatpacking workers (Kindy, 2020).

This story repeats all over the globe in chicken, pork, beef, and fish processing facilities. Tönnies, a leading European meat factory based in Guetersloh, Germany, had more than 1,500 people tested positive for COVID-19 since the beginning of the pandemic until June 2020 (Soric, 2020). Approximately 60–80% of the workforce in German slaughterhouses are foreigners, primarily from Eastern Europe. Most of them are contractors (Huber, 2020). In Ireland, around 60% of the meat processing workforce are migrants. A reported 20% of meat plant workers have contracted COVID-19 nationwide as of February 2021 (Bowers, 2020). Such a high infection rate would have impeded any other industry from continuing its activities. Still, governments seem to be opting to save one sector over all others.

The Migrant Rights Centre Ireland (MRCI) interviewed 150 meat plant workers from 21 different nationalities regarding the working conditions in the Irish meat industry. Sixty per cent of the respondents had sustained injuries at work or developed occupational diseases, ranging from regular bruises and repetitive strain injuries to bone fractures and finger or limb amputations. Verbal bullying and psychological bullying were reported by 43% and 35% of the workers, respectively. Only 4% of the cases reported to the employer were addressed with corrective actions. Insufficient training (62%), absence of occupational sick pay schemes (90%), and no payment for extra work (27%) were common complaints. None of the workers felt valued at work. With an hourly pay rate of EUR 10.85 to EUR 13.56, some claimed to work more than 55 hours in a week. Here are a few statements about their work conditions (MRCI, 2020):

- 'When the factory needs to kill 500 cows, if people don't show up we still have to meet that target—they get in inexperienced people and there's loads of accidents as a result.'
- 'We suffer racism and discrimination, all of the time we are forced to do the worst and heaviest work because we are migrants and don't have access to rights.'
- 'Management has zero respect for us—they want people's production to be beyond their capacity. We only get a few 10-minute breaks maximum across the day, and when people go to the toilet you can see them running to get back to their station.'
- 'My colleague called me monkey and made comments about the colour of my skin. I decided to speak to management. The manager decided to warn my work colleague but said, sarcastically, why did I stress myself with it seeing that I wasn't so black (so dark), then he smirked.'

And things are no different in Brazil. The award-winning documentary *Carne e Osso* (2011), directed by Caio Cavechini and Carlos Juliano Barros, exposed the work conditions in poultry, cattle, and swine slaughterhouses in Southern and Midwestern Brazil. The incidence of occupational diseases and accidents in the sector is stratospheric. Deep cuts, amputations, tendonitis, epicondylitis, injuries to the spine, hands, wrists, shoulders, and neck, anxiety, depression, and other mental disorders are common among the employees. The low temperatures, long hours, and very few breaks during the day further impact their health. Forced to meet excessive targets, workers make up to 18 movements every 15 seconds with a sharp knife to debone chickens (Carne e Osso, 2011).

Data from the Brazilian Ministry of Social Security (MPS) reveal that the chance of developing tendonitis in a chicken deboning line is 743% higher than in any other industrial sector nationwide. The risk of cranial or abdominal trauma in beef slaughterhouses is three times higher than in other types of work . When the employees fall ill or get involved in labour accidents, which happens sooner or later, they are replaced by others. Some seriously injured workers may go months or years without getting a new job, while others become permanently disabled (Carne e Osso, 2011).

The social cost of occupational diseases is passed on to society. According to Juliana Varandas, an occupational therapist at the Brazilian National Institute of Social Security (INSS) based in the city of Chapecó (in the state of Santa Catarina), approximately 80% of the workers served in the region come from slaughterhouses. (Carne e Osso, 2011).

Not only the treatment of animals is hidden from the public, but so are human rights violations and COVID-19 outbreaks (Kendall, 2020). The livestock industry's *modus operandi* is the same across the globe: the lack of transparency and the third-party exploitation combined ensure the sector's prosperity. Consumers should ask themselves if this is the kind of industry they want to support.

You may be wondering if it would not be enough for companies to adopt superior labour standards. After all, introducing breaks during the day, reducing the workload, and hiring more people sound like great solutions. However, implementing these measures entails higher production costs, which are generally passed on to the consumer as most companies are unwilling to reduce their profit margins. Would you buy the same product for three times the price?

To explain why slaughterhouses are still operating amid the COVID-19 pandemic, policymakers claim these are essential services. Essential to whom? Not to me, for example. I have not eaten meat since 1999, and I can affirm that these products are far from being essential. Nor is the sector vital to economically vulnerable people worldwide, who often cannot afford animal products and rely on cereals and beans as their main protein sources. Most humans might well appreciate animal flesh, but it is not required for health maintenance—quite the opposite.

Who is profiting by keeping these places open? Not the staff, as we have seen. Probably not even small farmers, who struggle to produce ever more to supply big corporations. And certainly not the animals. Those who profit are a select elite that uses consumers of animal products as a manoeuvring force to keep animal agriculture alive and growing. For example, the annual exports of Irish beef and Brazilian beef are valued at EUR 3.9 billion and USD 13 billion, respectively (DAFM, 2019; Carne e Osso, 2011). This income is not shared with the rest of society. What spreads across regular citizens is, in reality, the environmental and social cost of an industry that depletes the planet and makes people sick.

For that matter, 75% of newly emerging diseases affecting people originated in animals. Scientists have been long warning about the dangers of exploiting farmed and wild animals. Among the zoonotic diseases that spread to humans through direct contact or via animal product consumption, many have the potential to become a new pandemic. The diseases with the most pandemic potential are considered to be Nipah virus infections (with bats being the primary reservoir), yellow fever, dengue and chikungunya (transmitted by mosquitoes, such as *Aedes vittatus*), Middle East respiratory syndrome or MERS (caused by MERS-CoV virus through animals, such as camels), swine flu (pigs), yellow fever (monkeys), Buruli ulcer (opossums), and bird flu (Devnath & Masud, 2021; Kushner, 2021a; Kushner, 2021b; Kushner, 2021c; Constable, 2021a; Constable, 2021b; Greger, 2020; Simpson et al., 2019).

Many zoonotic diseases keep re-emerging in the same place, spreading through varying distances, or returning as new (and more dangerous) pathogen variants. In all cases, the outbreaks lead to human mortality and economic fallout. Small countries dependent on tourism or livestock farming are especially affected (Haque & Haque, 2018). If we continue oblivious to the suffering of animals, the onset of a new pandemic is just a matter of time.

Animal agriculture is unsustainable

It is crucial to understand the following basic concept: milk, meat, and all other animal foods derive from the metabolic transformation of the nutrients ingested by the animal, following the principles of physics, such as Conservation of Mass and Thermodynamics. When we consume animal products, we have a "middleman" (the animal) converting plant nutrients to tissues and secretions instead of using those nutrients directly. As no machine (let alone living organisms) has a 100% conversion rate, consuming animal products is an excellent way to *waste* natural resources.

For example, if all grain fed to cattle were distributed to humans, we could feed a further 3.5 billion people, almost half of the current global population (Nellemann et al., 2009). This is extremely relevant as the world population is expected to reach 10 billion people by 2050. From a food engineering perspective, we will not be able to feed everyone the way we are eating now. Earth's resources are finite, and animal products are incredibly resource-intensive.

By deforesting ever-larger areas for animal farming, we are menacing local communities, especially Indigenous people. Shockingly, 36% of all cropland area on the planet is used to feed livestock, not to mention the space occupied by animal farms (Cassidy et al., 2013). In Brazil, fair-right president Jair Bolsonaro is relaxing vigilance against illegal deforestation in the Amazon forest. The already alarming levels of deforestation reached their peak in 2020. More than 90% of the Amazon tropical deforestation is attributable to grazing, cattle ranching, and soybean cultivation, which is primarily grown to feed cattle (Foer, 2019). Timber and mineral exploration are *secondary* sources of deforestation in terms of cleared area.

With confirmed coronavirus cases numbering 21,866,077 and nearly 609,573 deaths by the 9th of November 2021 (Johns Hopkins University & Medicine, 2021), Brazil has seen illegal loggers introducing the deadly virus into Indigenous communities. Native people are facing violence, the loss of demarcated land, and discrimination fuelled by the government's actions. By purchasing meat, dairy, eggs, and even fish, we are supporting these ethics because South American monoculture crops are vastly used in animal agriculture and aquaculture everywhere in the world. In addition, Brazil is one of the largest exporters of beef and poultry meat. But we will address that later.

It is worth stressing that Brazil is not the only country with an unsustainable economic agenda whatsoever. China and the US are the primary GHG emitters, with a 2020 CO_2 production of 10.43 billion tonnes and 4.6 billion tonnes, respectively. These two nations emitted 43.8% of the CO_2 produced globally in 2020. Brazil comes in at 14th in the list, with 433.8 million tonnes, after countries with varying population sizes and net national wealth levels, like India (3rd position, 2.4 billion tonnes), Germany (6th position, 637.4 million tonnes), Indonesia (9th position, 567.6 million tonnes), and Canada (11th position, 530.6 million tonnes) (World Population Review, 2021).

As these are the 2020 CO_2 emission sum totals by country, some might say these numbers penalise large, populated nations. If we discuss these figures in terms of *per capita* emissions, we have The Republic of Palau (66,408 tonnes), New Caledonia (49,291 tonnes), and Qatar (32,978 tonnes) as the top 3 countries. In comparison, the *per capita* emissions were 13,917 tonnes in the US, 7,246 tonnes in China, and only 2,027 tonnes in Brazil (World Population Review, 2021).

According to a comprehensive study published in the *Science* journal in 2018, even the least sustainable plant-based food (e.g., chocolate made from cocoa beans grown in deforested land) is less damaging to the environment than the most "sustainable" animal product (e.g., local grass-fed beef) (Poore & Nemecek, 2018). I used quotes in "sustainable" as people tend to believe (erroneously) that grass-fed meat and free-range eggs are more ecologically friendly. While these raising methods can be a bit more "humane" for animals (can exploitation be *humane*?), they can be less sustainable due to the greater amount of feed and larger pasture areas required. Moreover, transport accounts for a small fraction of the carbon emissions involved in animal food production and distribution—typically less than 1% for red meat and around 2% for poultry meat (Ritchie, 2020). *What* we eat is far more critical than *where* it comes from. I will discuss these issues carefully in the following chapters.

In 2012, science writer Tim De Chant produced a very insightful infographic showing the estimated amount of land needed to sustain the consumption patterns of seven billion people from different countries. The global hectare, defined as a biologically productive hectare with world-average productivity, was used in the calculations. Bangladesh would need only a small portion of the planet, while the US would require 3.9 Earths and Australia, 4.8 (McDonald, 2015). Such large footprint numbers are largely attributable to animal foods consumption.

There are vested interests in keeping people misinformed about the leading causes of climate change. While fossil fuels take the full blame, the livestock and fishing sectors continue unregulated, moving us closer to the 1.5°C warming limit established in the 2015 Paris Agreement, considering pre-industrial average global temperatures as a reference (UN, 2015; IPCC, 2015).

I am not whatsoever saying that we should dismiss fossil fuels' burning as one of the leading emitters of GHG. What I am defending is that we have a vital opportunity to curb climate change, which is through food producers and consumers. As a matter of fact, it will be *impossible* not to exceed the 1.5°C warming without addressing animal agriculture. For example, the meat and dairy sector alone is projected to emit 80% of the GHG emissions threshold by 2050 (IATP & GRAIN, 2018). In fact, a 2020 paper published in *Science* demonstrates that without major changes to current food systems, we will exceed the 1.5°C target even if fossil fuel emissions were immediately stopped (Clark et al., 2020).

Some people like to point out that overpopulation is the problem, not animal agriculture or other resource-intensive and polluting sectors. Overpopulation is definitely *a* problem, but it is not *the* leading problem. How we manage our resources can be much more important than the resources themselves. There are too many of us *doing the wrong thing*.

There are way too many resources and collateral effects involved in the production of animal foods. The Earth can sustain the 7.8 billion people living on it now (and even the 10 billion estimated in 2050), but clearly *not* the way we have been eating. Conversion rates are very low in animal products, with meat production typically requiring many kilograms of grains for each one kilogram of meat. It is true that people in poorer communities tend to have more children due to their cultural upbringing and education level. However, the amount of food (and other resources) consumed by a child in a developed country could sustain the life of *several* children in less industrialised countries within healthy limits.

Bottom line: we are all responsible for climate change and resource use, at both governmental *and* individual levels. Our principles form the values of our society.

Science will not save us

Many of us seem to believe there are technological fixes to everything. But look at where our current technology has brought us. While we certainly have made progress in many aspects of life thanks to science and technology, this has not necessarily translated into higher levels of happiness and fulfilment. As late Barbara Marx Hubbard says in her exceptional book *Conscious Evolution*, we cannot solve our problems in the same state of consciousness in which we created them (Hubbard, 2015). Furthermore, it is unwise to wait for engineers to develop a revolutionary energy source to fuel our industries and transport while we keep consuming unsustainable (and cruel) products.

It is time we put our intelligence *and* wisdom into effect in the material world. We are doing neither. All the crises we have been facing are more than a wake-up call—they are evolutionary catalysts. We are being asked to make higher choices, to choose another path towards a lucid, just, and prosperous society. Science alone will *not* do that. Science can only operate at the level of our awareness.

The flesh of dead, tortured creatures and their secretions can no longer be regarded as food given our current evolutionary stage. Raising animals as food is causing too much damage not only to animals, but also to our health, to the environment, to social equality, and to our humanity. We obviously cannot go on like this. Not if we care about the next generations, not if we wish to evolve morally, not if we want to live in a peaceful and humane world as we say we do.

With the advent of food technology, we can now replace *any* animal product with plant-based alternatives. Not that food must mimic the texture and flavour of animal products to be good, for that matter. But if the priority is the resemblance, that problem is quite solved too. We do not need to hurt feeling beings.

It is widely accepted in the scientific community that our food choices are an efficient and *essential* way of curbing the climate crisis and other environmental problems, especially in the short term (Rao, 2021; Clark et al., 2020; Willet et al., 2019; Springmann, 2019; Springmann, 2018a; Springmann, 2018b; Reisinger & Clark, 2018; Goodland & Anhang, 2009). And as far as global warming is concerned, the impact of animal agriculture can be explained by an energy balance. Please bear with me.

Climate research usually reports results in terms of CO_2 emissions or CO_2 equivalents (CO_{2eq}). However, this does not mean that CO_2 is the only GHG or even the most impactful. For example, the global warming potential of nitrous oxide (N_2O) and methane (NH_4) are, respectively, 296 and 28 times that of CO_2 (Koneswaran & Nierenberg, 2008). Animal agriculture is the primary emitter of both N_2O and NH_4. In addition, sulphate aerosols, which originate primarily from the burning of coal and oil, have a *cooling* effect (Rao, 2021; Reisinger & Clark, 2018). Therefore, for comparison purposes, emissions from different sources are computed as CO_{2eq}.

From an engineering analysis of anthropogenic emissions, Systems Engineer Dr. Sailesh Rao concluded that a total shutdown of the fossil fuel sector would result in a net *increase* in radiative forcing of 0.901 W/m^2 annually (Rao, 2021). In other words, a *warming* effect. In contrast, a total shutdown of animal agriculture would result in a net *decrease* of 0.104 W/m^2 annually, *cooling* the planet. Dr. Rao is Executive Director of Climate Healers, a US non-profit initiative.

But how on Earth has the UN Intergovernmental Panel on Climate Change (IPCC) disregarded these critical assumptions in their emission estimations? The underlying reason becomes evident when we consider that IPCC relies on data from the UN Food and Agriculture Organization (FAO)—and FAO has partnered with the International Dairy Federation and the International Meat Secretariat to promote intensive animal farming (Rao, 2021). FAO also collaborates regularly with international organisations such as the Pan-American Dairy Federation, the International Congress of Meat Science and Technology, the International Poultry Council, and the International Egg Commission (FAO, 2002; FAO, 2012; Rao, 2021).

The good news is that most of us eat several times a day, allowing us to directly participate in the climate crisis mitigation by removing animal products from our meals.

A higher path

In a spiritual and moral context, the highest path is the one that leads us to the highest good; that is, actions that will benefit ourselves and others while causing the least harm. When we talk about food production, the highest path is the one that results in higher yield, safety, and productivity, lower costs, better nutrition, and is more sustainable. On the social and political level, the highest path refers to decisions that benefit the greatest number of people rather than a privileged group. At the environmental level, the highest path is the one that sustains life while preserving Earth's resources and biodiversity. The exciting thing about ending animal agriculture is that it is indisputably the highest path at the moral, spiritual, engineering, social, political, *and* environmental levels.

You might be asking me about the livelihoods of animal farmers in a plant-based world, which is a fair question I will address later in more detail. But in summary, adaptation to a new market will be the key. State subsidies currently directed to animal agriculture can be repurposed. In the US, for example, taxpayers spend USD 38 billion every year to subsidise meat and dairy (Simon, 2013). And globally, USD 15–35 billion are used to finance fisheries (UNEP, 2008). Knowledge and material resources once used to produce animal foods can be diverted to the cultivation of plants, algae, and fungi, but also to cell agriculture, and varying food technologies. Dairy farmers who process the milk into yoghurt, cheese, and other derivatives can adapt their equipment and processes to a new ingredient base: plant milk.

The animal industry's reassessment will be inevitable in the coming years, whether we are fine with abusing animals or not, due to the escalating climate emergency and ecological crisis. An industry based on fundamentally unsustainable practices is destined to collapse sooner or later, simple as that. We have a decade to decide our fate.

I kindly invite farmers, with their unquestionable expertise and hard work, to be part of the progression of human values and seek to transition to a more sustainable and just food system. But governmental aid to animal farmers willing to switch market niches will hardly happen *unless* you and I make our food choices wisely. Our individual and collective ethics have always shaped the world's economy and industries. Every product we buy is a vote for the world we want to see.

Throughout history, we have seen a progression from hand production to industrial, large-scale manufacturing systems. Such systems were first fuelled by steam and later by fossil fuels, which are now being replaced, to some extent, with greener energy sources such as solar power and wind power. We have seen the progression of communication systems, from telegraphs to ADSL internet to high-speed, broadband Wi-Fi. As humanity evolves, technology progresses too, and people must adapt.

The need to change is constant throughout the history of humankind, as life is impermanent itself. Would you like to be the very same person throughout your life? Would you like to live as your grandparents lived, including the good and the bad parts? Would you have given up your smartphone with a high-resolution camera to continue supporting landline phone service providers? Industries can either use the opportunity to adapt to social and technological advances and reinvent themselves, or collapse.

Regarding livestock farming, people cling to the illusion that family farms with a dozen animals roaming freely supply the supermarkets with the meat, dairy, and eggs they buy. They also seem to believe that fisheries and aquaculture are efficient and sustainable ways of producing food. I can assure you that the animal exploitation industry does not resemble that in any way. You can also watch some documentaries if you like: *Earthlings* (2005), *Cowspiracy* (2014), *The Milk System* (2017), *Dominion* (2018), and *Seaspiracy* (2021) are a great start. These are some among numerous other documentaries on the industry's backstage.

And even if backyard animal farms and subsistence fishing dominated the animal exploitation industry, it would still be wrong: *animals are not ingredients*.

Worth mentioning, small farmers are being crushed by big food corporations that squeeze them for maximum yield at the lowest price. Rather than being returned to small farmers, profit margins fill the pockets of a powerful, wealthy elite. By fuelling the economy with animal product purchases, we support the entire culture behind this industry—animal abuse, human abuse, and environmental degradation. I profoundly hope more and more people choose to oppose violence in all its forms.

You make a difference

Some might say that we, as individuals, cannot control how things are produced and the machines are fuelled. But we can (and must) strive for a less impactful lifestyle, which includes choosing our products wisely. I cannot emphasise this enough: our food choices have a *profound* impact on our planet.

The problem with our current unsustainable food system is two-faced. On the one hand, the solid body of evidence on the ill effects of animal agriculture and fisheries does not reach far beyond scientists. On the other hand, those who have access to the information choose to conveniently ignore or deny it. What such denial says of regular people and high-level stakeholders is up to you to reflect upon. But I choose the higher path and kindly invite you to do the same.

The sum of individual actions drives fundamental changes. We will undoubtedly need governmental aid to regenerate our food system, including the promotion of plant-based diets. However, this change will never occur if it does not start with you and me. Besides, it is unwise to wait for governments and industries to take the first step, as they are profit-driven. Instead of waiting for others to act, why not start a bottom-to-top social revolution? Use your power as a consumer and citizen for the greatest good!

The average consumer genuinely has no idea where food comes from and how it is produced. And I do not blame them; our excessively urban lifestyle can promote this disconnection. That is why I will give you a glimpse into how animal foods are produced, ranging from themes such as process efficiency and sustainability to animal rights and social equality.

Chapters 1 to 4 describe the current practices in animal-based food production, including dairy, eggs, meat, and fish. I will also tackle economic, environmental, and health issues. In-depth analyses will be illustrated by real examples recorded during undercover investigations by Animal Equality, an animal rights organisation founded in 2006. Chapter 5 is an invitation for a happier, healthier, and more ethical way of living. I will provide you with concise yet practical guidance on how to live vegan and improve your physical, spiritual, and intellectual health.

If you are perfectly fine with everything I am reporting, go on with your old lifestyle. But if the invitation for a compassionate and wiser living resonates with you, that is your authentic self telling you to *choose the higher path*.

Whatever the case is, thank you for reading this book and reflecting on its profound message of love.

CHAPTER 1

DAIRY

Our diet represents way more than mere food choices based on nutritional requirements, cultural identity, and personal preferences. Food consumption is also a political, social, and environmental issue. We say we want to go in one direction, yet do everything that takes us towards the opposite. We want justice and peace, yet we buy products based on violence, injustice, and environmental depletion. That is why knowledge outreach is so important. Being aware of the backstage of industrial production makes us freer, as we only have *real* choice if we have *informed* choice. Otherwise, we are just living unconscious lives, like hand puppets controlled by external forces.

In the history of humankind, there have always been those who used wisdom to better the world and those who employed it to dominate others or profit at their expense. Secretive practices behind agribusiness are not news to anyone. We have been fooled many times, from the alleged biodegradability of Monsanto's herbicides to the purported health benefits of smoking by tobacco companies in the past decades. But as people awaken to values of justice and fairness, they no longer accept being deceived.

For decades, respected scientists, animal rights advocates, and environmental activists have been reporting what happens behind the scenes in the livestock industry. Unfortunately, public authorities and civil society have paid little attention, relegating fairness to the last place. But perhaps the coronavirus pandemic has given us back a bit of lucidity, showing that we must correct what is gone astray. Even if building a fairer society (strangely) is not a priority for many, perhaps our survival on this planet is. The climate crisis timebomb is ticking, and our diet has a major role in it.

In this chapter, let us examine ethics, process efficiency, sustainability, and the socio-economic aspects involved in dairy production.

Dairy farming practices

The US, India, and China are the world's largest producers of bovine milk, with a 2020 production of 91.6 million, 86.2 million, and 29.0 million tonnes, respectively. Other large countries, such as Russia and Brazil, produce more modest amounts: around 27–32 million tonnes per year (USDA, 2019; Embrapa, 2020).

Climatic issues in 2019 (e.g., drought in Europe and Australia, warmer temperatures in South America, and cold, wet spring in New Zealand) led to poor pasture conditions, impacting milk production. However, global production grew by 2% from 2019 to 2020, totalling almost 822 million tonnes (FAO, 2021). Such production growth is mainly attributed to the increased milk production per cow, achieved through husbandry techniques, optimised diet, and genetic improvement (USDA, 2019). Over the past 20 years, Brazilian milk production, for instance, has increased by around 80%, using almost the same number of cows (Embrapa, 2020).

While cows are manipulated to produce more and more, under the support of national and international authorities, the environmental impact of the dairy industry is growing. However, the sector is not being held accountable for the GHG emissions generated (IATP & GRAIN, 2018). I highly recommend the documentary *The Milk System* (2017), directed by Andreas Pichler, which uncovers secrets behind the dairy industry, based on interviews with farmers, dairy companies, politicians, and scientists.

For decades, government subsidies have forced milk prices below the production costs to boost consumption and, therefore, ensure the sector's growth (Simon, 2013). Noteworthy is that the dairy industry would not survive without these incentives. An industry that cannot stand on its own is economically unviable.

Milk production exceeds domestic consumption in many countries, such as the US, France, Australia, Ireland, and New Zealand (FAO, 2021). Therefore, the industry grows through the expansion to emerging markets. The European Union (EU) produces approximately 141 million tonnes of milk a year (USDA, 2019). Part of the surplus has been channelled to new markets, especially in Asia and Africa, destabilising small local producers (IATP, 2020).

For example, Australia's milk production exceeds the national market's demand by 30–60%, and the surplus is exported mainly to Asian countries, such as China, Hong Kong, Japan, Indonesia, Singapore, and Malaysia (Dairy Australia, 2019). Paradoxically, these markets buy Australian milk even though 70–100% of East Asians are lactose intolerant (NIH, 2020).

A similar trend is observed in India, the number 1 country in dairy cows' population—over 44 million. In 2020, India produced approximately 189 million tonnes of milk (nearly twice the US production), which are mainly consumed internally (Dairy Global, 2019). Ironically, over 74% of Indians are lactose intolerant (Sharda, 2015).

Lactose intolerance affects more than 70% of people of colour and over two-thirds of the world population (NIH, 2020). Those who have the enzyme lactase-phlorizin hydrolase in sufficient quantity to digest lactose in adulthood do so because of a genetic mutation. The allele associated with lactase persistence is typical in descendants of Northern and Central Europe (Burger et al., 2007).

Like most industries, the livestock sector was hit by the coronavirus pandemic. The industry was impacted by the reduced access to materials, inputs, services, and markets, restrictions in the workforce, and changes in demand. To protect animal agriculture, governments have been helping producers in different ways. In Italy, for example, the Ministry of Agricultural, Food and Forestry Policy (*Ministero delle Politiche Agricole Alimentari e Forestali*) has allocated EUR 6 million for the purchase of milk from Italian producers (FAO, 2020).

In a 2020 report on measures to contain the impact of COVID-19 on the livestock industry, FAO claims that *"protecting public health is the first priority"* and that the livestock sector is a *"key contributor to food security, nutrition, and livelihoods"* (FAO, 2020). However, the organisation curiously fails to mention that the animal exploitation industry is linked to nearly all pandemics and most of the current diseases, from cancer, cardiovascular conditions, metabolic syndrome, and diabetes to drug-resistant infections. We will discuss that later.

Investigation results published by the World Health Organization (WHO) in 2020 revealed that SARS-CoV-2 (the virus that caused the COVID-19 pandemic) migrated from bats to humans *through intermediaries*, possibly farmed animals. The report says:

> [...] all available evidence suggests that SARS-CoV-2 has a natural animal origin and is not a manipulated or constructed virus. The SARS-CoV-2 virus most probably has its ecological reservoir in bats.[...] it is thought that SARS-CoV-2 jumped the species barrier and initially infected humans from another animal host. This intermediate animal host could be a domestic animal, a wild animal, or a domesticated wild animal and, as of yet, has not been identified. (WHO, 2020a)

We want to colonise Mars but are still stuck in the Dark Ages, breastfeeding from cows, eating ovulatory secretions of birds, and craving the taste of animal flesh. We have mastered the craft of creating global pandemics through the exploitation of animals but still believe that being cruel is worth the price.

In 1983, veterinarian, researcher, and author Dr. Michael W. Fox published a comprehensive paper on animal welfare in the dairy industry (Fox, 1983). He reported how dairy cows were treated in farms, from extended confinement and mutilation to forced impregnation and early separation from calves. Dr. Fox has been granted Honor Roll status by the American Veterinary Medical Association for his exceptional work on animal rights. His publications include tens of books and papers on animal rights, sustainable agriculture, and bioethics. Despite some progress since his 1983 paper was published (e.g., lower incidence of tight stalls that restrict movement), current practices remain virtually the same, with minor improvements.

When a cow gives birth, her calf is taken away within as little as 24 hours so her milk will feed humans instead. Following separation, the mother will scream in agony for days, looking for her baby. If the calf is male, he will most likely be shot and tossed into a bin or sold to be raised for veal. Calves raised for veal are confined in crates to ensure a softer meat and are killed at approximately three months. A small percentage of male calves can also be used for breeding purposes. These can be kept in the farm, as long as the bull can inseminate a predetermined number of cows during a breeding season, or are sent to dairy breeding farms, which produce stocks for artificial insemination. If the offspring is female, she will be either mounted or artificially inseminated as of 12 months of age (or at 300 kg), following her mother's path into dairy production (Sartori, 2007).

Artisanal and some Protected Designation of Origin (PDO) cheeses such as Parmigiano Reggiano are traditionally made with animal-origin rennet, a complex of enzymes extracted from the lining of a calf's stomach.

After being separated from their mothers, baby cows that were not slaughtered because of the deemed commercial value will spend their first months of life confined in barren hutches, fed a milk replacer. Suckling is an innate behaviour in young mammals, and deprivation in calves affects digestion and satiety, in addition to hindering their growth and behavioural development (de Passillé, 2001; Meagher et al., 2019).

While cow-calf separation is always an emotional stressor, studies show that the maternal bond tightens over time (Meagher et al., 2019). This piece of evidence comes in handy to the dairy industry, which can justify cow-calf separation at 24 hours from birth or less as an alleged concern with "animal welfare".

Would you prefer to have your baby torn away from you right after birth or two months later? Can you see how the animal industry is missing the point?

Images recorded by the animal protection group Farmwatch showed farmers killing male calves at four days of age in New Zealand (Geary, 2015). Ireland faced a surplus of nearly *a million* calves in 2020 due to their record dairy exports. Day-aged animals are killed, exported long distances, or simply left to die—especially male calves, which are considered an unwelcome by-product. Farmers refer to this problem as "calf tsunami" (Kevany & Busby, 2020).

Also in Ireland, younger male calves are frequently sold to European veal farms, while older male and female animals are sold to more distant markets to be used as beef and dairy cattle. As the minimum age allowed for calf transportation is 14 days, some dairy farmers countrywide are registering calves as born earlier to get them off the farm as quickly as possible (Kevany & Busby, 2020).

Now take a moment to connect with the anguish and physical discomfort of these calves during transport, aggravated by the absence of the mother, fear, and lack of food and water during the journey. The same treatment would seem diabolic in the eyes of an average person in any country if we were talking about human infants. We cannot complain about not having peace in our lives as individuals and as a society if we are every day depriving other creatures of peace and dignity.

As cows get older, they usually undergo painful mutilations, including disbudding, marking with a hot iron, extra teat removal, and tail shortening with scissors or a blade. Generally, the use of anaesthetics and analgesics is encouraged, but not mandated by law. In the UK, for instance, teat removal can be done without anaesthetics if the calf is three months old or over (Animal Equality, 2020). Although disbudding is advised before eight weeks of age, many small farmers interviewed in a recent Australian study were found to disregard this recommendation. Besides, the animals were not always provided with pain relief (Beggs et al., 2019). Now imagine what the standard practices in less industrialised countries are.

Dairy cows give birth *every* year to maintain their milk output (pregnancy lasts nine months). If you are a female, imagine having only three months of respite a year, between one pregnancy and the next. Cows are forcibly impregnated shortly after calving, generally with sexed semen—semen capable of producing offspring of a specific sex, according to the producer's need. In other words, their natural cycles are manipulated so female calves are produced instead of "low value" males. This cruelty cycle will only end with their slaughter.

Animal farming advocates (including not only individual farmers and dairy farmer associations, but also scientists whose education aims at animal food production) like to claim that certain farming practices face controversies due to the excessive urban life most people lead. As if integrity was not enough to tell right from wrong. We are not stupid.

To induce or synchronise oestrus (heat), increase fertility, and boost milk production, genetic selection, nutritional management, and husbandry techniques are used, including the administration of reproductive hormones, both natural (e.g., prostaglandin, progesterone, and oestrogen) and synthetic (e.g., recombinant bovine somatotropin—rBST) (Dervilly-Pinel et al., 2014; Sartori, 2007). The use of rBST is banned in the EU, Canada, and other places, but is legal elsewhere, including in major milk-producing countries, such as the US, India, and Brazil. These efforts combined ensure that cows produce a huge volume of milk (dozens of litres in their udder, instead of an average of a few litres) for about ten months a year.

As the excessive weight in the udder hurts the tissue, causing ligament laxity, dairy farmers have been fitting cows with *bras*. The claim that the technique aims to bring comfort for the animals masks the primary motivation: to ensure the cows' longevity and productivity. Well-maintained machines give maximum output, right?

Overmilking, environmental agents (e.g., high temperature and humidity), and microbiological contamination result in the development of a painful inflammation of the mammary gland called mastitis. In Brazil, mastitis affects approximately 30% of dairy cows, with incidences of up to 88% in certain herds (da Silva et al., 2017). In the UK, more than 300 cases of mastitis are detected among 100 cows a year (AHDB, 2020). If you are someone who has ever had mastitis, you can resonate with the pain and discomfort these females experience as they are continually milked.

A growing number of dairy farms in Europe and the UK employ the *zero-grazing* system to boost productivity (Keaveny, 2020). Cows are fed grass or feed in a system that does not involve any time at pasture, which means they spend their entire lives indoors.

Poor ventilation and high levels of ammonia in confinement systems irritate the respiratory tract. The combination of these conditions with climatic factors (e.g., cold and humidity) and especially microbiological agents—*Mannheimia hemolytica, Pasteurella multocida, Arcanobacterium pyogenes*, bovine coronavirus (BCoV), bovine diarrhoea virus (BVD), etc.—cause a range of serious complications such as pneumonia, diarrhoea, and even death (Gulliksen et al., 2009).

Studies show that ammonia emissions affect residents of areas close to farms, worsening asthma and other lung diseases (Sigsgaard & Balmes, 2017). Imagine what they do to factory-farmed animals, exposed 24 hours a day to high concentrations.

Lameness is another frequent problem among cattle, especially those raised indoors in industrial systems. In England and Wales, up to 79.2% of dairy cows in a herd suffer from lameness, with a 36.8% national average (Barker et al., 2010). Lameness can be caused by physical injury, inadequate hoof trimming, digital dermatitis, sole ulcer, white line disease, and other health conditions resulting in limb pain. Poor rearing conditions (e.g., extreme confinement, high stocking density, low hygiene standards), metabolic and hormonal changes, and excessive milk production causing unbalanced weight bearing can also lead to claw lesions and lameness. Injuries and swelling of legs, knees, and hocks are more common in housed cattle because they lie on hard or rough surfaces (von Keyserlingk et al., 2012).

Acidosis is a disease of the rumen caused by inadequate diet, for example high acidity silage or grain overfeeding, the latter being increasingly common in factory farms (Lean et al., 2010; González et al., 2012). Acidosis, mastitis, and ovarian cysts, in turn, worsen laminitis, as ruminal acidity triggers the release of vasoconstrictor substances, disrupting the blood supply to the extremities (Melendez et al., 2003; Motamedi et al., 2018).

Although little discussed in the literature, bone fractures are common in farmed animals, especially females. Injuries to the lumbar, cervical, and sacral thoracic segments and even humerus fractures can occur for a multitude of reasons, such as slippery floors, fast growth, nutrient deficiency, milk overproduction, mounting by a heavier animal, and rough handling (Borges et al., 2003; Gibson et al., 2019). Cows that collapse frequently can receive extra doses of violence to stand up. Fractures to the coccygeal vertebrae are common when farmworkers twist the cows' tails to make them walk.

In fact, deliberate violence is common in dairy farms, where farmworkers hit, kick, torture, kill using painful methods, and sexually and verbally abuse both calves and cows. An overwhelming number of undercover investigations show that these behaviours are not at all isolated. For instance, an article by *The Independent* revealed the sickening treatment of animals in a dairy farm in Essex, England. The investigators witnessed repeated cases of tail twisting, punching, shouting, and swearing at the animals. Disturbing footage shows a worker touching the genitals of two cows on different occasions and even movements suggestive of masturbation. After declaring that "*We aim to uphold the highest standards of animal welfare and care*", the farm implemented disciplinary action and continued business as usual (Dalton, 2019).

People with sadistic inclinations are attracted to work in animal farms and abattoirs, where they can be cruel and violent without legal consequences. Based on multiple evidence, researchers have developed The Link theory, where animal abuse is a predictor of violent behaviour towards humans. Animal cruelty frequently co-occurs with weapon offences, sexual assault, arson, homicide, and other violent crimes (Hodges, 2008). Workers who do not enjoy hurting animals fight against empathy to survive in their jobs. Kinder World, a non-profit organisation dedicated to encouraging compassion for animals, shares hundreds of videos recorded in animal farms on their website. To see the reality behind dairy farming, please check www.kinderworld.org/videos/dairy-industry/.

In conversation with my colleague Dr. Renato Pulz, he reminded me of the suffering involved in dairy and beef expos, where farmers showcase and sell their animals. Dr. Pulz is a Brazilian veterinarian, lawyer, and animal welfare lecturer. He says that "the transport, change of environment, exposure to unfamiliar animals and people, special diet, grooming, and the contests themselves are extremely stressful to the animals." Traditionally, exhibitors compete for who "owns" the cow with the highest milk production. In the yearly held World Dairy Expo, dairy cows are milked in front of crowds of more than 60,000 people from all over the globe. Commonly, a dairy cow produces over 70 kilograms of milk in a few milking sections during the competition. In 2019, a Brazilian cow broke the world record for milk production, with 127.57 kilograms of milk in 3 milkings during a contest (Azevedo, 2020).

Cows are slaughtered when they are no longer able to produce the expected milk yields. Dairy cows have an estimated lifespan of 5 years instead of a typical life of 20–25 years, since the successive pregnancies and milking are exhausting.

The meat from "spent" dairy cows is lesser valued than meat from beef breeds but is still a source of revenue to the dairy producer. Around 10% of the cattle slaughtered in the US are culled dairy cows (USDA, 2015). In Ireland, 60% of the beef produced comes from the dairy herd, according to a chairperson of ICMSA Livestock Committee (The Irish Independent, 2020).

Transport to the slaughterhouse is a major source of stress for the animals, including fear of the unknown, thermal discomfort, overcrowding, injuries, thirst, hunger, fatigue, and other factors. I will elaborate on this in the next chapters. Upon arrival at the slaughterhouse, cows are forced into stunning chambers. There, they are shot in the head to render them unconscious during bleeding, which is called 'humane slaughter.' The very term is contradictory—is it *humane* to kill? Is it *humane* to breed animals to use them as machines? And worst of all, we do this in the name of products that harm human health.

Stunning is not always effective due to either equipment malfunction or human failure, since abattoir workers must keep a fast pace along the production line. Therefore, many cows are still conscious when their throats are slit. All of this happens within "animal welfare" standards, the popular term evoked by animal industry professionals to mask the inherent exploitation of animals in the sector.

Many cows are slaughtered *whilst pregnant*, as the maintenance costs do not justify keeping them for longer. A study involving 54 German slaughterhouses revealed that more than half of them slaughter pregnant cows. Of all slaughtered females, 15% are pregnant, 90% of which are in an advanced gestation (Riehn et al., 2010).

Another research also carried out in Germany reported the slaughter of pregnant cows in 77% of 55 slaughterhouses. Of these females, 51% were in the second or third trimester of pregnancy. Vital signs in foetuses whose mothers were killed were reported in 11% of cases, including umbilical cord pulsation and independent movement of the offspring (Maurer et al., 2016). Germany is considered a model country in terms of animal welfare. Think about it.

India has the largest dairy herd in the world, with over 56 million cows as of 2020 (Shahbandeh, 2021a). More than 16 million cows are exploited for their milk in Brazil, while the US has around 9 million dairy cows (Embrapa, 2020; Shahbandeh, 2021a). In the UK alone, which occupies a fraction of the area of India, Brazil, or the US, two million cows are used in milk production (Animal Equality, 2020).

In a day, a cow consumes 50 to 140 litres of freshwater and up to 70 kilograms of feed (Kavanagh, 2016; USDA, 2019). The numbers vary depending on climate, season, production system, type of feed, milk production, and body condition. The average 1 litre of liquid manure obtained for every 3 litres of milk produced generates 52 million tonnes of manure only in the EU each year (USDA, 2019).

Massive daily amounts of manure are polluting the soil and water bodies, impacting soil health and marine life. Significant quantities of methane, carbon dioxide, and nitrous oxide are being emitted into the atmosphere, worsening the climate crisis. To better illustrate the idea, bovine milk production emits up to three times more CO_{2eq} than soy milk, from farm to fork (Poore & Nemecek, 2018).

In industrialised countries, though, we have several more sustainable and cruelty-free alternatives to dairy milk at hand. Making plant-based milk at home with common ingredients (e.g., water and oats) is cheap, fast, and does not require any special equipment. A new reality where intelligence and empathy are the norm is right ahead of us; why not take one step further?

I do not blame consumers for not knowing the facts behind dairy farming. The system operates in such a way to keep us disconnected from primary truths sustaining our consumption patterns. The misinformation about animal farming practices is common among the very professionals working with food.

Having studied food engineering myself, we were never taught what happened *before* the milk left the farm. As food engineers, we deal with processing techniques, product and packaging development, optimal transport conditions, and shelf-life extension, but often ignore how the resource (the milk) is obtained. When it comes to livestock practices, our subjects cover slaughter methods and some zoonotic diseases, as these impact food safety and the quality of the final product. Animals represent little more than their role as commodities.

Nowadays, I honestly doubt that our course modules exclude animal farming practices for lack of relevance. The educational system, corporations, and governments are not interested in getting the facts out to the very consumers who keep the industry alive. However, as the truth is unveiled, we have the chance to opt out of products that sustain animal abuse and environmental degradation at the expense of our physical, emotional, and spiritual health.

As everything we buy creates demand, our food choices shape the world. Now, dear reader, you know the truth. You have just been empowered to define *who you are* in relation to the violence against our fellow animals.

Many believe that animal exploitation can happen within ethical standards or in a compassionate way. It cannot. The terms 'ethical exploitation' and 'compassionate exploitation' do not even make sense. People also commonly assume that violence against animals is limited to poor, undeveloped countries. They could not be further from the truth. Is there a kind way of breaking the bond between a mother and her baby? Is there a compassionate way of confining, raping, and killing a living being?

As we have seen from the facts so far depicted, cruelty is standard practice in dairy farms worldwide. This does not mean that individual dairy farmers and farmworkers are heartless people. However, both farmers and consumers are embedded in a system based on the false assumption that animals are here to meet our needs and whims. They may not *be* heartless people, but they *act* as such.

We are paying for the infliction of pain and death on an industrial level while ravaging the planet and our own health. And we are doing this in the name of an *ingredient*. What does this say about us?

Is the problem restricted to factory farming?

The ethos of specific farms and the moral values of individual staff can better or worsen the life quality of farmed animals. However, it is important to emphasise that they are exploited for economic purposes, regardless of differences between farms.

Ethical treatment is not related to herd size or production system, as animal farming itself relies on animals being treated as a source of income and materials. Calves are still separated from their mothers so their milk will feed humans, *despite the farm size*. Cows are repeatedly inseminated to keep producing milk and are milked against their will in any production system. When their productivity declines, they are discarded like a broken machine, whether it is a family farm or a mega-dairy. On both small and large farms, male calves are considered an inconvenience, a by-product.

Furthermore, the assumption that family-run, small farms are more sustainable and humane than factory farms does not necessarily hold. First and foremost, there is no moral justification for using sentient creatures. Moreover, research shows that grass-fed and pasture-free cattle (most common in small farms) can use more resources and emit more GHGs than zero-pasture confined ones (more frequent in larger farms) (Clark & Tilman, 2017).

While cows allowed to graze outdoors are theoretically "happier" than those confined, they exercise more and hence require additional feed. Besides, a faster metabolism results in increased methane and nitrous oxide emissions, important GHGs. In other words, greater resource use and a higher environmental impact.

One of the main reasons why producers are increasingly opting for indoor cattle rearing is that by minimising exercise and movement, they increase the energy the animals put into milk (or meat or eggs) production. Also, as outdoor grazing is reduced, the required farming area is massively diminished. Therefore, confining the animals in restricted, indoor spaces can be economically advantageous.

On the other hand, raising animals indoors means higher use of fossil fuels to grow and transport cereal feeds. The cultivation of commodity crops, such as soy and maise, in the quantities required by animal farming entails using fertilisers and pesticides, taking a further toll on the environment (USDA, 2014). Commonly, crops used to feed cattle are grown in poorer countries, where deforestation rises each year along with the soaring demand for animal products. In addition, overfeeding grain can result in higher rates of diseases such as acidosis and laminitis, which translate into lower animal welfare levels (Lean et al., 2010). In summary, confinement is beneficial for the farmer but terrible for the animals.

Nonetheless, we must consider that climate and weather conditions do not always enable farmers to feed a cow purely on outdoor grass. Besides, cows seek shelter from thunderstorms, snow, and intense sunlight. Winter-housed production systems, in which the cows are fed on hay, silage, or grains, can be essential protection against adverse weather conditions in colder countries, especially for young, pregnant, and convalescing animals. In warmer countries, protecting the animals from the sun and heat can be the issue instead.

Researchers from the University of Melbourne, Australia, conducted animal welfare assessments on 50 pasture-based dairy farms of varying herd sizes in Western Victoria (Beggs et al., 2019). The 2009 Welfare Quality dairy cattle assessment protocol (WQ-ME) (Table 1) was used to evaluate whether welfare levels depend on herd size. The farms were ranked as small (<300 cows), medium-sized (300–500 cows), large (501–750 cows), and very large (>751 cows).

In the study, very large farms tended to provide better drinking water quality, with 90% offering water that was suitable for human consumption versus 36–39% in smaller farms. Electronic identification for the monitoring of individual cows was present in 90% of large and very large farms, while 40% identification and 23% drafting were found in medium-sized herds, and 0% in small farms (Beggs et al.,

Table 1. Indicators used in the 2009 Welfare Quality Dairy Cattle Assessment protocol (WQ-ME)

Welfare criteria	WQ-ME indicators
Good feeding	
1. Absence of prolonged hunger	Body condition scores
2. Absence of prolonged thirst	Water provision, cleanliness of water points, number of animals using water points
Good housing	
3. Comfort around resting	Time needed to lay down, cleanliness of the animals
4. Thermal comfort	As yet, no measure
5. Ease of movement	Pen features according to live weight, access to outdoor loading area or pasture
Good health	
6. Absence of injuries	Lameness, integument alterations
7. Absence of disease	Coughing, nasal discharge, ocular discharge, hampered respiration, diarrhoea, bloated rumen, mortality
8. Absence of pain induced by management procedures	Disbudding/dehorning, tail docking, castration
Appropriate behavior	
9. Expression of social behaviours	Agonistic behaviours, cohesive behaviours
10. Expression of other behaviours	Access to pasture

Source: Beggs et al. (2019).

2019). Electronic monitoring enables the early detection of diseases and, therefore, can be translated into better welfare levels.

Anaesthesia, sedation, and analgesia during or following husbandry procedures (e.g., disbudding, tail docking, and castration) were more common in larger farms (73%) than in other herd sizes (40%) (Beggs et al., 2019). For obvious reasons, *all* painful procedures should be accompanied by anaesthesia and pain relief in a humane industry. Not to mention that a humane industry would not be using sentient beings in the first place.

Thermal comfort strategies were found in all farms, ranging from shade points to field sprinklers (these were more common in larger herds). Although a 5% incidence of mastitis within 30 days of calving is recommended by the dairy industry, around 30% of the herds exceeded the target (Lean et al., 2010; Beggs et al., 2019). In larger farms, cows could be approached by an unfamiliar investigator at shorter distances, which might be related to lesser fear (Beggs et al., 2019).

The results of this study corroborated the finding of Canadian animal welfare researchers, who reported there is little evidence of any straightforward relationship between herd size and animal welfare (Robbins et al., 2016). In summary, the evidence points to the fact that larger farms can either improve or worsen the animals' quality of life.

But *even* if the animal welfare levels were considerably higher in smaller farms (which is not true), we could not honestly claim we only purchase dairy products from local, small farms. Most of us dine out and buy processed food; therefore, it is impossible to guarantee we are not supporting cruel farming practices *unless* we give up on dairy products.

Over 10 million dairy cows in the EU are zero-grazed or housed in tie stalls with little time outdoors. Depending on the EU country, up to 90% of the dairy cows are not grazed. In 2012, the top 3 EU countries with zero-grazing systems were Italy (90%), Spain (87%), and Greece (85%) (CIWF, 2012).

In Brazil, Australia, and India, the climatic conditions allow for extensive rearing of cattle. However, herd size and intensive farming are growing in these countries too.

Animal welfare: for animals or profit?

The largest dairy farm in Brazil, Fazenda Colorado, has a herd of around 1,500 cows. The highly mechanised production system associated with the qualified team of veterinarians and other staff supports a daily milk production of 70,000 litres—more than 45 litres per cow each day. The farm grows part of the feed and uses manure as a natural fertiliser. A semi-confinement regimen is in place, where cows spend most of their lives inside a two hectare-sized pen with temperature control. Temperature is controlled to ensure milk yield is not reduced by the energy demand for thermal balancing in the animals' bodies (Fazenda Colorado, 2020).

To ensure optimal herd size, the farm applies genetically manipulated semen to impregnate cows and ensure a high probability of female descendants. Cows are inseminated by inserting a long stick with semen into their vagina during the heat. The procedure is repeated twice within a certain period to guarantee the cows get pregnant. Soon after birth, the newborns are separated from their mothers and the female ones are put into a small pen, where they will receive veterinary care and grow into adult lactating cows themselves (Fazenda Colorado, 2020).

In the first days after calving, bovine colostrum is collected from the cows so a veterinarian can feed their offspring. Colostrum is the fluid produced in the first four days after birth, which contains nutrients, antibodies, and growth factors for the calves. Over the following days, the calves will be fed a milk formula while their mother's milk is destined for human consumption. All this information is publicly available on the company's website (Fazenda Colorado, 2020).

"Welfare" standards are high in Fazenda Colorado, where cows receive quality feed and fresh water according to their needs, have a whole team of experts at the site, cutting-edge automated milking facilities, and clean individual pens. Nonetheless, as discussed earlier, animal welfare and animal exploitation are two opposing concepts. The cows are sexually violated throughout their lifespan, have their babies torn apart from them soon after birth, and their milk stolen several times a day, nearly all year long. Baby calves only have the chance to taste their mothers' milk as colostrum is essential for their health and good development—after all, they are deemed machines that need to be efficiently fuelled to ensure high yield.

An efficient measure of our ethics (or lack of it) towards animals is to imagine these procedures being applied to humans. Please re-read the previous paragraph replacing the terms 'cow' with 'woman' and 'calf' with 'infant'.

The most modern farming system cannot give these animals back their dignity as sentient beings who have a body and a consciousness, like you and me. As a matter of fact, simply having a body capable of feeling pain is a moral imperative for us to respect them, regardless of their level of consciousness or intelligence. Or are people in a vegetative state or with a lower intelligence quotient (IQ) unworthy of consideration?

We all know how much energy is put into pregnancy and lactation. Now imagine having a period of merely 60 days between one pregnancy and the next. Try and picture how you would feel after consecutive pregnancies, births, and multiple daily milking sessions.

Worth noting is that the short break before calving and the next gestation, called dry period, is not borne out of the farmer's mercy towards their cows. The udder tissue must regenerate so the milk production of expected levels is obtained in the subsequent lactation (FAWEC, 2015).

According to the Farm Animal Welfare Education Centre (FAWEC) from the Autonomous University of Barcelona, 'the dry period is critically important for the welfare of dairy cows and their production in the following lactation. The main welfare problems during the dry period are an increased risk of intramammary infections, pain and discomfort due to udder engorgement, feed and water restriction, and aggressive interactions between cows' (FAWEC, 2015). Nonetheless, a study by Finnish researchers based on data from 715 dairy farms (13.2% of all farms in the country) revealed that 9% of them did not use any dry cow therapy (DCT) (Vilar et al., 2018). DCT is the treatment of cows with antibiotics against intra-mammary infections.

Going back to the argument regarding herd size and ethical treatment, let us bear in mind that small farmers must compete with intensive farms for their market share. This means increasing production, cutting costs, or both, which generally reduce animal welfare levels. Cost-saving techniques might mean smaller worker-to-animal ratios, lower quality feed, lesser access to healthcare, further overcrowding, and other measures that negatively affect the animals' health and wellbeing. Therefore, our rising demand for dairy products pushes for further abuse and suffering, whether we are talking about extensive or intensive farming.

China alone imported 14.5 million tonnes of fresh milk in 2018 (FAO, 2019). We clearly cannot supply the global demand for milk and dairy if all farms use free-pasture systems and operate on smaller scale.

A literature review published in the *Journal of Dairy Science* showed a growing interest in the treatment of farmed animals in scientific papers in the last three decades (von Keyserlingk & Weary, 2017). Nevertheless, most studies focus on the relationship between animal welfare and *production*, with little or no emphasis on the cows' perspective of wellbeing (e.g., being free from exploitation).

While investigations on the management of health conditions that reduce milk yield and fertility rates (e.g., mastitis and metritis) improve the life of dairy cows, they remain treated as commodities. Their *economic* value outweighs their intrinsic value as individuals.

Animal welfare, both physical and mental, impacts productivity—the biggest priority to the dairy industry. As negative animal–human interactions lead to lower fertility and milk yields, training on animal handling is encouraged. An additional justification for undertaking animal handling courses is that cows are a major cause of injuries to livestock handlers (Lindahl et al., 2016). Rough handling and aggressiveness on the farmer's side (e.g., kicks, punches, slaps, and shouts) induce fear of humans and anxiety, reducing the milk yield and lengthening the milking sections. An estimated 30% of the variability in annual milk production between farms is ascribed to differences in fearfulness levels (FAWEC, 2014).

Therefore, when farmers strive to create a clean and "comfortable" environment for their dairy cattle, who are confined, forcibly impregnated, and milked continuously over the years, they are obviously not focusing on animal welfare *per se*. They are focusing on the minimum standards by which animals can be enslaved and exploited to generate the expected profit. True animal welfare means animal *liberation*, not exploitation.

Organic, free-range, grass-fed, and other labels

Depending on the country, climate conditions, location (lowland versus upland), and farming system, dairy cows can never make it to the pasture. But even when the animals do have access to outdoor spaces, this does not mean the time they spend grazing and roaming compares to the amount of time they spend indoors. This is especially true in wetter and colder regions, such as in the UK, where 20% of cows spend their entire lives confined and 80% are fed outdoors on pasture for *some* time, but not all (The Courtyard Dairy, 2017).

'Grass-fed,' 'pasture-fed,' and 'free-range' labels can be much more of a marketing strategy than a concern about the animals' welfare. In the UK, grass-fed cattle must be fed on grass for at least 51% of their lifetime. In other words, the animals can spend the other half of their lives indoors, being fed on cereals, grains, and supplements. Furthermore, the term 'grass-fed' includes *all forms of grass*—grass out at pasture, grass stored in the form of silage (cut, fermented grass), or hay (cut, dried grass). Therefore, 'grass-fed animals' are not necessarily animals that eat fresh grass outside all year, as we might have imagined (The Courtyard Dairy, 2017).

Still in the UK, the animals must be fed 100% on grass only if the farmer subscribes to a private scheme called 'Pasture for Life.' The 'Pasture Promise' label, for its turn, ensures that the animals had access to pasture for at least 180 days a year. The 'free-range' label has no legal standing regarding milk (The Courtyard Dairy, 2017).

'Organic milk' is another label that might be misleading, giving a false impression of happy cows living an idyllic life. The term 'organic' refers to various aspects, from the time spent at pasture to authorised medical treatments. In the US, the Department of Agriculture (USDA) supervises the National Organic Program (NOP), which regulates organically produced crops and animal-based foods. Organic milk US producers must, therefore, meet USDA and NOP standards, as well as observe the regulations of the Food and Drug Administration (FDA) on food labelling.

To have their milk certified as organic, farmers must provide cows with outdoor access all year round; 30% of their diet must include pasture grass; a minimum of 120 days per year on pasture must be assured; antibiotics, growth hormones, and genetically modified organisms (GMOs) are not permitted; feed must be 100% organic and contain no hormones, plastic pellets for roughage, or food by-products from other mammals. Dairy cows cannot be treated with antibiotics during their whole producing life—solely natural and homoeopathic remedies are allowed (USDA, 2016).

If treated with antibiotics, the animal can never return to organic production. Nonetheless, average mastitis rates in the US are around 25%, while the prevalence in Europe ranges between 32% and 71% (García-Fernández, 2019; European Commission, 2017). Therefore, it is reasonable to imagine that many cows with mastitis may not be given antibiotics so that the farmer does not lose capital. In fact, bacterial counts compatible with the causative agents of mastitis (e.g., *Streptococcus* spp., *E. coli*, and *Mycoplasma bovis*) are commonly found in milk, organic or not, indicating that the cow was in pain during milking (Lima et al., 2018).

Mastitis is obviously not the only disease that affects cows and requires allopathic medication. For example, respiratory infections, metritis, and laminitis are common health conditions that happen in larger or smaller numbers depending on several factors, such as udder hygiene, feedlots cleanliness levels, stocking densities, and feed quality. Can we be sure there is thorough and frequent monitoring of the herd by authorities to detect whether organic farmers are withholding treatment to sick cows? Are 100% of producers honest? The significant incidence of fraud in the dairy sector raises these questions (Tibola et al., 2018; Moore et al., 2012).

Many courses in Europe and worldwide are offered on mastitis control in organic dairy farms, since mastitis has a profound economic impact. Again, the industry's concern is with the proper functioning of the "dairy machines." So, if you consume organic dairy with the true purpose of improving the life of dairy cows, please re-examine your choices.

Now, if your motivation for changing diet is health-wise, legitimate organic milk indeed contains no chemicals, hormones, GMOs, or antibiotics—but they do contain oestrogen, cholesterol, saturated fat, casein and other animal proteins, lactose, and other unhealthy components.

This argument reminds me of the dairy industry's claim that conventional, non-organic milk has no synthetic hormones, antibiotic residues, or other unwanted substances in significant concentrations. If that were true, why would organic milk production happen under such strict regulations? Think about it.

Worth noting is that organic dairy farming is generally more profitable than its conventional counterpart. A study published by the European Commission in 2013 on crop and milk farms in European countries showed that raw organic milk is sold at significantly higher prices than conventional milk in France, Germany, and Austria. While labour per hectare increases significantly and yield decreases in organic farms, profit margins are generally compensated by the higher market value. The report also states that the income of organic milk farmers is commonly complemented by other activities, such as beef production and agritourism (European Commission, 2013).

In Brazil, dairy processors pay 60% more per litre of organic milk. The sale price covers the cost of organic production, which can be 40% higher than conventional farming (Canal Rural, 2017). Importantly, subsidies for sustainable production and animal welfare also leverage the income of organic producers in different parts of the world (European Commission, 2013).

I end this section by commenting on another term related to cow's milk that is on the rise in the Western world: 'Ahimsa milk.' Ahimsa is a Sanskrit word dating back nearly 4,000 years ago that means respect and avoidance of violence towards all living beings. Although Ahimsa is a key principle in many religions, such as Hinduism, Buddhism, Jainism, and Sikhism, humans have conveniently overlooked it. As enslaving, killing, eating, and using animals is wrong, people found a way to use a beautiful term to justify their deeming females as milk sources.

'Ahimsa milk' means the females are not forcibly inseminated and the cows and calves are not slaughtered, but instead protected for life. Sounds good, right? However, when we buy the concept of 'Ahimsa milk,' we forget that mammals need to get pregnant to give milk. Therefore, 'Ahimsa milk' farms (mostly Hare Krishna communities) must keep cows and bulls too. Well, keeping cows and bulls means their herd is growing steadily. As feeding, sheltering, and providing care for large animals is obviously expensive and requires large areas, these farms end up giving calves away to other farms, which not necessarily will follow nonviolence. 'Ahimsa milk' farms also often use bulls as oxen, claiming working and serving humanity is their dharma.

Another important aspect is, where do these animals come from in the first place? Do 'Ahimsa milk' farms rescue cows and bulls exclusively to give them a good life or take them in to obtain milk and labour in return? And is only "excess milk" taken from the females?

Also, these farms usually sell milk and cheese. How do they ensure a constant supply throughout the year? Or are their sells only seasonal?

'Ahimsa milk' might give people precisely what they want to hear: it is okay to use animals as long as they are "well-treated." We have seen that before with the claim that organic milk and locally produced milk are humane. They are not. If all ambivalent consumers switched to 'Ahimsa milk,' we soon would have a market for this kind of product. And with economic exploitation comes animal suffering.

We should not complicate what is simple: living Ahimsa means not seeing animals as a source of products or services of any kind. Following Ahimsa requires living vegan.

Pro-dairy propaganda: who benefits?

We cannot deny that the dairy sector is concerned with the rising awareness of animal rights. Television commercials showing bucolic scenes, attractive packaging with graceful images, and labels that evoke health and sustainability terms try to convey the impression that dairy products are obtained from happy cows who kindly donated their bodies, calves, and freedom for the good of humanity. Cows are even taken to studios to be photographed or filmed in green scenarios, which by no means represent the actual farming practices.

The truth is that the dairy industry exploits the reproductive cycle of our fellow animals and takes a massive toll on the environment, while insisting on the archaic, unscientific statement that milk is healthy for humans. What is more, mega-dairies are crushing small farmers with unbeatable production rates, which force milk prices down.

Over decades, overproduction and government subsidies have pushed milk prices below the production costs, benefiting powerful dairy companies while leading small farms to bankruptcy. According to the USDA, the number of US dairy farms dropped from more than 83,000 to 40,000 between 2000 and 2018. In the EU, 4 out of 5 dairy farms disappeared between 1981 and 2013 (IATP, 2020). In Brazil, the crisis has also accentuated among small producers, who increasingly give up on dairy farming because the price paid per litre does not cover the production costs. Every day, 45 producers quit dairy farming nationwide, according to Embrapa, the Brazilian Agricultural Research Agency. According to Rosangela Zocca, a researcher and zootechnician at the institution, small producers have two alternatives: expand production or leave the sector (Canal Rural, 2018).

Nonetheless, milk production continues to rise as small family dairies are replaced by factory farms and new markets are found around the world. Countries with little or no tradition in dairy consumption are the target of main producing countries, whose domestic consumption is falling.

The China Study (2005), the most comprehensive study on diet and health ever published, demonstrated the close association between the consumption of animal products and health conditions such as metabolic disease, diabetes, heart disease, stroke, breast cancer, prostate cancer, bowel cancer, and autoimmune diseases (Campbell & Campbell II, 2005). The research, which lasted about 20 years and involved 880 *million* people in 2,400 Chinese cities, was conducted by renowned nutrition scientist Dr. Thomas C. Campbell and his son Dr. Thomas M. Campbell II, a medical doctor, in partnership with researchers from the University of Oxford (England) and the Chinese Academy of Preventive Medicine.

Born and raised on a dairy farm in the 1930s, Dr. Campbell initially intended to help people improve their health by advocating for the consumption of animal products. As a matter of fact, his PhD research at Cornell (where he is a Professor Emeritus of Nutritional Biochemistry) was dedicated to increasing growth rates in sheep and cows. Little did he know that *The China Study* would point him in a completely different direction. For instance, he discovered that casein (87% of cow's milk protein) promoted all stages of the cancer process; that animal protein stimulates the production of hepatic insulin-like growth factor-1 (IGF-l) in the body, prompting cancer development; and that animal protein and excessive calcium found in milk suppresses the production of "supercharged" vitamin D, which is associated with cancer, autoimmune diseases, and osteoporosis (Campbell & Campbell II, 2005).

In 2020, Dr. Walter C. Willett and Dr. David Ludwig, both physicians and professors at Harvard University, published an extensive literature review on the health implications of milk consumption in the *New England Journal of Medicine*. The consequences range from cardiovascular disease, obesity, and diabetes to allergies and various types of cancer (Willett & Ludwig, 2020). The review included more than one hundred scientific articles. The results of these surveys are consistently supported by numerous studies by different researchers across the world.

But what about the nutrients found in milk? And what about calcium intake? Quitting dairy can be dangerous—you may be thinking. While milk does contain significant amounts of protein, vitamins, and minerals (including calcium), it is also high in saturated fat, cholesterol, lactose, D-galactose, casein, growth factors, and hormones. As we have seen, these substances have pro-inflammatory, carcinogenic, and neurodegenerative properties; accelerate ageing; decrease the immune response; increase blood pressure; cause heart disease and diabetes; and affect the hormonal balance in the human body (Barnard, 2020; Willett & Ludwig, 2020; Shwe et al., 2018). Furthermore, research shows *inconsistent results* regarding cow's milk consumption and bone health (Byberg & Lemming, 2020; Weaver, 2014; Michaëlsson et al., 2014; Bischoff-Ferrari et al., 2011; Kanis et al., 2005). In fact, findings suggest an unaltered and even reduced risk of osteoporosis in people following a long-term plant-based diet (Hsu, 2020).

An analysis conducted with 61,433 women and 45,339 men in Switzerland indicated higher mortality rates, greater incidence of fractures, and superior levels of oxidative stress and inflammation in individuals who consumed more milk (Michaëlsson et al., 2014). Another study, carried out with data from 39,563 people (69% women) from different European countries, found no significant associations between milk consumption and protection against fractures. The investigation comprised bone fractures in general, osteoporosis-related fractures, and hip fractures (Kanis et al., 2005). In a meta-analysis involving 195,102 women and 75,149 men from different countries, no clear relationship was found between milk intake and hip fractures (Bischoff-Ferrari et al., 2011).

In other words, the dairy industry's propaganda, heavily based on the supposed relationship between milk consumption and strong bones, is *not* supported by scientific evidence. In addition, plants can also be rich in calcium: for instance, cabbage, figs, beans, soybeans, broccoli, oranges, spinach, almonds, and many others. Milk is unnecessary and detrimental to human health, unless you are a baby drinking your own mother's milk.

Even organic milk, which by law must be free of synthetic hormones or drug residues, naturally contains substances that are harmful to the human body, such as the aforementioned casein, lactose, cholesterol, saturated fat, and hormones. Note that oestrogen, for example, is an *unavoidable* compound in milk, since milk secretion by the mammary glands depends on hormones present at specific concentrations during pregnancy and lactation. The onset of breast, ovarian, and prostate cancer has been associated with dairy consumption as a consequence of the disturbance in our levels of oestrogen and testosterone, among other factors (Barnard, 2020; Campbell & Campbell II, 2005; Qin et al., 2004; Chan et al., 1998; Outwater et al., 1997).

Nutrients, growth factors, and hormones (e.g., oestrogen) from dairy also trigger the secretion of insulin, growth hormone (GH), and IGF-1 (Aghasi et al., 2019; Juhl et al., 2018; Rich-Edwards et al., 2007). IGF-1 is not only associated with cancer and type 2 diabetes, but also with acne occurrence in children and young adults and precocious puberty in girls (Rich-Edwards et al., 2007; Gaskins et al., 2017; Harrison et al., 2017; Brouwer-Brolsma et al., 2018). Early menarche and changed breast composition may influence hormone-related cancers and metabolic syndrome in adulthood (Gaskins et al., 2017), explaining the additional benefits of excluding dairy from our diet.

Furthermore, the combination of high levels of oestrogen with growth factors present in milk is linked to the incidence of premenstrual syndrome, worsening of menopausal symptoms, endometriosis, polycystic ovary syndrome, and even difficulty conceiving (Barnard, 2020).

Milk proteins are also linked to the incidence of type 1 diabetes. β-casein, for example, can trigger immune reactions that destroy insulin-producing cells in the pancreas (Chia et al., 2017; Campbell & Campbell II, 2005). Studies also show that dairy and other animal products play an important role in asthma and autoimmune diseases, such as rheumatoid arthritis, multiple sclerosis, rheumatic heart disease, lupus, and others (Alwarith et al., 2020; Alwarith, 2019; Campbell & Campbell II, 2005). According to Campbell & Campbell II (2005), "[...] almost no indication of the dietary connection to these diseases has reached public awareness," despite the strength of the evidence.

Milk consumption is also related to other gastrointestinal disorders, in addition to those of lactose intolerance, such as irritable bowel syndrome (Willett & Ludwig, 2020; Campbell & Campbell II, 2005). Many people face abdominal bloating, gas, belching, and stomach pain their entire lives without ever making the connection to dairy consumption. Allergy to milk proteins is another relevant problem, which affects 4% of children, but also adults. These are immunological reactions, against what the human body understands as harmful substances, which can lead to fatal or near-fatal anaphylactic shock (Archila et al., 2017).

As if that were not enough, the consumption of cow's milk is associated with faecal occult blood and anaemia in children, as milk proteins are potent inhibitors of iron absorption. For this reason, many food and nutrition bodies around the world discourage feeding babies with cow's milk (Leung et al., 2003).

Dr. Milton Mills, a physician and expert in plant-based diets who graduated from Stanford University in 1991, spoke on many occasions about the racial and ethnic bias behind the USDA Dietary Guidelines. He says the consumption of dairy products is highly encouraged despite the widespread incidence of lactose intolerance among Afro-American, Latinx, and Asian people, who make up a significant part of the US population. This type of institutional racism occurs in countless countries with a mixed population.

Dairy products are also sources of microbiological and chemical risks. Milk can be contaminated with microorganisms from the animals (including enteric bacteria) and from handlers (e.g., farmers and workers in processing plants). Typical pathogens in unpasteurised milk include *Mycobacterium* spp., *Bacillus cereus*, *Listeria monocytogenes*, *Yersinia enterocolitica*, *Salmonella* spp., *E. coli* O157: H7, *Campylobacter jejuni*, among others.

As dairy products are extremely perishable for their nutrient composition and microbial load, sanitary production practices, energy-intensive heat processing, and refrigeration are required to reduce foodborne diseases, such as tuberculosis, brucellosis, and typhoid fever (Cremonesi et al., 2020).

According to the World Health Organization (WHO), an estimated 10 million people fell ill with tuberculosis and 1.4 million died in 2019 (WHO, 2020b). Although it is unclear how many of these cases are related to bovine tuberculosis, studies show that human tuberculosis is linked to the bacteria *Mycobacterium bovis* in up to 15% of cases, with incidence varying among countries (Bolaños et al., 2017). The bacteria can infect people by direct contact with sick animals (e.g., farmers), by ingestion of animal-derived foods (including milk), and through the environment.

Prevalence rates of multi-drug resistant tuberculosis (MDR-TB) are increasing each year. The WHO registered a 10% rise between 2018 and 2019, with MDR-TB affecting 206,030 people globally in 2019 (WHO, 2020b). In fact, the leading cause of the development of super-resistant pathogenic microorganisms is the indiscriminate use of antibiotics in farmed animals to improve health and boost growth and productivity. Astonishingly, around 84,648 tonnes of antibiotics are used every year in animal agriculture (Tiseo et al., 2020). In the US and other countries, some 80% of the antibiotics are used in livestock (WHO, 2017).

Chemical risks associated with milk consumption include residues from antibiotics, dioxins, heavy metals (such as lead and mercury), synthetic hormones (e.g., rBST), mycotoxins, pesticides, GMOs, and feed and forage contaminants (e.g., polychlorinated biphenyls), especially those derived from intensive non-organic agriculture. These substances pose serious health risks due to their pro-inflammatory action and carcinogenic potential, among other undesirable effects (Benkerroum, 2016; Nero et al., 2007). By the way, one of the leading global causes of fertiliser and pesticide use is animal agriculture, as livestock feed is highly dependent on monoculture crops (USDA, 2014).

The ingestion of antibiotic residues from animal products is a serious public health threat, with consequences ranging from imbalances in the gut flora, malformations, and allergic reactions (including anaphylaxis) to cancer and antibiotic resistance (Sachi et al., 2019). Antibiotic resistance occurs through the selection of resistant bacterial strains, which can arise from repeated exposure to low doses of antimicrobials (Tiseo et al., 2020; Nero et al., 2007).

Studies have repeatedly shown that milk, eggs, fish, seafood, and meat, especially organ meats (e.g., liver), contain significant concentrations of antibiotics, such as amoxicillin, ciprofloxacin, oxytetracycline, chloramphenicol, and many others (Hassan et al., 2021). A research study carried out in 4 Brazilian states (Rio Grande do Sul, Paraná, São Paulo, and Minas Gerais) indicated that 11.4% of the 210 raw milk samples collected contained antibiotic residues in levels above the legislation limit (Nero et al., 2007). In 2010, the largest shares of global antibiotic consumption in animal agriculture were attributed to China (23%), the US (13%), Brazil (9%), India (3%), and Germany (3%). By 2030, this ranking is projected to change to China (30%), the US (10%), Brazil (8%), India (4%), and Mexico (2%) (Boeckel et al.).

Speaking of chemical residues, milk is one of the food items with the highest adulteration rates worldwide (Nagraik et al., 2021; Tibola et al., 2018). Antibiotics, for example, can be used to reduce the microbial load in raw milk. Adulterants such as water, vegetable fat, animal fat, and varied chemicals (e.g., melamine, urea, formalin, detergents, ammonium sulphate, boric acid, caustic soda, benzoic acid, formalin, salicylic acid, hydrogen peroxide) can also be added to milk to mask defects, improve fat and protein content, enhance technological and sensory aspects, or increase yield (Nagraik et al., 2021; Bolaños et al., 2017; Nero et al., 2007). Fraud in the milk supply chain can occur in all farming systems and tiers (farms, processing plants, and retail).

A 2019 Dutch study including 20 dairy farms, 4 milk processors, and 4 retailers revealed that organic farms were slightly more prone to food fraud than non-organic farms, especially due to opportunities for adulteration and weaker control (Yang et al., 2019). Some organic farmers interviewed said it is easy to mix conventional and organic milk, which is rarely detected (Yang et al., 2019). The Netherlands is considered a country with low corruption levels. Can you imagine what the reality in other countries is?

Earlier in this section, we saw that the milk production of leading producing countries exceeds domestic consumption. In addition to the expansion into foreign markets to ensure the continuity of the dairy sector, another strategy to give way to excess milk is to direct it to canteens, cafeterias, hospitals, and other public institutions. The distribution occurs through government programs, which aim to guarantee dairy farmers a minimum economic return for their production. In Brazil and elsewhere, many agricultural cooperatives that purchase large volumes of milk directly from producers to save them from financial failure have exceeded their storage capacity (Canal Rural, 2015).

As political and economic measures spur the global consumption of milk, mega-dairies and multinational pharmaceutical companies become increasingly powerful. This is not a conspiracy theory; this is a fact—the pharmaceutical industry profits from the incidence of *diseases*. The US-based Institute for Agriculture and Trade Policy (IATP) also details how corporate and government strategies are fuelling a rural crisis by driving the expansion of the dairy sector:

> EU [...] dairy corporations remain competitive in the global market by paying EU farmers below the cost of production and dumping "cheap" dairy exports into developing country markets. In the US, 93% of family farms have shuttered since the 1970s. Yet, overall dairy production in the US continues to rise due to new or expanding mega-dairies. These are often funded by outside investors and propped up by several Farm Bill programmes. New Zealand exports 95% of its milk, primarily through Fonterra, the world's second largest dairy processor. [...] Every major dairy-producing region (Europe, North America, New Zealand and India) has increased production, while indebtedness and farm loss in rural communities have increased. Five out of the 13 largest dairy corporations are headquartered in the EU, plus Nestlé in Switzerland. Nearly all of them benefitted from low farm prices [from 2018 to 2020] to boost their production. Such expansion happened while dairy farmers across Europe lost their farms or were on the verge of bankruptcy. (IATP, 2020)

Why not redirect these incentives to help small farmers transition to plant-based milk production? After all, healthy, tasty, affordable, and ethical foods that yield financial returns for producers benefit everyone. That is what numerous non-profit organisations are encouraging, for instance the Agriculture Fairness Alliance (US), ProVeg (which works in four continents with eight offices), The Good Food Institute (based in several countries), and Mercy For Animals through the Transfarmation Program. These organisations work with farmers, food industries, decision-making bodies, governments, investors, the media, and the general public to accelerate the transition to a fair, cruelty-free, and environmentally sustainable production system.

In fact, the plant-based industry, which moved USD 35.6 billion globally in 2020, has grown at large rates (IMARC, 2021). In Europe, the sector grew 49% between 2018 and 2020 (Smart Protein Project, 2021). In the US, the niche grew 43% in the same period (The Good Food Institute, 2021).

The consumption of plant-based milk is rapidly gaining ground. With excellent sensory properties and countless health benefits, plant-based milk is already sold at prices close to traditional milk in many countries, thanks to the increasing demand by more conscious consumers. In the US, 15% of the milk purchased in the retail market in 2020 was plant-based (The Good Food Institute, 2021). European sales of plant-based milk showed a 36% growth from 2018 to 2020, reaching EUR 1.6 billion in 2020 (Smart Protein Project, 2021). And in the UK, the sales of plant-based milk and cheese grew 85% and 154%, respectively, from 2018 to 2020 (Smart Protein Project, 2021). People do not want to consume products that promote disease anymore.

Nonetheless, while the global market for plant-based milk is estimated to reach USD 38 billion by 2024, dairy milk sales will reach USD 1.03 trillion (Shahbandeh, 2021b; ProVeg, 2019). As mentioned before, such growth in the dairy sector is mainly attributable to the expansion of large dairy companies into distant markets, which ends up expelling local producers from the activity as a side effect.

> In Sub-Saharan Africa, EU's milk powder exports increased by 20% between 2007-2017, with countries such as Mali, Cameroon and Nigeria particularly hard hit. This destabilises local dairy markets and rural communities heavily dependent on dairy animals in these countries. For example, Misereor, a German Catholic charity, found that milk powder imported from the EU was two to four times less than the price of local milk procured in Burkina Faso from the Fulani, a pastoralist ethnic group, dependent on livestock and dairy farming. Companies such as Arla, Friesland Campina and Danone have all made expansion into Sub-Saharan Africa a priority for their economic growth plans. [...] Fonterra spent the greater part of the last two decades expanding into global markets. Its export-led strategy has not only led to rising emissions, but also an economic crisis for New Zealand's dairy producers. Fonterra's decade long corporate expansion into China, Latin America and Australia, for instance, resulted in one-third of its suppliers unable to pay their bank debts. Its nearly 10,000 farmer shareholders incurred huge losses last year [2019], calling into question Fonterra's corporate structure and investment strategy. (IATP, 2020)

No wonder why industry and mainstream media try to defame animal rights activists, using derogatory language and objecting serious arguments with irrational claims. What would be of the current structures if people woke up regarding animal agriculture? Compassion for animals is a *threat* to supremacist-based systems.

Animal agriculture is so afraid of evolving societal values that it uses questionable strategies to shield industry practices from public scrutiny. Even though it is true that a great part of our society is perfectly fine with treating animals as objects, another part wants to do good for the animals and the planet. People of integrity would do better if only they knew how animals are exploited in the livestock sector and how this is depleting the environment.

Ag-gag laws are one of the many strategies used by animal agriculture to perpetuate ignorance among society. These anti-whistleblower laws criminalise undercover investigations of slaughterhouses and farms, so people remain oblivious to animal abuse and, of course, keep consuming their products. Ag-gag laws originated in the US in the 2000s and nowadays also feature in such countries as France and Australia. By silencing animal rights activists, these laws also prevent public disclosure of food safety problems and violation of labour rights, exacerbating the lack of transparency in the agri-food sector.

While farming practices are prevented from going mainstream, dairy and meat lobbies champion the use of marketing and misleading information to promote their products, to the detriment of ordinary citizens. Public authorities curiously use ag-gag laws to prevent the dissemination of legitimate and important information, to which consumers are entitled, but does not prohibit the transmission of obsolete, incomplete, or anti-science data on diet and health.

Fortunately, food and health authorities are beginning to make decisions based on the growing evidence about diseases caused or exacerbated by the consumption of animal products. In a 2019 survey conducted by ProVeg, 23 out of 86 countries had already given the green light to replace animal milk with plant-based milk in their official nutritional recommendations (ProVeg, 2019). For example, Canada's Food Authority emphasises the importance of plant-based foods and removed dairy as a food group in its 2019 dietary recommendations, placing it among numerous other protein sources. Harvard University (US) went one step further, removing milk entirely from its Healthy Eating Plate food guide. Based on scientific evidence and their commitment to public health, specialists in nutrition and medicine kicked all other animal products to the curb. Will animal agriculture lobbyists continue to censure similar progress elsewhere in the world?

We can continue digging our own grave or wake up. What do you choose?

Environmental impact of dairy production

Animal sentience is enough of a reason for us to shift away from animal products. However, the environmental impact of animal food production is also exceptionally concerning. Animal agriculture is considered a leading cause of the climate and ecological crises not only for its major carbon footprint, but also for the soil depletion, water pollution, biodiversity loss, and deforestation it entails.

The production of meat, dairy, eggs, and fish is massively resource-intensive and polluting. But before we proceed, it is useful to introduce the concept of sustainability within agri-food production.

Production systems and foods that require large amounts of water, land, energy, and other resources to provide similar nutritional value or function are inefficient and, thus, unsustainable. Animal products are much more resource-intensive than plant-based foods. Comparative studies confront nutrient-equivalent foods or equivalent portions: for instance, dairy milk versus plant-based milk or legumes versus meat. For example, the volume of water required in the production of bovine milk is 2–20 times greater compared to plant-based milk (Poore & Nemecek, 2018).

Important to note is that the resources embedded in food production are not limited to the actual ingredients in the food formulation. Let us retake the example of water, the most vital resource in Nature. All water used from farm or ocean to fork is accounted for in the water footprint of a food item. Animal foods have large water footprints for three main reasons:

1. animals drink water;
2. immense volumes of water are necessary to fill up or refresh fish tanks and clean facilities, instruments, and equipment—especially in slaughterhouses;
3. much water is required in food manufacturing and processing. The livestock sector is among the primary consumers of freshwater globally (Bustillo-Lecompte & Mehrvar, 2017).

The energy requirements are also substantially higher in animal-origin food production because:

1. intensive-reared animals are kept in pens or tanks with regulated temperatures to optimise growth and productivity;
2. fossil fuels are needed to transport animals between farms (for example, when a beef farm purchases male dairy calves) and from farm to the slaughterhouse;
3. animal products are perishable and must be kept refrigerated from farm or ocean to fork;
4. food processing requires energy. Unlike most plant foods, which can be consumed raw, meat, fish, eggs, and milk are hardly ever consumed without processing for safety reasons.

Milk is a great showcase to illustrate these points. "Dry cows," as non-lactating cows are termed, drink an average of 50 litres of water per day, a volume that can reach up to 140 litres during lactation depending on milk yield, feed's solid content, and weather (Kavanagh, 2016). Please note that milk has on average 87% of water in its composition, which comes from the animal's diet. Dairy cows can be fed pasture, grains, cereals, feed, and supplements, which carry embedded resources such as energy, soil, arable land, and water. In addition, grains and feed are processed by energy-demanding methods, such as drying.

Sheds, milking parlours, milking equipment, and the cow's udder must be sanitised continually to avoid contamination and the spread of diseases. Approximately 25 litres of water are used to clean every 1 m^2 of the milking facilities (Picinin, 2010). The culling of "unprofitable" cows and calves require large volumes of water to wash facilities and instruments in the abattoir. The slaughter requires approximately 1,500 litres of water per animal (Picinin, 2010).

Thermal regulation plays a vital role in cattle metabolism, and thermal discomfort translates into production losses. For this reason, energy is used to maintain a controlled temperature in barns and milking parlours. Additional energy is used to transport raw milk to dairy processing plants in climatised tanker trucks. As nearly all animal products, milk must be processed to ensure food safety—microbial counts and other microbial, chemical, and physical standards are set and controlled by legislation.

Milk is classically treated thermally, either by pasteurisation or sterilisation. Differently from sterilisation, which destroys all viable microorganisms in milk, pasteurisation destroys pathogenic bacteria but not deteriorative microorganisms. Therefore, pasteurised milk needs to be refrigerated until the final point of consumption (e.g., restaurant, household, hotel). If special types of milk are considered, such as lactose-free milk or skim milk, *further* energy, labour, and equipment are needed. For example, cholesterol and saturated fats can be removed from dairy milk using supercritical fluid extraction with CO_2 and solid-liquid extraction with activated charcoal as adsorbent (Sieber et al., 2011). Processing raises both the production cost and the environmental impact of a food item.

A comprehensive study published in *Science* magazine (Poore & Nemecek, 2018) showed that dairy milk production uses up to 20 times more water compared to different types of plant-based milk. The research comprised data from 38,700 farms and 1,600 processors in 119 countries and contemplated 5 environmental impact indicators—land use, freshwater withdrawals weighted by local water scarcity, and GHG, acidifying and eutrophying emissions.

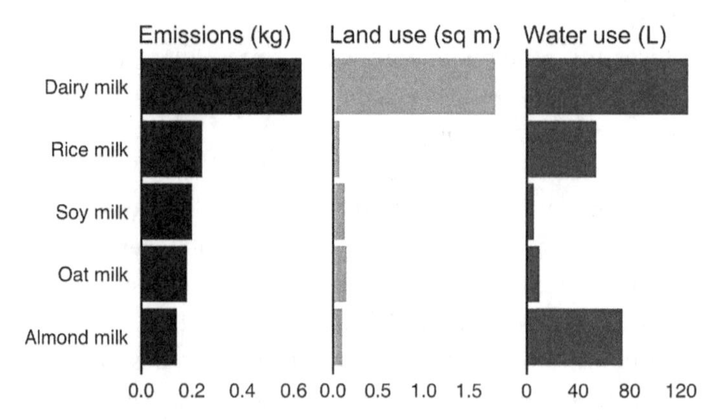

Figure 1. Environmental impact of one glass (200 mL) of different milks.
Source: Poore & Nemecek (2018), adapted by Guibourg & Briggs (2019).

The results showed that not only water consumption, but also land use and carbon emissions are significantly higher in dairy milk production (Figure 1). In addition, soil acidification and eutrophication are up to ten times higher in bovine milk compared to plant-based milk production.

Another critical factor when it comes to sustainability in agri-food production is the pollution generated. The livestock industry is one of the most polluting due to the large volumes of wastewater, GHG (mainly methane and nitrous oxide, but also carbon dioxide), and solid waste produced (Bandaw & Herago, 2017).

The famous 'carbon footprint' refers to GHGs attributed to an individual, organisation, industry, or product. These emissions derive predominantly from the burning of fossil fuels (through machines and vehicles), deforestation, enteric fermentation of ruminants, and fermentation of manure.

Deforestation is the main reason for the high carbon footprint of animal products, as the livestock sector requires vast areas of pasture and crop cultivation to feed farmed animals. The replacement of native and abundant vegetation with grass or agricultural crops generates two main effects: (1) a reduction of the plants' carbon absorption potential that prevents it from being emitted into the atmosphere; and (2) all carbon stored in trees for decades or even centuries is released into the atmosphere.

Solid, gas, and liquid waste from animal farms, slaughterhouses, and processing plants are rich in organic and inorganic matter, pathogens, and chemical compounds. Typical examples are animal and human excrements, blood, animal trimmings, gut contents, hair, horns, pathogenic microorganisms (i.e., *Salmonella* spp., *Brucella* spp., *E. coli*, etc.), parasites (i.e., *Giardia* spores), natural and synthetic hormones, feed and antibiotic residues, milk, meat, and egg from processing losses, wash-down water containing food residues, detergents and disinfectants, nitrate, phosphorus, cooling water and heating water, in addition to odours and carbon emissions—again, mainly CO_2, NH_4, and N_2O (Bandaw & Herago, 2017). Therefore, farm and industrial waste must be treated before disposal on soil and water bodies to minimise soil degradation and water pollution, both a threat to biodiversity.

The content, temperature, and pH of effluents can change the amount of oxygen dissolved in the water and contaminate surface and groundwater, affecting the planet's natural balance. Every litre of processed milk generates approximately 2.5 litres of wastewater, which once treated generate up to 10 litres of effluent (Raghunath et al., 2016). These figures show that the treatment of wastewater is also resource-intensive.

A 450 kg dairy cow in lactation produces an average of 48 kg of manure per day, which represents more than 17 tonnes per year *per animal* (University of Massachusetts, 2002). You can now picture the mess we are doing by raising billions of animals for food every year, including one billion cattle, of which a quarter are dairy cows (CIWF, 2020).

You might be wondering: why not use manure to fertilise the land? After all, fertile soil is a good thing, right? Well, manure is rich in nutrients like nitrogen, phosphorus, and potassium. When applied to the soil in excessive amounts, these nutrients disrupt the nitrogen cycle and leach down to the groundwater, causing severe environmental problems. Furthermore, nitrous oxide and methane emitted by manure heat the atmosphere, and nitrous oxide depletes the ozone layer, attenuating the barrier against the sun's ultraviolet radiation. As *nothing ever disappears* in Nature (it only changes form), overapplying animal manure to croplands is not always beneficial.

In climate change assessments, different GHG emissions are converted into CO_2 equivalents based on their heat-trapping ability. This methodology enables the comparison between different activities, for example, fossil fuel burning versus animal agriculture. Cattle emit CH_4 due to enteric fermentation, which is a potent GHG with 28 times the heat-trapping power of CO_2 over 100 years (Shindell et al., 2009). Animal agriculture contributes 37% of global methane emissions (Gerber et al., 2013). Animal manure and soil fertilisers (increasingly used to grow livestock feed crops) are major sources of N_2O, a GHG with around 300 times the warming potential of CO_2 over a 100-year timescale (Koneswaran & Nierenberg, 2008; Rao, 2021).

A 2020 study by IATP revealed that the 13 biggest dairy farms in the world emit as much GHG as the UK, one of the strongest economies worldwide. The group comprises Amul (GCMMF), Le Groupe Lactalis, Saputo Inc., Danone SA, Fonterra, Yili Group, Dairy Farmers of America, DMK Deutsches Milchkontor GmbH, California Dairies, Friesland Campina, Arla Foods, Nestlé SA, and Dean Foods. In 2017, their GHG emissions combined surpassed that of US-based ConocoPhillips and Australian-based BHP, two major oil companies (IATP, 2020).

A thorough study on the dietary role in climate targets published in *Science of the Total Environment* assessed 313 food-system scenarios (Theurl et al., 2020). A change towards plant-based diets proved to be the most efficient way to reduce global deforestation rates and increase carbon uptake through vegetation regrowth. The research also concluded that a vegan diet is significantly more effective than a vegetarian diet, given the GHG emissions ascribed to milk production. A similar conclusion was reached in the 2018 study by Poore and Nemecek, which showed that a vegan diet is the most powerful way to curb our dietary environmental impact (Poore & Nemecek, 2018).

As more and more studies emerge on the association of dairy production with the climate and ecological crises, the livestock sector insists that milk is produced within sustainable standards, offering ludicrous solutions or using vague rhetoric. When confronted by a thorough report by IATP on the dairy industry's carbon emissions, the International Dairy Federation's president, Judith Bryans, and the Global Dairy Platform's executive director, Donald Moore, continued to give empty statements: 'The dairy sector is committed to producing nutritious foods in environmentally sound and responsible ways' (Carrington, 2020).

To be noted is that few, if any, of the mega-dairies worldwide are reporting their carbon emission levels (IATP, 2020). Although the role of animal agriculture in the climate crisis is addressed in the 2015 Paris Agreement, governments do not require animal-based food companies to measure or publish their GHG emissions. And interestingly, the Montreal Protocol on Substances that Deplete the Ozone Layer does not include N_2O (a potent depleter whose primary source is manure production).

To make things worse, governments have been traditionally investing in marketing to generate further demand for dairy products. The US Congress created USDA-sponsored public service announcements (PSAs) that have been promoting milk to the public for decades. Bord Bia, the Irish Food Board, has been sponsoring posts on social media on the alleged environmental and health benefits of dairy consumption.

In fact, we cannot reasonably say the decision-making process is unbiased. Animal agriculture lobbyists have a regular presence in Brussels, Belgium, where the EU food and agriculture legislation is discussed and voted on. Powerful corporations such as COPA-COGECA (Committee of Professional Agricultural Organisations-General Confederation of Agricultural Cooperatives) are based in Brussels. Not uncommonly, members of the European Commission and Parliament are directly involved with the meat and dairy industries as shareholders. Similar scenarios repeat in many other countries.

In the US, the animal industry lobby moves tens of millions of dollars annually to influence agricultural policy. The dairy sector invests approximately USD 7 million a year in lobbying (Open Secrets, 2020). In Brazil, government decisions are consistently influenced by the legendary Bancada Ruralista, a powerful lobby group within the National Congress linked to agribusiness. Their presence in Congress greatly affects economic, labour, and environmental legislation.

Two years after the 2015 Paris Climate agreement, the 13 biggest dairy farms already cited exhibited an 11% emissions increase. The GHG emissions of Amul, Le Group Lactalis, and Saputo increased by 40%, 30%, and 25%, respectively, from 2015 to 2017. These companies increased their production by extending to new markets from South America to East Europe (IATP, 2020). What little effort is being made by these big dairies to lower their carbon footprint is mainly through a reduction in processing plants, transport, off-site (e.g., electricity generation), and office emissions. Nonetheless, 90% of the carbon emissions associated with dairy production are related to the *animals' enteric metabolism* (IATP, 2020). That is like covering an open fracture with a band-aid.

None of the dairies cited in the IATP report considered reducing their production to meet the climate targets. Only five of them are reporting their emissions, from which three are addressing supply/production emissions: Arla (Denmark), Nestlé (Switzerland), and Danone SA (France) (IATP, 2020).

In the face of increasing consumer awareness around the environmental impact of dairy production, some dairies came up with a cunning strategy: to report emission reductions as "emission intensity reduction." Well, *emission intensity* refers to emissions per litre of milk. It becomes clear this is an utterly inefficient solution when you consider that the dairy industry *emission intensity* lowered by 11% between 2005 and 2015, while their *overall emissions* increased by 18% in the same period (IATP, 2020). Such negligence (to say the least) by policymakers, governments, and dairy companies is suicidal.

According to IATP and the international non-profit GRAIN, we must reduce GHG emissions by 34.5 million gigatonnes to avoid exceeding the 1.5°C limit. However, if the meat and dairy sector grows as projected, it could consume 80% of this volume by 2050 (IATP & GRAIN, 2018). To highlight the urgency of the issue, a global average increase in temperature of 1 to 3°C could reduce African agricultural yields by 50% (IPCC, 2019). In fact, a planet 2 to 3°C warmer would be *uninhabitable*.

And let me emphasise this: those in the Global South will probably suffer earlier and the most, as richer countries will be more able to protect themselves. It is childish at best to cling to our food preferences when the planet is collapsing.

Given the critical situation, a growing number of researchers advocate for the increase in milk yield per cow to reduce GHG emissions (Van Middelaar et al., 2015; Knapp et al., 2014; Zehetmeier et al., 2012). Such strategy bases on the assumption that a smaller number of cattle is necessary to produce the same amount of milk when milk yield per cow is greater.

As methane emissions by cattle and other ruminants are a result of the activity of methanogenic microorganisms belonging to the *Archaea* domain, methods to reduce GHG emissions in dairy farms include even vaccination against rumen archaea (Hristov et al., 2013; Van Middelaar et al., 2015).

But this is not the most ingenious solution suggested by the animal exploitation industry. The British group Zero Emission Livestock Project (ZELP) signed a partnership with Cargill and other companies in the agricultural sector to develop masks for cattle that neutralise methane emissions. According to the company, 90–95% of the methane is exhaled through the nostrils and mouths of ruminants. The masks will be available to European farmers in 2022. Imagine spending your entire life with a mask on your face.

Methods of reducing emissions or increasing productivity per animal often lead to *more suffering*. The production of unnatural amounts of milk by cows result in several welfare consequences, including increased strain on the mammary tissues, more milking sessions per day, higher incidence of udder inflammation (i.e., mastitis), leg problems, metabolic disorders, and reduced immune defence (Oltenacu & Broom, 2010).

Reducing GHG emissions by pushing animals to their biological limit (or vaccinating them against natural microorganisms in their body, or even covering their nose and mouth with masks) is clearly unethical, but also reckless, since all the other environmental problems associated with milk production remain—e.g., manure overproduction, soil degradation, water pollution, resource overuse, deforestation, and so on.

Dairy cows currently produce at least twice the milk they produced some decades ago. An average Swedish dairy cow, for instance, produced 4,200 kg of milk in 1957.

This figure raised to 9,500 kg in 2013 (Oltenacu & Broom, 2010; Agri Benchmark, 2013). Notably, a beef cow produces only 1,000–2,000 kg of milk (Webster, 1993). As mentioned before, the distinction between beef and dairy cows is human-made and was achieved by selective breeding for particular traits. All mammals produce milk according to the needs of their offspring.

Fossil evidence dates the emergence of the first modern humans, *Homo sapiens,* 200,000 to 350,000 years back, a tiny fraction of Earth's 4.54 billion years of age (Hublin et al., 2017). In other words, if the Earth were 24 hours old, humans would have existed for less than 7 seconds! That said, do we honestly believe Nature created all the other species for our convenience?

Even during the 2021 UN Climate Change Conference, also known as COP26, there were not enough mentions of the role of animal agriculture in the climate crisis in the media. And ironically, the event's menu featured a wide variety of animal products, demonstrating a disgraceful lack of understanding (or concern) about the leading causes of global warming. On the bright side, over 100 countries have pledged to cut methane emission levels by 30% by 2030 (*BBC News*, 2021). That is clearly not enough, but it is a good start.

Nations can curb emissions *if* they want to. The coronavirus pandemic is a clear example of the managing power of governments amid a life-threatening situation. Despite controversies, most of us changed our habits overnight to follow social distancing, lockdown, and other guidelines. Politicians must treat the climate crisis and animal agriculture with the same urgency they are managing the pandemic if they are serious about our future as humankind.

Political and economic strategies to help animal farmers transition to other livelihoods would facilitate the reform of our food system. Fiscal incentives benefiting powerful and polluting dairy corporations could be reallocated to crop production. Business grants and tax breaks could be granted to plant-based farmers and companies. Dairy farmers and industries could use their know-how and existing infrastructure to produce plant-based dairy products. Former dairy factories could be used as part of the new supply chain, and the equipment could be repurposed or retrofitted. Governments could highlight the economic opportunities for farmers to move into the new business. And finally, health, food, and agricultural organisations could redirect marketing and educational campaigning funds to promote plant-based diets.

Small farmers and local food companies would greatly benefit from a free market instead of the current subsidised and manipulated one. Animals would be freed from exploitation, and humans would gain physical, emotional, and spiritual health. This transition would reduce the burden on public health systems, global warming, and deforestation, allowing vegetation to grow back (Theurl et al., 2020).

Clearly, governments must start taking responsibility for both the climate and rural crises associated with the dairy sector (and the animal exploitation industry in general). Nevertheless, we can always reject products that are harming the animals, the planet, and other humans. We have the power to frame moral issues.

All social achievements in history started with a few people fighting for justice. Be one of them.

INVESTIGATIONS

Now, let us examine some examples of the reality of dairy farming, registered by Animal Equality's investigators. I confess I sobbed every time I researched the pictures and descriptions kindly provided by them. I humbly ask animals for forgiveness for taking so long to go from vegetarian to vegan in the past. How I wish I knew these things back then! I did not have this luck, but you do.

I also must say I profoundly admire the work of undercover investigators, who have the courage to witness these animals' suffering and pretend they are fine with it, only to give them a chance of a dignified life in the future. To them, my deepest gratitude.

INVESTIGATION 1 – BABYBEL SUPPLIER (US)

This investigation was conducted in 2019 at the Summit Calf Ranch, owned by Tuls Dairies, a farm that houses thousands of calves in Nebraska, US. Summit Calf Ranch is a supplier of Bel Brands, producer of the famous BabyBel and Laughing Cow cheeses (Animal Equality, 2019).

Shocking scenes of animal neglect and cruelty, including calves with frozen limbs due to extreme winter temperatures, contradict the official narrative on the Tuls Dairies' website:

> Superior milk cows grow from healthy, happy calves. [...] The ranch cares for approximately 6,000 calves from newborn to five months old. All newborn calves are bottle-fed for their first two months and great attention is spent to meet the health and wellbeing needs of each calf. Summit Calf Ranch provides excellent care for calves' first formative months. (Summit Calf Ranch, 2022)

Similar narratives are found on the websites of every dairy farm in the world.

The undercover footage revealed the prolonged suffering of dozens of calves, aged 1 to 150 days old, freezing to death in temperatures as low as -6°C (Figure 2). Many died of hypothermia and pneumonia.

The plastic barns obviously do not offer enough protection against the cold and wind (Figure 3). Water and milk froze with the below-zero temperatures, leaving the calves hungry and thirsty. Please note that these barns and feeding systems are commonly used in dairies all over the world.

The reality exposed in these pictures contradicts the welfare rhetoric on Summit Calf Ranch's website: '[We have] 3,200 hutches for newborn calves. When calves move from a diet of milk to a specially balanced nutrition plan, they live in mono-sloped barns' (Summit Calf Ranch, 2022).

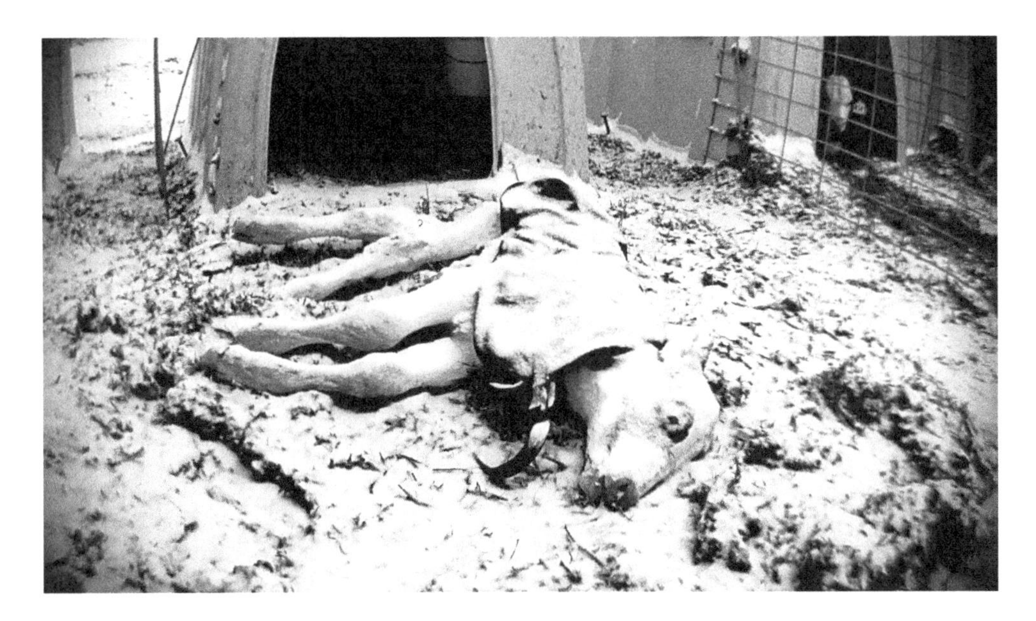

Figures 2a and 2b. Calves frozen to death at Summit Calf Ranch, Nebraska (US).

Figures 3a to 3c. Shelter and food and water supply for calves at Summit Calf Ranch, Nebraska (US).

Repeated exposure to extreme temperatures resulted in frozen limbs and the separation of the hoof from the leg of several calves (Figure 4).

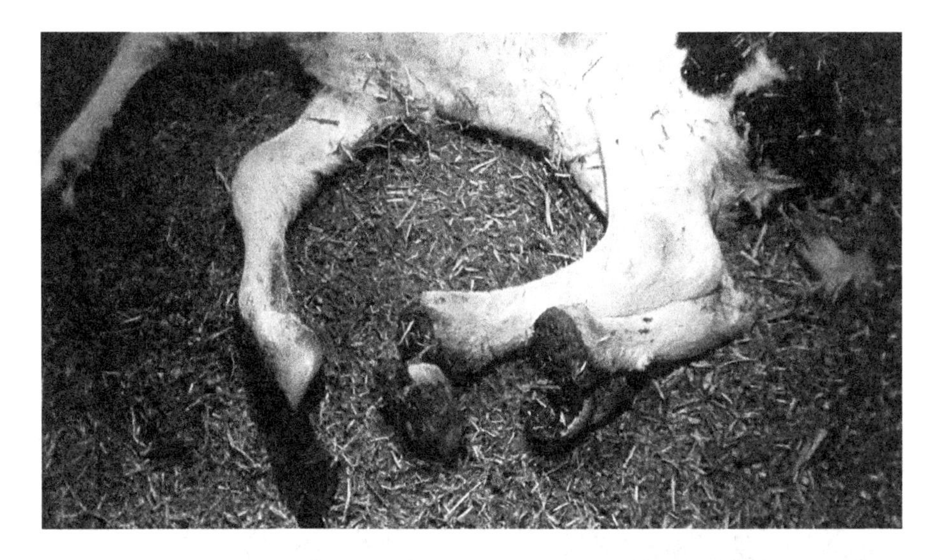

Figures 4a and 4b. Calves with severe injuries caused by extreme cold temperatures at Summit Calf Ranch, Nebraska (US).

Sick and injured calves received minimal or no veterinary care (Figure 5). Others were left to die by negligence next to dead calves (Figure 6). Scours, a disease that entails fluid loss and dehydration, was the number one cause of death at the farm, as proper medical treatment was not offered.

According to Sean Thomas, international director of investigations for Animal Equality: 'We documented sick newborns surrounded by their already dead pen mates in an area referred to as the "hospital" where they were left to suffer for days and in many cases finally die' (Animal Equality, 2019).

Figures 5a and 5b. Negligence towards sick and injured calves at Summit Calf Ranch, Nebraska (US).

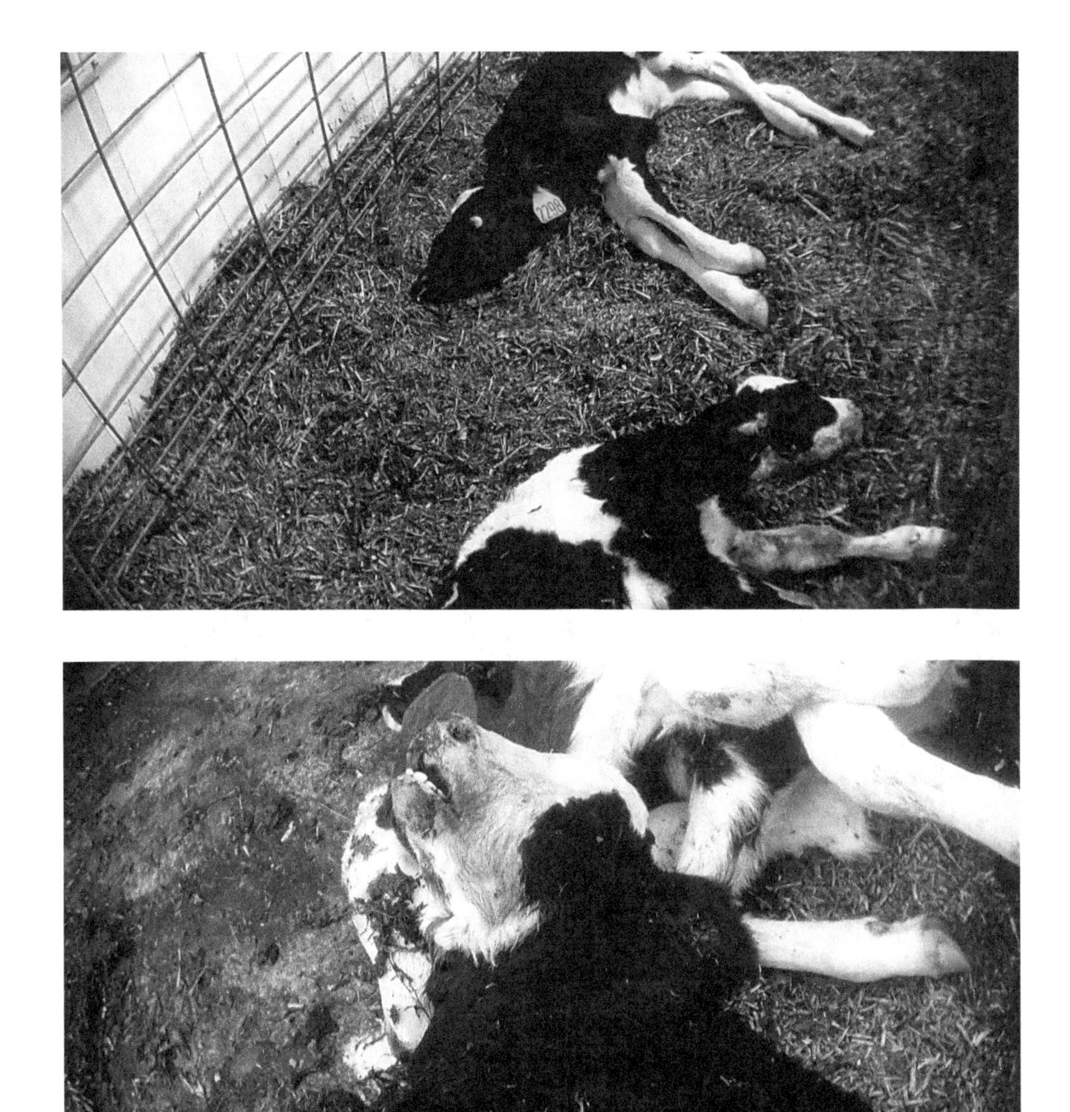

Figures 6a and 6b. Sick newborns left with dead calves at the Summit Calf Ranch, Nebraska (US).

Disbudding, castration, and other painful procedures were performed without pain medication, which is usual in livestock farms. Rough handling of animals was also common. In Figure 7a, you can see a worker restraining a calf while the other removes the horn with a hot iron. Figure 7b shows a male calf being castrated by banding, a brutal procedure that cuts the blood supply to the testicles by placing a rubber band around them until they fall off.

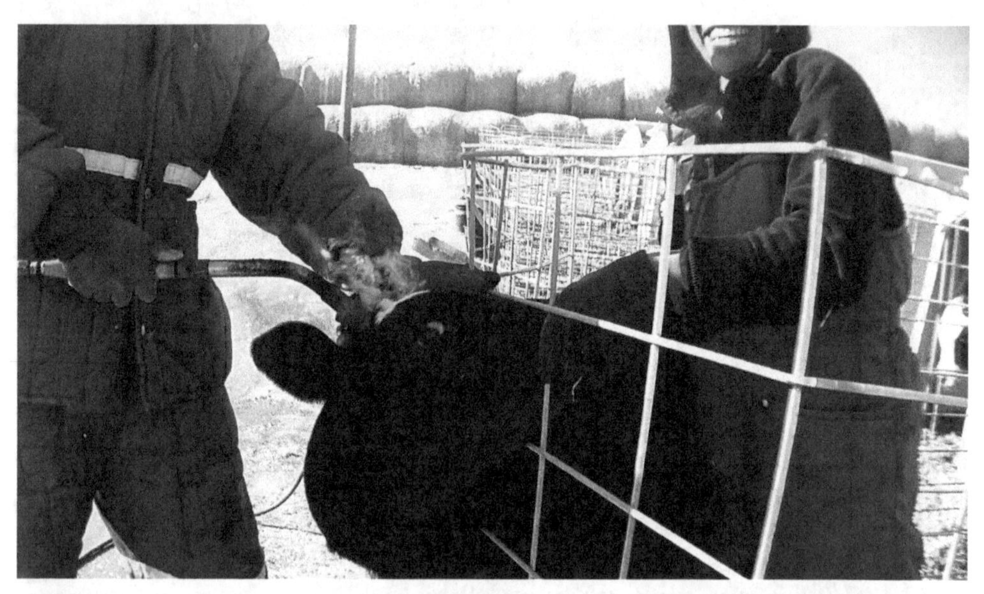

Figures 7a and 7b. Painful procedures without anaesthetics at Summit Calf Ranch, Nebraska (US).

Calves are separated from their mothers soon after birth and must fend for themselves without maternal care. Those who are deemed of commercial value are tube-fed while the others are left to die (Figure 8).

Figures 8a to 8c. Mother–calf early separation at Summit Calf Ranch, Nebraska (US).

It is impossible not to sense the fear of the calves being transported in a truck in Figure 9.

Figure 9. Calves during transport from Summit Calf Ranch, Nebraska (US).

Among the cruelty witnessed, the investigators found workers shoving, jabbing, and hitting calves with sorting sticks and hut rods as the manager's dog lunged and bit the backs of their legs. Newborn calves with raw legs from lying on the concrete, weak calves with wet umbilical cord trying to stand in the pen, and rough handling of calves during loading onto trucks were witnessed by the investigators *on a daily basis*. Shame on us.

INVESTIGATION 2 – BUFFALO MOZZARELLA FARM (ITALY)

Italian buffalo mozzarella cheese is widely known for its prime quality. However, the reality captured by Animal Equality's investigators on buffalo farms includes animal overcrowding, poor hygiene conditions, and environmental pollution. Figure 10 shows animals confined in small spaces and living in unsanitary conditions, covered in mud and faeces.

Figures 10a to 10d. Degrading life conditions of buffalos milked for mozzarella cheese production (Italy).

Figure 11 shows the poor hygiene conditions during the milking of buffaloes, and the inadequate infrastructure of the farm. From a food engineering standpoint, this is *unacceptable* for the health hazards associated with milk contamination. No food technology can compensate for high levels of microbial load in raw materials without compromising nutritional quality and increasing processing costs.

Figure 11. Poor sanitary conditions in buffalo milking (Italy).

Other sad life conditions were photographed, such as a dead animal hidden under a pile of mud, faeces, and straw (Figure 12) and baby males torn apart from their mothers, left to die on their own by negligence (Figure 13).

Figure 12. Dead buffalo found in a pile of mud, faeces, and straw (Italy).

Figure 13. Baby male buffalo treated as undesirable by-product (Italy).

The investigators also documented cases of waste disposal directly into nearby waterways, affecting the province of Napoli and Caserta area.

As we continue to buy dairy products, the food industry understands we agree with their ethics. The *only* way to stop supporting this atrocity is by going vegan.

CHAPTER 2

RED MEAT

I was a 19-year-old Food Engineering undergrad when this story happened. Our course description included a module called 'Meat Processing Technology.' I remember reading the course title for the first time and rolling my eyes thinking of how we have trivialised (and modernised) violence in our society. But then a day came when our lecturer gave us a practical assignment—a visit to a beef abattoir and processing plant. I knew the visit would be challenging for me, but I wanted to see up close the processes used in my future profession. And, who knows, maybe the reality was not as terrible as I had pictured... Well, it was a thousand times worse.

It was a well-known, modern meat processing facility in Southern Brazil. In the first minutes of the visit, it hurt me to see those overcrowded animals waiting for death, treated as mere objects. One of them had a bleeding bud. The look on their faces as they approached the slaughterhouse was heart-breaking. As they approached death, they started to feel agitated and widened their eyes. Many of them had tears rolling down their faces. After being shot in the head with a captive bolt gun and hung up by the hind legs, the animals had their throats cut to bleed to death.

Being in a slaughterhouse is like visiting hell—brutality, flailing animals, agonised screams, the noise of colossal iron chains carrying heavy carcasses, their skin being peeled off, the smell of blood, the cold, the *contempt for life*. But the part that struck me the most was something unexpected.

My colleagues and I were watching the evisceration stage, where carcasses hanging upside down are sliced open to eliminate the internal organs. It is a physically strenuous job that requires the staff to stab one animal after the other with quick movements. Each animal's organs are collected in a vessel, and the fast-paced process continues. As we watched a worker eviscerate one more animal, I was appalled to see a perfectly shaped foetus falling out of the carcass' belly, just to be discarded along with the viscera. It was a perfect, well-formed calf, whose size indicated the mother was in an advanced stage of pregnancy. It was not profitable to incur the costs of feeding and housing the cow and her offspring, so why keep them alive? The vessel contained tens of other unborn calves. I sobbed the entire visit.

Every corner of that place revealed a scene of utterly unnecessary violence, but nobody else seemed to see it. Everyone was watching the same things, but no one seemed to *perceive their meaning* other than me. At the end of the visit, we were "awarded" with a sensory analysis section. My colleagues and lecturer looked incredibly happy while tasting the factory's meat products, in a ritual of complete disconnectedness from what they had just witnessed.

It is in this very same state that the vast majority of us are living. Lovelessness and cognitive dissonance explain why the world is facing so much chaos. We have all been sleeping for too long, and now is the time for a great awakening. We *need* to take this civilising step.

In this chapter, we will discuss red meat production and consumption, and their multidimensional implications. Although the term 'meat' refers to the flesh of any animal (land, flying, and aquatic), I am splitting different types of meat into three chapters for didactic reasons: red meat in Chapter 2, poultry (and eggs) in Chapter 3, and fish in Chapter 4.

Market overview

The world produces 4 times more meat than in the 1960s, totalling 309 million tonnes in 2018. Beef (meat from cattle and buffalo) and pork are the two most consumed red meats in the world. There are about 1 billion cattle and 700 million pigs on the planet (Shahbandeh, 2021; Cook, 2020; Ritchie & Roser, 2019). Of these, nearly 450 million pigs are found in China. In terms of cattle population, Brazil ranks first on the globe, with an estimated 214 million animals in 2018—versus 94 million in the US (Ritchie & Roser, 2019).

Approximately 63 million tonnes of beef are produced every year globally, with the US (11.1 million tonnes per year), Brazil (9.0 million), and China (5.9 million) accounting for the largest shares. The EU produces an additional 7.8 million tonnes (Buchholz, 2020; Ritchie & Roser, 2019). The global beef production meant the slaughter of 302 million animals in 2018 alone (Ritchie & Roser, 2019).

The global production of pork far exceeds that of beef: almost 109 million tonnes per year. The leading pork producers are China (with a hefty 49.9 million tonnes per year) and the US (10.8 million). Germany (4.9 million), Spain (4.1 million), Brazil (3.4 million), and Russia (3.4 million) have a prominent place in global production but are far behind the top two producers. Around 1.5 billion pigs were killed in 2018 (Ritchie & Roser, 2019).

Lamb and sheep meat are also popular in many countries. Still, the global numbers of cattle and pigs are unparalleled. For example, the 2014 global production of sheep meat amounted to 8.7 million tonnes. Therefore, I will focus on beef and pork in this chapter.

The three major consumers of pork meat *per capita* in 2019 were Portugal (42 kg/person), China (38 kg/person), and the US (30 kg/person). In the same year, beef *per capita* consumption was higher in Argentina (54 kg/person), Brazil, and the US (both with 37 kg/person). Figure 14 illustrates how meat consumption has grown over the decades.

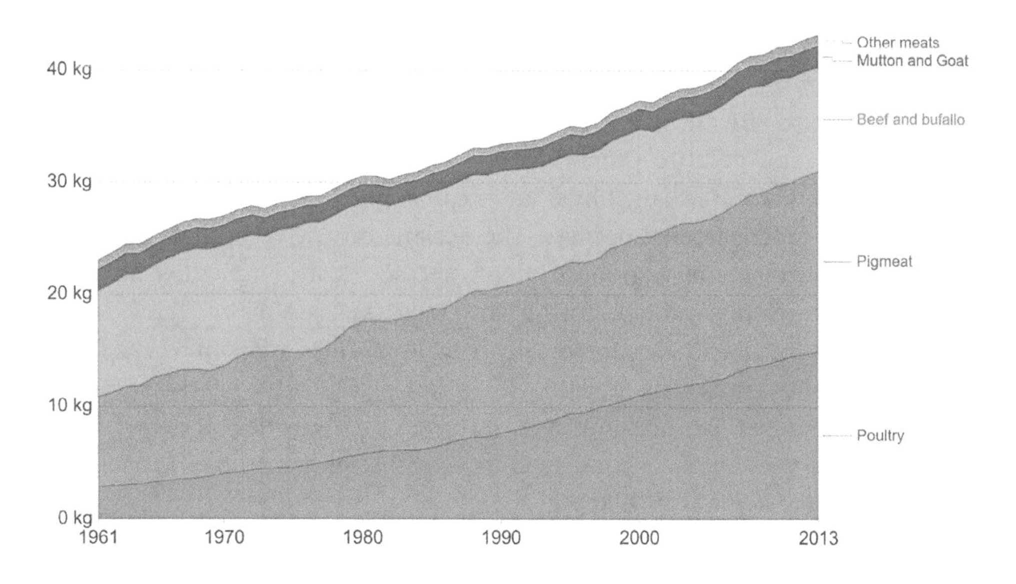

Figure 14. Global meat consumption (kg/person, 1961–2013). Source: Ritchie & Roser (2019).

The appetite for animal flesh is accompanying the income rise in many countries, such as China, whose *per capita* consumption has increased 15 times in the last 50 years (Ritchie & Roser, 2019). Therefore, it is not surprising that the incidence of cancer is 2.5 times higher in countries with a superior human development index (HDI). The occurrence of rectal cancer, for example, approximately quadrupled in countries with a higher HDI, while the incidence of colon cancer is more than 5.4 times higher. This data, which comprises the incidence of 36 types of cancer in 185 countries, is used by the World Cancer Research Fund itself (Bray et al., 2018).

In fact, the WHO and the International Agency for Research on Cancer (IARC) advised people on the carcinogenic potential of processed and unprocessed red meat in a scientific report dated 2015. Processed meat (e.g., sausage, salami, and ham) is considered a *Group 1 carcinogen*, the same category as asbestos and tobacco (WHO & IARC, 2015). More specifically, the consumption of processed meat causes colorectal cancer. Red meat (unprocessed) is considered *probably carcinogenic* to humans, in Group 2B. These conclusions are based on 800 studies on the association between meat consumption and several types of cancer across different countries.

Numerous other health agencies worldwide recommend limiting meat intake. As we continue to disregard animal rights, our health deteriorates.

Red meat farming and processing

Like all other animals, cows and pigs are sentient beings. This means they feel pain, fear, sadness, affection, and anxiety and can process what is happening to and around them. The 2012 Cambridge Declaration on Consciousness formally communicated that mammals, birds, and cephalopods possess the neurological substrates that generate consciousness. The declaration was a result of a series of systematic experiments by a group of renowned scientists, employing advanced techniques for assessing and monitoring consciousness in animals.

Bulls and cows, in particular, are socially complex animals who appreciate music, choose best friends within the herd, have long-lasting affiliative relationships, and grieve when separated. The bonding between cows or bulls, widely known by those in close contact with them, is confirmed in several scientific studies (McLennan, 2012; Reinhardt & Reinhardt, 1981).

Pigs are curious, smart, and playful animals who have individual personalities and love to socialise. Researchers demonstrate that pigs are highly intelligent, establish relationships between facts, emit emotional responses to different situations, and are capable of solving complex problems just like dogs (Broom et al., 2009; Imfeld-Mueller et al., 2011; Fraga et al., 2021).

It is funny, though, that we need scientists to state the obvious. Or do we honestly think animals are inanimate objects?

The documentary *The Emotional World of Farm Animals* shows the rich emotional lives of farmed animals as best-selling author Jeffrey Masson visits sanctuaries across the US. This heart-warming documentary is suitable for all ages—and every school-age child should watch it.

Although animals have been extensively proven to feel pain, experience complex emotions, and possess intelligence, 80 billion land animals are exploited and slaughtered each year for human consumption (Ritchie & Roser, 2019). If we include aquatic animals, then *trillions* of animals are killed annually—it is even difficult to imagine this number (Mood, 2010).

Painful mutilations are common in the meat industry (and in animal agriculture in general). Male calves and pigs are often castrated *without anaesthesia*. Hot iron marking and dehorning are frequently carried out with no anaesthesia, sedation, or painkiller administration, as regulations vary from country to country and law enforcement is commonly absent. Females are forcibly impregnated to generate the maximum number of descendants throughout the few years they are entitled to live. Successive pregnancies exhaust their bodies and minds, in a strenuous hormonal rise and fall. When their productivity declines, they are slaughtered to become second-graded meat. Babies are separated from their mothers, without respecting their affective bonds and biological needs. Typically, they are taken away at 21–28 days for piglets and from 6 months of age for calves. Genetic selection, balanced diet, drug administration, and other practices are employed to boost growth so the animals become meat sooner. The animals are transported between farms, to slaughterhouses, exhibitions, and even international markets, which is a major source of stress, injury, and death in extreme cases.

The quality of life of farmed animals varies depending on the production system, husbandry techniques, level of training among farmworkers, and their personal values. But even in the best-case scenario, farmed animals are *income sources*. The treatment given to them is a function of the economic return they represent, never of their inherent value as feeling beings.

Beef and pig farms operate under three main rearing systems: (1) extensive, in which the animals are kept predominantly free, feeding on pasture, agricultural waste, and food leftovers; (2) intensive, in which they are confined for life and kept on a balanced diet; and (3) semi-intensive, which combines aspects of intensive and extensive farming, that is, the animals have access to pasture, but are kept in pens for certain periods.

The confinement of pigs and cattle is a source of emotional and physical suffering: respiratory problems due to high levels of ammonia, gastrointestinal disturbances ascribed to excessive grain consumption, boredom, depression, and anxiety leading to obsessive behaviours, aggressiveness, and apathy, foot injuries because of hard and rough floors, and so on. In pig farming, females perhaps suffer the most, as they can spend nearly all their lives in individual cages or pens, being impregnated, giving birth, and nursing piglets.

In addition to the husbandry method, rearing systems can also be classified as conventional or organic. In any country, there are farms of all sizes and rearing systems, from "backyard" production to mega properties with thousands of confined animals. Extensive systems are typical of subsistence farms. Intensive production requires more technicality and capital but generates greater profitability. Importantly, most beef and pork consumed worldwide come from factory farms.

The animals do not necessarily spend all stages of their productive life in the same location. For example, in veal production, calves can be born on the beef farm or be purchased from dairy farms to be fattened. In pig farming, rearing comprises several stages: gestation (reproduction and pregnancy), farrowing (birth and suckling—around three weeks), nursery (first care of the weaned piglets, until they reach a certain size—approximately 6 to 8 weeks), and growth and finishing phase (an average of 16 to 17 weeks, when the weaned pigs are fattened until the slaughter weight). These stages can take place on different properties, with many farms dedicating exclusively to the finishing phase. In addition, some farms are specialised in breeding "replacement" pigs, females and males that will replace culled and slaughtered animals and add specific genetic characteristics to commercial herds (Embrapa & ABCS, 2011). Slaughter of both cattle and swine is most commonly carried out in slaughterhouses, except for very small farms.

Transport is a major cause of stress and injury in animal agriculture. Farmed animals can travel in cramped spaces, without food and water, covered in their own excrements. Traffic accidents involving livestock transport are more common than we think, and can result in minor or serious injuries, exposed fractures, death, and even animals left abandoned, agonising in the street (Miranda-de la Lama et al., 2011).

Intentional animal abuse, beyond the inherent violence animal agriculture entails, is common in livestock farms. This includes rough handling, kicks, punches, verbal abuse, and other forms of aggression. Even though ag-gag laws try to contain undercover photo shootings in animal farms, animal rights organisations frequently document cases of severe negligence and unjustified violence against the animals. Visit www.animalequality.org and www.kinderworld.org to see for yourself.

Many people do not mind causing suffering to others. They are perfectly capable of watching a video showing the horrors of animal agriculture and then turn away to unabashedly order a bacon sandwich. These are sad and hopeless cases. But other people seem to believe that farmed animals live a peaceful life until the day they gladly die to serve humans as their divine mission. However, the reality is entirely different.

First, the bodies and lives of other sentient beings are not goods we can avail of. Second, the slaughter is the *last* violence farmed animals face after a lifetime of repeated emotional and physical suffering. Third, 'humane slaughtering' is an oxymoron. Not only is there no way to kindly kill a feeling being who does not want to die, the so-called "humane" methods are not that humane. Or is shooting a cow in the head and cutting their jugular so they bleed to death what we consider humane?

On top of that, clandestine slaughtering, problems in the processing line, and even religious slaughtering methods (e.g., halal and kosher) mean many animals have their throats slit while still conscious.

Let us now examine some details associated with beef and pork production separately.

Beef and veal

Differently from the dairy sector, male calves and bulls have the highest commercial value in the beef industry. Cows are mainly raised for breeding stock, to ensure herd size is maintained or increased. Male calves are either castrated (becoming steers) or kept as bulls for reproduction. Steers are fattened until they are slaughtered for beef at the age of 12–24 months. Considering that a cow, bull, or steer has an average life of 20 years, these animals live a maximum of 10% of their lifespan. A minority of castrated males can be used as haulage oxen. The use of non-castrated bulls as haulage oxen is avoided to prevent unwanted mating and aggressiveness.

Castration of male cattle can be done by the application of rubber rings around the neck of the scrotum, surgical removal of the testicles, or the use of a tool called burdizzo that crushes the vessels carrying blood to the testicles. One person performs the procedure while at least another one restrains the animal.

Although these methods are extremely painful, they are broadly carried out *without* anaesthesia or analgesia. Commonly, the animals faint due to the immense pain, especially in burdizzo castration. Complications can occur after castration, including inflammation, tetanus, infection, maggot infestation, sepsis, suppression of immune function, and haemorrhage. In Ireland, for example, cattle can be castrated without anaesthesia or pain relief using a rubber ring until eight days of age or using a burdizzo until *six months* of age. The meat industry refers to these methods as "humane castration" of cattle (Teagasc, 2020).

Veal (or calf beef) is obtained from male or female calves aged 1 (yes, one!) to 32 weeks. Most veal comes from young dairy calves who are not used for breeding purposes. Veal is basically a product the dairy and meat industries came up with to turn dairy male "valueless" calves into an income source.

I take the opportunity to highlight the *direct* connection between the dairy and meat industries. Many of us spend years, if not decades, being vegetarian for ethics, believing we are keeping violence off our plates by saying no to animal flesh. But as we have seen in Chapter 1, the dairy industry is terribly cruel, in addition to being closely associated with the meat sector. And you will see in the next chapters that the chicken, egg, and fish industries are not far behind.

I also want to emphasise that the problem is not the *treatment* of animals, but rather *using* them. Subjugating and using sentient creatures is morally unjustifiable.

Beef from female animals is deemed less valuable as they produce less muscle than males. Therefore, sexed semen can be used in the meat industry to ensure the predominance of male calves. Semen can be bought in or collected from the farm's bulls and stored in liquid nitrogen. Artificial insemination is preferred because it is a means to select for certain genetic features and, thus, improve the quality of the herd. In addition, handling bulls poses safety issues to the workers, not to mention the extra costs incurred in housing, feeding, and providing veterinary care to larger animals.

There are several semen marketing companies in the world, such as Semex, Genex, Cattle Genie, and Select Sires. Semen can be collected by electroejaculation or using an artificial vagina. The first method, which is more common, comprises the electrical stimulation of the seminal vesicles of the bull. Bulls are confined to a squeeze chute, have their heads locked in with a headgate, and the rectum is emptied. Then, a probe with linear banded electrodes connected to a voltage source is inserted in the anus, so the electrical stimulation causes ejaculation (Statham, 2015; Bedford-Guaus, 2014). After collection, the semen is either frozen in liquid nitrogen or used to inseminate a female on the same day. Several pulses or a higher voltage may be necessary to induce ejaculation in certain bulls.

Cattle ranchers will certainly claim these procedures are painless (which I doubt)—but are they *ethical*? Would you allow this to be done to your male cat or dog?

Cows and heifers (young cows) in heat are artificially impregnated using an insemination gun, in a process that can involve dozens of females a day. As the semen is stored at -196°C, the straw containing semen is thawed into warm water. The farmer or technician raises the animal's tail, inserts one arm in the rectum, while the other reaches inside the cow to insert the instrument into her vagina. The fingers of one hand are used to feel for the tip of the gun through the rectal wall. The semen must be deposited specifically into the cow's uterus, which is not straightforward as the cervix contains several projections that can somewhat block the canal. This means injuries can happen if the worker is not professionally trained or experienced (or is in a bad mood). Anaesthesia is not required in the procedure. Each insemination has a 50–60% success rate (Statham, 2015). If a cow comes back into heat around three weeks following insemination, it means the impregnation was not successful, and therefore the procedure must be repeated.

As I was writing this chapter, Dr. Renato Pulz related to me two disturbing facts: (1) the rectal palpation for insemination and diagnosis of pregnant cows commonly results in anal fissures, leaving the veterinarian's glove covered in blood; (2) in artificial insemination programs, sterile males can be used to detect females in oestrus, stimulate oestrus and ovulation in adult females, and accelerate puberty in heifers by mounting them. To prevent copulation, surgeries are performed that *deviate the penis* or expose the prepuce, causing pain and even illness. Females that have been androgenised can also be used for this purpose—they are given regular doses of testosterone propionate or enanthate to induce bulling (Beef Point, 2006).

In farms that follow natural breeding, fertile bulls are a very desirable commodity. Bulls become fertile from 12 months of age, when 35% of them produce good-quality semen. At 16 months old, the rate increases to 95% (Barth, 2000). The way to test their libido and fertility is empirical and based on the number of cows "serviced" (a term used by farmers) and impregnated in a given period. This means several cows are restrained in a certain space to allow for impregnation by a bull at fertility age (Hamilton, 2007).

As libido and mating ability must be evaluated periodically to ensure rapid herd growth, females are exposed to these tests repeatedly throughout their lifetime (Perry & Walker, 2008). Cows that exhibit low fertility are culled, as they are deemed uneconomical.

I pause here to invite you to a couple more reflections:

1. How can women fight for their rights when they allow other females to be dominated, forcibly impregnated, have their offspring kidnapped, and their milk or eggs stolen?
2. And how can we oppose sexual abuse when we submit males and females of other species to rape, genital mutilation, and androgenisation? Think about it.

As bulls are inherently less docile than castrated animals, they are set apart when their services are not required. Commonly, they are restrained to a pole by a rope attached to a nose ring. The metal rings are pierced into their septum, sometimes without anaesthesia, and used to control the animals. The pain and discomfort when manipulated by the ring ends up taming the animals.

Many agricultural shows require bulls to have a nose ring. The stress these animals face in these events is massive. They are submitted to environment change, stressful transport conditions, proximity to other animals, grooming, confinement, and exposure to humans they are not familiar with.

Some might say the farming practices vary widely among farms. However, when we consume animal products, we cannot guarantee their origin—unless we raise and kill the animals ourselves and *never* eat out. Do you remember the BabyBel cheese supplier from the previous chapter? If a large US farm that supplies milk to an international dairy brand treats its animals that way, imagine what happens in smaller, less supervised businesses. And what will be the reality in countries with vague or even non-existent animal protection laws?

As in dairy farming, calves are separated from their mothers earlier than natural, at 5–10 months old (Enríquez et al., 2011; Haley et al., 2005). Natural or no management-imposed separation occurs at 7–14 months, with a gradual reduction in maternal care and suckling (Reinhardt & Reinhardt, 1981; Enríquez et al., 2011).

Enforced weaning is a significant stressor for calves and dams, resulting in decreased immunity for the calves and lower rearing performance for the dams. Low immunity in calves is associated with coccidiosis (a digestive infection by the protozoan parasite *Eimeria* spp, causing diarrhoea), diarrhoea caused by *E. coli*, *Cryptosporidium parvum*, rotavirus, coronavirus, diarrhoea due to mineral deficiencies, and respiratory diseases (including pneumonia), leading to respiratory distress, fever, nasal discharge, loss of appetite, and even death (Lorenz et al., 2011a; Lorenz et al., 2011b).

Long-term studies by Reinhardt and co-workers (Reinhardt, 1983a; Reinhardt, 1983b; Reinhardt & Reinhardt, 1981) with semi-wild cattle showed that the affection between mother and calf remains after weaning. Suckling can persist in weaned calves, even after their mothers are lactating her new offspring. In addition, male calves never mounted their mothers, even if they started copulating with other females, at the age of 16 months or older. Affiliative bonds endure for many years, giving place to families within the herd.

In a review paper on weaning methods in beef calves, Enríquez et al. (2011) state:

> The mother–young bond has been defined as a preferential mutual, emotional attachment, of relatively long duration, and that resists temporary separations. It is characterised by affiliative behaviours such as licking, provision of food, warmth and protection, rest in company, synchronisation of activities and maintenance of closeness and body care. The willingness to establish such an intimate social relationship begins before birth and is reinforced by the contact between mother and young during the first hours after birth. In the dam, this phenomenon involves several physical and physiological changes that occur

> during pregnancy, parturition and the first contacts with the offspring. [...] In cattle, which are characteristically gregarious, the moments before birth are among the few in which the cow seeks to depart from the herd. Although domestic cows seem to adapt their calving strategy to ecological factors, cattle are usually regarded as hiders. It has been shown that cows isolate themselves from the rest of the herd to calve; this relative isolation allows the association with the newborn during the period of highest sensitivity and predisposition for the establishment of a mother–young bond.

As welfare in animal agriculture is closely (if not exclusively) related to productivity targets, different weaning methods exist aiming at reducing the detrimental effects of cow–calf bond-breaking. Weaning of beef calves follows three main methods: (1) abrupt separation; (2) fenceline weaning; and (3) two-stage weaning, by preventing the calves from nursing their dam before separation.

Abrupt weaning is the most common method. Calves are simply drafted from their mothers and moved as far as possible. The emotional consequences for the animals are visible: they spend from days to weeks vocalising repeatedly, walking in search of the other, in extreme restlessness. Some animals even break through fences to reach each other. As suckling is a natural behaviour in calves, privation leads to further anguish and compulsory substitute sucking. The calves try to suck on objects, suck the farmer's finger, or anything else that might offset the instinctive behaviour. The calf's distress is worsened by the need to adapt to a new physical and social environment and a different diet. A move from milk and grass to silage and feed commonly leads to digestive problems (Enríquez et al., 2011). However, the financial return is greater, as milk is more expensive than solid foods.

If it seems heartless to ignore the pain of a woman who has lost her newly born child, there is no reason to disregard the suffering of cows and calves separated against their will. The bonds between mothers and babies are sacred, no matter the species.

The second method, called fenceline weaning, involves separating cow and offspring into adjoining paddocks while allowing visual contact through a fence. The cows are moved to a distant paddock after 4–14 days, and older calves will help the calf through their new life. While this can sound like a more "humane" weaning method, studies show contradictory results in terms of trauma indicators.

Price et al. (2003) found higher daily weight gain and lesser vocalisation and searching for the dam in fenceline weaning, compared to abruptly separated calves. In contrast, Enríquez et al. (2010) observed that calves exhibited lower growth rates, as well as more frequent and persistent vocalisation when taken away after 14 days of fenceline separation. In any case, such limiting factors as longer weaning time, larger pasture area, and extra cattle management still render abrupt weaning the preferred method.

A third method is the gradual weaning by preventing the calf from suckling, using nose-flaps or calf-weaning rings (Haley et al., 2005). These plastic rings are not pierced, only clipped on the nose. However, they have spikes that hurt the mother, causing her to push the calf away when they try to suckle. Although the typical distress responses are reduced compared to the abrupt separation, the young faces frustration for not being able to access the udder, resulting in slower weight gain (Riggs et al., 2019).

It is of utmost importance to realise how flexible the truth is in animal agriculture. *Meat* scientists and ranchers defend the calves must be separated from their moms when they are around six months old. They claim an earlier and abrupt separation is detrimental to the animals' welfare and overall health, while natural weaning is uneconomical. At the same time, *dairy* scientists and dairy farmers claim that cow–calf separation must be conducted as soon as possible (a.k.a. one day from birth) to avoid animal suffering. Well, regardless of the breed, we are talking about the same animals (cows and calves). How come animal agriculture's arguments are so conflicting? Because 'animal welfare' serves a major purpose in animal agriculture: to increase profits (and, as a bonus, to build consumer trust).

In the meat industry, care from the dam is crucial to ensure the calves grow strong and fast to be slaughtered for beef sooner. In the dairy industry, however, allowing cows to feed and raise their babies means a lesser milk production. Thus, it is better to separate them soon after birth. Furthermore, male dairy calves are unwanted by-products, which is why the welfare argument of weaning calves at six months of age, widely used on beef farms, is quickly discarded on dairy farms. A flexible truth is nothing but ploy.

As mentioned earlier, transport is particularly stressful for farmed animals. Even though livestock transport is regulated by national guidelines in most countries, calves and adult cattle may be transported short and long distances under weather extremes, sometimes without food or water. Careless handling, tight spaces, manure accumulation, rough driving, and traffic accidents aggravate the situation.

An investigation of traffic accidents involving the transport of cattle for human consumption was published by Miranda-de la Lama et al. in 2011. The research included cases that occurred in Spain over ten years. The researchers reported on the pain and stress the animals are submitted to in cases of traffic accidents, which can result in survivors being killed *in situ* or severely injured ones being reloaded and transported to the slaughterhouse. The leading cause of accidents was attributed to low training of the drivers, who are mostly self-employed hauliers. Nearly a quarter of the accidents involved other vehicles, killing 41 people (Miranda-de la Lama et al., 2011).

Livestock can also be transported internationally by sea or air. In 2019, 3,800 cattle ran out of food onboard a ship denied docking in Cartagena, Spain. The animals were on their way from Brazil to Turkey, and a fodder and water loading was to happen at the port after exhausting 24 days of travelling (Badcock, 2019). Exports of Brazilian live cattle can generate incomes 25% higher than commercialisation in the internal market. In 2017, over 400,000 cattle were exported, a 42% increase compared to the previous year (Selistre, 2018).

Overcrowding in ship transport is common, which entails deprivation of movement, excrement accumulation, thermal discomfort, hierarchical disputes between animals, faecal contamination of water and food, trauma, and injury. The new diet commonly causes digestive discomfort and illnesses, including diarrhoea and ruminal tympany or "frothy bloat" (Pulz, 2020). Ruminal tympany is characterised by an excessive volume of gas in the rumen, leading to rumen-reticular distension.

In a day, cows and bulls produce 20–40 kg of faeces and 10–20 litres of urine. An atmosphere rich in ammonia from the manure can cause airway irritation and respiratory diseases. In addition, lameness and dermatitis are frequent consequences of prolonged contact with faeces and urine, not to mention the discomfort of being covered in excrement for days. Noise pollution, seasickness, and merely finding themselves in a different social and physical environment are sources of intense stress (Pulz, 2020).

In Brazil, in 2018, a judge granted an injunction prohibiting the shipment of 27,800 cattle bound for Turkey after animal rights activists filed an action in federal court requiring the suspension of livestock exports. After one week docked at Santos, in the state of São Paulo, the injunction was revoked and boarding allowed. The reports by two veterinarians who examined the conditions inside the ship (one of them nominated by the Federal Public Prosecutor's Office) were clearly contradictory. While the nominated professional stated the presence of perfect animal welfare conditions, the independent veterinarian registered scenes of extremely poor hygiene and neglect, including lack of feed and water and intense smell of ammonia from the animals' urine. The injunction was suspended on the grounds of damage to the country's economy and reliability in the international market. Important to highlight is that the cattle had reached the Santos port from different regions of Brazil, after several days of land or river transport (Pulz, 2020).

Livestock transport is a big market opportunity for cattle ranchers worldwide. In 2015, the European Court of Justice ruled that any transport of live animals from the EU to other countries must ensure that EU animal welfare standards are met. In theory, a veterinary practitioner must accompany the load and unload of the animals all the way until destination. Nonetheless, even when the guidelines are followed, it is impossible to control what happens to the animals *after* they reach destination.

In 2017, animals exported from such EU countries as Ireland, France, and Poland to the Middle East were reported to arrive in gruesome conditions and be slaughtered inhumanely. Videos and pictures recorded by Animals International, an animal rights organisation, showed 'animals hoisted by one hind leg, spinning on a chain as a man with a knife makes several slashes at their necks. Other images show an animal being stabbed in the eyes, having leg tendons cut while others have their heads and necks restrained by devices covered in the blood from earlier killings' (O'Brien, 2017).

Turkey, Lebanon, Bulgaria, and Middle Eastern countries are valuable beef outlet markets to Ireland, for example, especially in view of the colossal number of unwanted male dairy calves resulting from the growing herd (Claffey, 2018). Ireland exported 13,122 cattle to Libya in 2019 and more than 5,000 in the first semester of 2020, despite the chaotic war times in the destination country, which can jeopardise the animals' safety. The country earned more than EUR 450 million with live animal exports in 2019 (Kevany, 2020). EU exports of live bovines, chicken, sheep, pigs, goats, and horses grew nearly 63% from 2014 to 2017 (Daragahi, 2018).

Nonetheless, a ban on EU live animal export to third countries has been under discussion due to rising awareness of the conditions the animals endure at the Turkey–Bulgaria border. The border is a passageway for livestock from EU to Middle Eastern countries, where the demand for fresh beef is on the rise. After arriving in appalling conditions, the animals are sometimes held in the border for days inside metallic containers under the scorching sun, with little fresh water and feed, while paperwork and fees are resolved (Daragahi, 2018).

Lesley Moffat, the founder of Dutch-based charity Eyes on Animals, accompanied a case of cattle and sheep EU export stuck at the Kapikule border. 'Many of the cows are crammed into quarters so tight that they are unable to lie down. One cow was bleeding from a raw stump where her horn caught in the railing of the truck and was ripped off,' she told *The Guardian* (Daragahi, 2018). This is only one single case of animal cruelty related to transport among thousands across the world every year.

In 2017, Australia exported almost three million farmed animals to countries such as Indonesia, Qatar, and Kuwait (Daragahi, 2018). Cattle exported from Australia were reported to face extreme violence in Indonesia in 2019. Footage by Animals Australia documented cattle dragged by ropes in a car park, before being slaughtered *in situ* without prior stunning. A total of 172 complaints have been received by the Australian Exporter Supply Chain Assurance System between 2012 and 2019 (Wahlquist, 2019).

When faced with this kind of complaints, government agencies across the globe simply insist that the transport and slaughter of animals follow high welfare levels.

As long as most of us tolerate institutionalised violence, there will not be significant civilising progress in any sector of society. The world outside us is only a reflection of our inner selves.

Pig meat

Back in 2018, I remember having a chat with a fellow scientist at the Irish Food and Agriculture Development Authority (Teagasc), where I used to work as a project manager. She was preparing a presentation to non-scientists on sausage production. As her audience would include both adults and kids, she had tailored her speech accordingly and prepared an illustrative example of the ingredients that go into the sausages. Conveniently, she would not mention the way the ingredients are sourced.

Before her presentation, she suggested that I pretend I was part of the audience and ask her one question. Naturally, I asked her about the use of animals in the meat industry. She gave me an evasive, memorised answer—something like: 'The Irish meat industry is known as one with the highest animal welfare standards in the world.'

Many people would feel entirely satisfied with that answer, as it puts their consciousness at ease. To me, though, this reminds me of people on a diet who deliberately underestimate their food portions so they can eat more. Who are they cheating?

The percentage of US pig farms numbering more than 5,000 animals rose from 5% in 1982 to 73% in 2017 (Held, 2020). An estimated 98% of pig meat produced countrywide comes from factory farms (Anthis, 2019). In Spain, the second largest pig producer in the EU, the number of farms dropped to a third between 1999 and 2013, while the average herd size quadrupled (EPRS, 2020). In Brazil, over 70% of pigs are raised in intensive farms (Fiebrig et al., 2020). Although there are farms of all types and sizes globally, the trend is that industrial-scale farms, heavily reliant on technology and with herds of thousands of animals, replace small and medium-sized properties.

Pigs live around 15 years in Nature. They love to forage, sunbathe, and socialise, and spend more than half of their time roaming and exploring. Contrary to their reputation, pigs are clean animals that dislike urinating and defecating near where they eat and sleep. As they have low natural protection against the sunlight and do not have sweat glands, they like to roll in the mud to keep their skin protected and an ideal body temperature. The mothers often make nests for their young, using straw, leaves, and other materials. These animals are denied their basic rights in factory farming.

In swine farming, pigs are slaughtered on average at 6 months old, when they reach 100–130 kg. Sows live 3–5 years, as their primary function is to increase the herd size. They have just over two litters of around 10–15 piglets a year, with each pregnancy lasting 114 days on average. As the consecutive pregnancies and life conditions deplete the sows' health, they are sent to slaughter after a few years.

In extensive and semi-intensive farms, pregnant sows can be kept free, in collective pens or in mixed systems of individual cages and collective pens. However, intensive farmed sows are frequently kept in sow stalls and farrowing crates.

Sow stalls are cages measuring around 2 m long, 60 cm wide, and 1 m high, to which sows are confined throughout part or all of their pregnancy. This means spending two-thirds of their lives in extreme confinement. In these crates, sows have space to either lie down or stand up, but not turn around. Note that adult sows weigh an average of 200–250 kg and may be even heavier depending on the breed. The industry claims that gestation cages prevent aggressive behaviours and reduce injury and miscarriage rates.

To facilitate excrement collection and disinfection, sow stalls are usually paved with concrete, metal, or slate stone, without bedding materials. These types of flooring cause thermal discomfort, feet injuries, slipping, and noise, stressing the animals. Pregnant sows are given controlled amounts of feed to ensure milk production and a certain longevity, which would otherwise be compromised by the obesity caused by regular feeding under severe sedentarism (Bergeron et al., 2008). The lack of exercise and the genetic manipulation commonly result in bone fragility and other health issues, such as weakness of knees, arched legs, metatarsus varus, urinary infections, and cardiovascular diseases. In addition, the constant friction against the bars causes skin lesions. Privation from freedom and innate behaviours leads to severe psychological trauma and depression, resulting in obsessive bar-biting and head-weaving (Zhang et al., 2017).

Sow stalls are *widely* used around the world, although a movement is underway to ban them in several countries. In the EU, confinement of sows in cages for the entire length of pregnancy has been prohibited since 2013, but use is allowed for up to four weeks after mating. In Europe, only Sweden, Norway, and the UK had completely banned sow stalls by 2020. Other European countries are still passing regulations to ban these cages permanently (CIWF, 2020a). The four largest Brazilian meat companies (which process 60% of the country's pigs) promised to phase out sow stalls in their supply chain by 2026 (Azevedo, 2019). In the US, three out of four sows are kept in sow stalls, although several states have officially banned them in the last decade, and dozens of brands have pledged to abandon the practice (World Animal Protection, 2020).

Notably, many pig farmers, meat processors, and meat farmer associations see the ban as a marketing strategy to increase global competitiveness, under the guise of "animal-friendly brand." After all, ambivalent consumers are more comfortable eating animals whose mothers did not spend pregnancy in a cage. Furthermore, trade barriers related to the levels of animal welfare end up forcing certain countries to adapt to the standards of import markets. This is the case of Brazil, which exports over 568,000 tonnes of pork per year, equivalent to 15% of its production (Guimarães et al., 2017).

A few days before giving birth, sows raised in intensive farms are commonly moved from sow stalls to farrowing crates. Farrowing crates are approximately the size of sow stalls but have an attached enclosure from which the piglets can nurse through metal bars. The crates can also be integrated to heat pads, pens heated with incandescent light bulbs or other heat sources. The purpose of the heat pads is to ensure thermal comfort to the piglets, so that they gain weight faster, increasing the farm's productivity. Again, the mothers can only lie down or stand up.

Farrowing crates are illegal in Sweden, Norway, and Switzerland, but are widely present in pig farms in the UK, the US, Brazil, and China. Even in places where sow stalls are banned, such as the UK, farrowing crates are broadly used. In 2016, 60% of all 350,000–400,000 sows in Britain farms were kept in farrowing crates from a few days before birth until the piglets are taken away (Viva!, 2016). This means spending around three months per year in extreme confinement.

Pig farmers sustain that sow stalls are crucial to avoid aggressive behaviour among pregnant sows and towards the staff, while farrowing crates ensure a safe environment for mother and piglets, reducing competition for food with other pigs and avoiding that the sow smashes her babies to death. Nevertheless, pig farmers disregard that crushing happens primarily because mother and piglets are confined into tight crates in the first place. The livestock industry also chooses to ignore that aggressive behaviours are largely the result of artificial grouping of animals, social behaviour disturbance, and intense and prolonged stress.

Sows are prevented from performing their most basic natural behaviours, such as building a nest for their piglets in a more secluded area using straw, leaves, and other materials; providing proper maternal care to the offspring, including close contact; exploring, grazing, and roaming under the sunlight; toilet away from their food and water; mental stimulation; and socialising. Piglets are usually weaned after three to five weeks from birth and transferred to a nursery area. Meanwhile, their mother is injected with hormones (e.g., equine chorionic gonadotropin, human chorionic gonadotropin, and gonadotropin-releasing hormone) to induce oestrus and ovulation, preparing her body for a new gestation.

Both sow stalls and farrowing crates remain a standard practice worldwide in intensive farms, countries with national bans being the exception (Thring, 2012). The use of these crates is sustained by the global demand for pig meat, especially for *cheap* pig meat.

Importantly, the scientific literature states mortality rates are the same whether farrowing crates are used or not (KilBride et al., 2012). The number of piglets accidentally suffocated to death offsets deaths caused by other reasons, such as stillborn piglets due to stress. The true reason behind the broad use of these appalling crates is *profit*. The crates maximise the number of animals housed in a certain space, herd management is facilitated, and human labour is reduced. Additionally, animals that exercise less require less feed.

According to the US National Pork Producers Council vice president, veterinarian Paul Sundberg: 'Farmers treat their animals well because that's just good business. [...] Science tells us that she [a sow] doesn't even seem to know that she can't turn. She wants to eat and feel safe, and she can do that very well in individual stalls' (Kauffman, 2001). That is the level of ethics in animal agriculture.

Notwithstanding, animal scientists recognise that most welfare problems in the EU pork production are related to intense confinement, early weaning, intensive breeding, and failures during stunning and slaughter (Pedersen, 2018; EFSA, 2020). In other words, a food system based on animal exploitation constitutes a *generalised* animal welfare problem.

In the chapter 'Sow welfare in the farrowing crates and alternatives,' published in the scientific book *Advances in Pig Welfare*, the authors say: 'The key to success has been recognising that allowing the display of species-typical behaviours contributes to the biological fitness of the animal, which encompasses important economic performance parameters including number of offspring produced, viability of offspring, and maternal rearing ability' (Baxter et al., 2018).

Semi-intensive farms that offer outdoor access benefit from the better productive performance of sows that can manifest their characteristic behaviours. For example, females can share farrowing pens, where they can build nests for their offspring. However, the piglets are weaned at the same age as in intensive farms and the sows are impregnated again shortly thereafter. "Happy exploitation" does not exist.

Mutilations of various types are common on swine farms, regardless of the rearing system or herd size. For example, day-old piglets have their teeth clipped, or ground with pliers or rotating porous stones *without* anaesthesia. Tooth clipping aims to minimise biting injuries among piglets and to the sow's teats, which reduce the farm's productivity. Tooth clipping may involve the last third of the tooth or most of it, right up to the gums. The procedure causes severe pain, acute or chronic, due to the exposure of the dental pulp, and can lead to infections. Injuries to the tongue, lips, gums, and the tooth itself can also occur, especially when a larger portion is clipped and when the team is not well-trained. Moreover, being handled to have their teeth clipped is itself a significant source of stress for piglets.

Bite injuries occur mainly because the sow cannot get away from her piglets when they compete for her milk. In other words, the underlying problem is the confinement combined with excessively large litters promoted by genetic improvement, and not the presence of teeth in piglets, which obviously have an important physiological function. Tooth clipping is a *standard procedure* on commercial farms all over the world.

In Brazil, Directive 195/2018 of the Ministry of Agriculture, Livestock, and Supply prohibits clipping piglets' teeth with pliers but tolerates other instruments. In Brazil, US, and many other countries, there are no laws that completely prohibit the procedure. Private schemes, such as Certified Humane, discourage the routine teeth clipping, but still allow it if the need is justified. EU Directive 91/630/ECC of 2001 prohibits tooth clipping as a *standard* measure in pig farming. Producers should resort to other management techniques before using tooth grinding/cutting (e.g., reducing stocking densities), but the practice is still legal if a veterinarian recommends it. In reality, tooth clipping is standard practice in most countries, including the EU and the UK.

Tail docking is another common mutilation that aims to discourage injuries caused by "cannibalism" or tail biting. The final third of the tail (or even larger portions) is cut using a blade, pliers, scissors, or an electric welder. Alternatively, the tail can be *crushed* with blunt pliers or scissors. The crush will cause the tail to fall off without the need for cauterisation. Despite being terribly painful, the procedure is *not accompanied by anaesthesia or analgesics* if the piglets are younger than seven days old, whatever the method used. Improper procedures can become infected, resulting in abscesses in the legs and spine, causing severe pain. Secondary infections can occur in the lungs, kidneys, and other parts of the body (Viva!, 2016).

Tail biting is an economic problem for pig producers because the resulting injuries reduce the integrity of the carcasses. Once an animal has its tail injured to the point of bleeding, cannibalism escalates within the herd. As the upper portion of the tail is more sensitive to pain, pigs with a clipped tail escape faster when other animals bite them, reducing injury rates (Viva!, 2016). Therefore, tail docking is done in the vast majority of commercial pig farms worldwide. In the UK, for example, 81% of piglets have their tails docked (Rivera, 2017).

Tail biting is a typical sign of stress and boredom, especially in factory farms, since pigs are deprived of exploring and rooting, may suffer from thermal discomfort, are mixed with unfamiliar animals, inhale high levels of ammonia, and are forced to eat, defecate, and rest in the same place. Bites can also occur due to competition for food and lack of nutrients, as nutritional deficiencies lead to gastrointestinal disorders, which stress the pigs (Rivera, 2017; Viva!, 2016). In the wild, pigs do not bite each other's tails (Viva!, 2016).

The surgical castration of male piglets, a common practice in most countries, is done within seven to ten days of being born, typically *without* anaesthetics or pain relief. The main reason for castrating male piglets is to avoid the undesirable pork flavour called 'boar taint,' which is caused by testosterone and androsterone. Besides, castrated males exhibit less competition, aggressiveness, and sexual behaviour. The procedure is done by cutting off the piglet's tests with a scalpel (CIWF, 2020b). After castration, piglets show behavioural changes indicative of pain, such as prostration, trembling, spasms, stiffness, scratching the bottom, and avoiding certain postures.

In Europe, 70% of all males are castrated, accounting for approximately 90 million piglets every year. Only a few countries use both anaesthetics and analgesics, for instance, Lithuania and Sweden. Castration is banned in Switzerland since 2010. Other countries prefer to slaughter their pigs at a younger age than castrating them—for example, the UK, Ireland, and the Netherlands (CIWF, 2020b).

In the UK, surgical sterilisation is prohibited, but chemical castration (or immunocastration) is permitted. For example, a vaccine called IMPROVAC is administered to immunise the animal against gonadotropin-releasing hormone (Viva!, 2016). Immunocastration is also approved by the Brazilian legislation, although surgical castration is more common (Embrapa & ABCS, 2011).

In the first days of life, piglets are also identified by ear tagging, ear notching, tattooing, or microchipping. Notching, for example, consists of perforating the edges and tips of the ears with pliers. Each notch represents an identifying number. These procedures are performed *without* anaesthesia or analgesia, although they involve acute pain due to tissue damage (Viva!, 2016).

Nose ringing is another common practice in pigs allowed outdoor access to suppress rooting and digging, innate behaviours that destroy the fields. The procedure is normally carried out *without* anaesthesia or analgesia, in two main ways: (1) metal rings with sharp edges are clipped to the rim of the nose; or (2) metal rings are perforated between the subcutaneous fibrous tissue and the nasal septum cartilage, like the rings used in bulls. In addition to being physically painful, these practices are sources of emotional suffering (Viva!, 2016; Horrell et al., 2001).

Owing to selective breeding, manipulated diet, and artificial life conditions (e.g., absent or insufficient sun exposure), many piglets are born with deformities and congenital diseases, such as cleaved palate, hermaphroditism, inverted nipples, absence of anus, deformed legs, and hernias. Additionally, most of them are immunocompromised. In the first days postpartum, piglets can be supplemented to outweigh the poor milk quality resulting from intensive breeding and undernutrition. Sows are especially prone to mastitis and metritis during the peripartum, which may affect milk production negatively (Fangman & Amass, 2007). Iron supplementation of piglets by injection or oral route is standard on intensive farms, since breast milk supplies only 10–20% of their daily needs (Embrapa & ABCS, 2011).

As piglets are forcibly weaned much earlier than natural (3–5 weeks versus 15–17), they receive treatment to prevent diarrhoea caused by digesting solid foods.

Pre-starter feed is administered from six days of age, either in dry or soft form. The purpose of the pre-starter feed is to accustom the piglet's palate to the flavour of the feed, in addition to activating the early secretion of digestive enzymes. In addition to intestinal disorders caused by the artificial diet, piglets are very vulnerable to enteric diseases caused by microbiological contamination of the pre-starter feed (which is highly perishable) or of the premises. The main microorganisms involved are *E. coli, Clostridium perfringens,* and *S. suis,* which cause colibacillosis, clostridiosis, and coccidiosis, respectively (Embrapa & ABCS, 2011).

Deliberate overcrowding ensures low calorie use and, thus, higher feed-to-meat conversion rates. High stocking densities, poor ventilation, artificial diet, hard, slippery, and excrement-soaked floors, lack of sun and exercise, and other typical living conditions on commercial farms render pigs susceptible to various health problems. Infectious agents causing respiratory diseases in pigs include *Mycoplasma hyopneumoniae* (Mhp), influenza A virus (IAV-S), porcine reproductive and respiratory syndrome (PRRS), and porcine circovirus type 2 (PCV2). Suckling pigs are mainly affected by *Streptococcos suis* infection (which is transmitted to humans), colibacillosis, and navel infections (USDA, 2016a). *S. suis* infects the animal's tonsils and causes meningitis, which can lead to death (Viva!, 2016).

Some pigs are so fat for their size they have difficulty standing up. Lameness is one of the main problems in sows and finishing pigs, whether due to congenital diseases, fast growth, varying injuries, bacterial infection of the feet, inflammation due to laminitis, or excess weight, all of them causing pain and discomfort in the limbs (Viva!, 2016).

During their short lives, production pigs are given antibiotics, hormones, and feed supplements to ensure they survive and fatten up fast to the desired weight. In many leading producing countries, antimicrobials classified as growth promoters are widely administered through the feed, in low dosage over long periods, to improve the zootechnical performance of pigs (USDA, 2016a; Embrapa & ABCS, 2011). The antimicrobials not only treat infections, but also prevent diseases and protect against the production of toxins in the gastrointestinal tract, resulting in increased feed conversion and growth.

A 2016 USDA report details the use of antibiotics in swine farms nationwide (USDA, 2016a). The report includes data collected in 2006 and 2012 from pig production sites of all types: breeding, nursery, grower/finisher, and wean-to-finish. Here are some conclusions:

Nursery sites

- Over 80% had administered *injectable* antibiotics during the previous six months;
- Almost 90% had administered *feed* antibiotics during the previous six months;
- Two-thirds had administered *water-soluble* antibiotics during the previous six months;

Grower/finisher sites

- More than 80% had administered *injectable* antibiotics during the previous six months;
- Over 90% had administered *feed* antibiotics during the previous six months;
- Almost 75% had administered *water-soluble* antibiotics during the previous six months;
- Around half used antibiotics in feed for *growth promotion.*

Wean-to-finish sites

- More than 90% had administered *injectable* antibiotics during the previous six months;
- Almost 98% had administered *feed* antibiotics during the previous six months;
- Over 85% had administered *water-soluble* antibiotics during the previous six months;
- Approximately 40% used antibiotics in feed for *growth promotion.*

The EU-wide ban on the use of antibiotics as growth promoters came into effect in 2006. However, they are still widely used under the pretext of disease prevention (Viva!, 2016). In the UK, half of the antibiotics sold are consumed by the livestock sector, with approximately 60% of these used in pigs (Viva!, 2016).

To be noted is that both small farms and factory farms use cruel practices routinely, as the animals are used as *production units*. Small-scale and even subsistence finishing farms will buy animals from other properties at some point, and these generally raise the animals under extreme confinement. And obviously, no one consumes animal products exclusively from "backyard" farms. Using feeling beings as food when there are abundant healthy food sources is unnecessary violence, whether we are talking about small farms or industrial-sized properties.

Worth noting, large-scale pig farms are predominant all over the world. The UK, for instance, has nothing less than 800 mega-farms—each with warehouses containing more than 750 breeding sows, 2,000 pigs, or 40,000 birds (Wasley et al., 2017). In China, intensive hog production is increasing, with farms as large as 7 hectares housing 10,000 sows and producing 280,000 piglets a year (Patton, 2018).

Insemination methods in swine farming do not differ significantly from that of cattle farming. Females are forcibly impregnated shortly after farrowing, either by mounting or artificial insemination. Natural breeding is still used, but artificial insemination is preferred due to the higher insemination rates and herd genetic improvement (Knox, 2016). In intensive farms, which constitute the majority of pig farms worldwide, the females are mounted by a male or are artificially inseminated inside their own crates. The females include weaned sows, sows that aborted or were not successfully inseminated, and gilts (young sows), who are inseminated from 150 days of age (Embrapa & ABCS, 2011).

The boars, or breeding males, are kept in individual pens for weeks during the breeding period. They frequently undergo several clinical examinations and andrological tests to verify their reproductive potential. In artificial insemination programs, boars are trained to mount a dummy sow to have their semen collected periodically.

As in cattle reproduction, sterile male pigs can be used to both stimulate and synchronise the oestrus among females in farms that use artificial insemination. The oestrus can also be synchronised within the herd through the administration of sexual hormones. Some of them (for instance, prostaglandin F2 alpha) are deliberately prescribed to induce abortion so all sows get pregnant at the same time, facilitating husbandry procedures (May & Bates, 2007).

Generally, sows are inseminated with a pipette containing semen twice a day during heat. It is not uncommon for physical injuries and infections to the genital tract to occur as a result of mishandling and microbiological contamination. A special diet is administered throughout all stages (oestrus, mounting, pre- and post-insemination, and gestation) to guarantee the maximum performance of the sows (Embrapa & ABCS, 2011). Periodically, gilts replace spent sows.

When they reach 7–8 kg, at 21–28 days, the piglets are weaned and transferred from the farrowing crates to the nursery area. Diet, ambience, and other factors are controlled. The goal is to smooth the transition so that weight gain and feed-to-meat conversion are optimised. But even using proper husbandry, the babies experience physical and psychological suffering. Piglets are separated from their mothers and introduced into new groups, resulting in fights and injuries; they must adapt to a diet based mainly on corn and soy; they may suffer from dehydration, as they no longer have breast milk (composed of approximately 80% water) and must learn to use drinkers and feeders (Embrapa & ABCS, 2011).

After about 42 days in the nursery stage, piglets weighing 18–25 kg move on to the growth phase, where they are fattened up to 50–60 kg, at just over 2 months old. In intensive systems, they will be raised in sheds with about 20 pigs per pen to allow for monitoring of sick animals. Finally comes the finishing stage, where the animals are housed in sheds containing tens to hundreds of other pigs until the final slaughter weight (100–130 kg). The finishing phase takes an average of 112–119 days. Diet is one of the most important factors at this stage, corresponding to 70–80% of breeding costs (Embrapa & ABCS, 2011). If the piglets have not been surgically castrated in the first week of life, immunocastration is applied in the growing or finishing phases, between 70 and 100 kg.

While most pork and bacon come from six-month-old pigs, suckling pigs are slaughtered between two and six weeks old for their tender meat.

Pre-slaughter is another critical stage because pigs are easily stressed during handling and transport. As stress can lead to physiological changes (e.g., increased heart rate, blood pressure, and body temperature, panting, and liberation of hormones into the bloodstream) that affect meat quality, scientists examine ways to increase animal welfare levels during pre-slaughter (Velarde & Dalmau, 2018a). Again, genuine animal welfare is not the focus, but rather the quality and safety of the final product.

In real life, rough handling and loud noises are commonplace during loading onto transport trucks and ships. Psychological stress due to social and physical environment changes, hunger, thirst, and fear are aggravators. Pigs are fastened for at least 12 hours before transport to reduce mortality from suffocation when regurgitating food, decrease waste production, and prevent contamination of the carcass with intestinal or stomach contents during slaughter and evisceration (EFSA, 2020).

The animals are crammed into lorries or ships, where they can compete for space and suffer from fatigue, high body temperatures, and water and food deprivation. Thermal comfort can be severely affected not only by the adverse climatic conditions (ventilation, humidity, temperature) and the high loading density, but also the impossibility of lying down, a natural thermoregulatory behaviour in pigs. Sudden death from heart or respiratory failure is widely known in pig transport. Pre-existing problems, such as pericarditis and endocarditis, common due to excess weight, contribute to heart malfunction. Fractures and varied injuries are also frequent. Losses of up to 1% are observed during transport of pigs (Dalla Costa et al., 2019).

Worth mentioning, the transport to the slaughterhouse is not the only time the pigs travel. Piglets can be transferred to finishing farms, while sows and boars can be transported from breeding farms to commercial farms.

Traffic accidents are frequent in livestock transport. In 2019, a truck carrying 238 pigs from Taber to Langley, in Canada, was involved in an accident. The driver and two more people tried saving as many animals as they could by moving them to pasture. Seventy pigs died, 60 of which suffocated to death due to overcrowding in the vehicle, and the other 10 were put down at the scene. The pigs had been loaded in Taber on a Thursday morning and were reloaded onto another lorry on Friday afternoon (Potenteau, 2019).

In 2020, a truck carrying 170 pigs on their way to an abattoir in Cuenca, Spain, overturned near Madrid. While many of them died immediately, others were seriously injured. Despite their obvious suffering, the wounded animals were completely ignored by the transport company's workers. They only opened the truck five hours after the accident. Some wounded animals were slaughtered *in situ* (Animal Equality, 2020). Animal Equality was present at the scene:

> Deafening shrieks could be heard for two hours after the crash before finally going silent as the hurt pigs succumbed to their injuries. It took hours for transport company officials to open the truck to check on the pigs, with the neglect displayed by those officials frustrating our team on the ground. We pleaded for someone to help the injured pigs, but those requests were ignored. Sadly, we filmed dead pigs—sometimes two-by-two—unceremoniously hoisted from the wreckage by crane and dumped into another truck. We also filmed one injured pig who had escaped, languishing tied-up by the side of the road. Our team watched in horror as the surviving pigs were ushered into a new transport truck, struck and kicked by workers to force them inside, and again taken to be slaughtered. (Animal Equality, 2020)

At the abattoir, pigs must be rested in lairage pens with access to clean water for two hours, where they can settle from the stress and activity—and ensure optimal meat quality (Velarde & Dalmau, 2018a). In the words of meat scientists Santé-Lhoutellier and Monin (2014), 'the best meat quality is achieved when the pigs are calm when they are being herded and stunned.' This reminded me that the only time we discussed animal welfare during my degree in Food Engineering was when we studied quality defects derived from animal stress, a.k.a. PSE (pale, soft, exudative) and DFD (dark, firm, dry) meat.

Lairage also ensures faster and more efficient slaughter, as the processing line can operate at a steady animal supply. Mortality during transport and lairage is widely known among pig meat farmers/processors and livestock scientists. 'Pigs should not be mixed with unfamiliar animals during transport and in lairages to minimise fighting and bruising. Apart from the direct loss associated with death during transport, bruising can result in increased carcass trimming or carcass condemnation' (Channon, 2014).

Suffering continues all the way between the lairage pen and the stunning area. 'Most of the time, animals are forced to move quickly in the last metres prior to stunning to maintain the chain speed. Vocalisation is associated with electric prod use, excessive pressure from a restraint device, stunning problems, or slipping on the floor' (Velarde & Dalmau, 2018a). Staff yelling and improper handling increase their fear and pain.

Pigs are stunned before slaughter to render them unconscious while they are hung by the hind legs and exsanguinated to death. The stunning stage is part of the "humane slaughter." Electrical stunning and CO_2 stunning are mostly used. To promote cardiac arrest, the animals can be stunned exclusively in the head, or alternatively pass an electrical current (EFSA, 2020).

High-speed processing lines reliant on underpaid, long-hour workers commonly result in improper stunning, with the animals remaining conscious and in extreme pain while they bleed to death. A recent report by the European Food Safety Authority (EFSA) stated that 29 out of 30 animal welfare issues during the slaughter of pigs are ascribed to inadequate training and fatigue by the workers. Wrong placement of head-on electrodes and short exposure time are typical sources of failed stunning (EFSA, 2020).

In CO_2 stunning, inadequate gas concentration, short exposure, and chamber overloading are common causes of unsuccessful loss of consciousness. According to meat scientists Matarneh et al. (2017), 'Most stunning methods will allow for the animal to regain consciousness if not exsanguinated shortly thereafter.' After stunning, pigs are usually bled to death by chest sticking. Basically, a knife is inserted into the animal's thoracic cavity, and the carotid artery and jugular vein are severed (Velarde & Dalmau, 2018a).

Slaughter can be done *without previous stunning* in religious meat production and on-farm slaughter, not to mention in clandestine abattoirs (Velarde & Dalmau, 2018b). Mechanical stunning by brain concussion, which is common in small slaughterhouses, is done with a captive bolt, free projectiles, or a blow to the head with a hard object. Due to their anatomical characteristics, pigs are amongst the most challenging animals to shoot in a way they lose consciousness (EFSA, 2020).

After bleeding, the carcasses undergo scalding, dehairing, and singeing. Evisceration takes place after these steps.

Just like in the dairy and beef industries, the slaughter of pregnant pigs is typical when keeping them alive is no longer deemed economically viable. EFSA's report entitled 'The animal welfare aspects in respect of the slaughter or killing of pregnant livestock animals' (EFSA, 2017) sets regulations for rendering unjustifiable violence a bit less emotionally and physically painful to the mother and her young. A complete paradox.

Organic meat, free-range meat, stall-free pork, and other labels

Organic meat

Although regulations for organic foodstuffs can vary geographically, organic accreditation encompasses the absence of antibiotics, chemical fertilisers, herbicides, pesticides, and GMOs. Animal welfare and environmental requirements also apply, such as outdoor access, adequate stocking densities and housing conditions, prohibition of some painful husbandry procedures, waste disposal, pasture management techniques, and others.

In the UK, organic meat refers to meat from animals that were provided outdoor access throughout the whole year, are fed on pasture at least 120 days per year, are given 100% organic feed, containing no GMOs, and have not been administered with growth promoters and antibiotics. Tail docking, nose ringing, and farrowing crates are banned by the non-profit Soil Association, which provides organic farming certification. Pigs must be allowed to root and dig, both outdoors or when housed, and therefore clean rooting substrates such as hay, grass, and alfalfa must be provided at all times. Calves cannot be kept in individual pens after seven days from birth. These and other housing, transport, handling, and slaughter standards must be met (Soil Association, 2020).

As livestock treated with certain substances (e.g., antibiotics) cannot be slaughtered for organic meat, the farmers must put preventive measures in place to ensure the animals are healthy. In the US, there are three additional requirements: animals must be raised as organic at least from the last third of gestation; a minimum of 30% of their feed must come from pasture; and temporary confinement is allowed only in the case of documented environmental or health issues (USDA, 2016b).

In Brazil, producers cannot use chemical fertilisers anywhere on organic farms and are prohibited from using fire in pasture management. Reproduction must occur by natural breeding or artificial insemination, with the use of embryo transfer and *in vitro* fertilisation being forbidden (Soares et al., 2014).

The EU organic standards further prohibit the administration of synthetic amino acids and animal cloning. In addition, suckling mammals must be fed preferably with maternal milk and the feed must come from the farm region whenever possible. Other requirements also apply for EU organic accreditation, such as trained personnel and a limited number of animals to minimise soil degradation from manure and overgrazing.

In practical terms, accredited organic meat comes from animals that lived under more humane conditions. Nevertheless, the animals are still enslaved, forcibly impregnated, and subjected to unavoidable stressful and traumatic experiences in an essentially *exploitation system*. Painful husbandry procedures (e.g., tooth clipping), early weaning, transport, and the slaughter itself are part of the routine of organic farms as well.

To be noted is that ranchers are attracted to organic farming for the economic return, as organic production tends to be more profitable (Soares et al., 2014).

Free-range pork

Free-range pork comes from pigs that were allowed free outdoor access for at least part of the day and were not confined to sow stalls or farrowing crates throughout their lives. Behavioural and physiological needs of the animals are better met because they can engage in natural behaviours, such as nesting, exploring, roaming, and grazing. However, free-range is not a legal term for pork, meaning that *the amount of time* the animals spend outdoors, how big the outdoor area is, and the livestock density in indoor pens are not controlled. Usually, husbandry practices are not regulated, and the animals are still subjected to cruel practices, such as castration of males without anaesthesia, culling of females with low productivity and health problems, ear notching and nose ringing without pain management, using androgenised or sexually mutilated males to induce oestrous, and so on.

In the US, the USDA free-range regulations apply exclusively to poultry. In Brazil, free-range farming is more common in small family properties, many of them aimed at subsistence. As in the US, the term 'free-range' is not subjected to government regulations.

In Australia and the UK, free-range pork is regulated by the Royal Society for the Prevention of Cruelty to Animals (RSPCA) Approved Farming Scheme (RSPCA Australia, 2018; RSPCA Assured, 2020). Nonetheless, meat carrying the RSPCA Assured label does not necessarily come from free-range animals as consumers understand the term. While the use of farrowing crates and sow stalls is prohibited, the animals can still be housed indoors, within restrictive spaces. Total minimum lying areas for growing pigs are 0.22 m^2 for 10 kg live weight and 1.03 m^2 for 100 kg. For reproductive females, minimum lying areas are 1.6 m^2 for gilts and dry sows and 4.3 m^2 for farrowing and lactating sows (RSPCA Australia, 2018). This means the animals can still live under confinement part or most of their lifetimes.

Furthermore, tail docking and teeth clipping are permitted by RSPCA, if not practiced routinely, and piglets are still weaned as early as 3–4 weeks of age, so the sow can be impregnated again. Surgical castration is not permitted, but chemical castration is (RSPCA Australia, 2018; Viva!, 2016).

Worth mentioning, the standards behind free-range pork labels vary between countries and the real practices may differ between farms. Besides, law enforcement is nonexistent, as this label has no legal standing.

Sow stall-free pork

In theory, sow stall-free pork refers to meat from pigs born to sows in group housing. In practice, however, sows can still be kept in cages for a certain period, which varies between countries and producers. The Australian industry, for example, defines as 'stall-free' sows confined in 'mating stalls' for five days from last mating to one week before giving birth. Mating stalls are claimed to prevent injury or abortion due to aggressive behaviour between sows. The use of such stalls can mean up to three months of confinement *each gestation*. Note that mating stalls have the same dimensions and function as gestation stalls—they are simply named differently (RSPCA Australia, 2020).

In addition, stall-free pork does not mean *farrowing crates* are not used. The sows can be confined in farrowing crates for up to five weeks, from about a week before giving birth until weaning. Sow stall-free pork is one of the labels accredited by RSPCA in Australia (RSPCA Australia, 2020).

In Brazil, the Certified Humane label prohibits sow stalls. Sows must remain in a 1.8 m by 2.4 m farrowing area for at least 28 days after the offspring are born. The area must be clean and contain such bedding material as straw (Certified Humane Brasil, 2021). However, mothers and piglets remain confined.

In 2020, Australian supermarket chain Coles had been advertising their 'Own Brand' sow stall-free pork, conveying a flawed idea that ethical problems related to pork-eating are over. Cole's Own Brand also bans artificial growth promoters. Although these are, in fact, important advancements, especially for a major supermarket chain, they pale into insignificance compared to the whole suffering entailed in pig meat production (and animal use in general). While it is true that sows will be group-housed in sheds instead of stalls during their pregnancy, sow stalls *can still be used* for five days between weaning and mating each lactation. Besides, sows and piglets are still confined to a farrowing crate from birth to weaning (Humane Choice, 2020).

Grass-fed meat

Grass-fed meat generally requires that pigs and cattle are fed on grass for more than half their lifetime (The Courtyard Dairy, 2017). The remainder feed can be cereals and grains. But here is the ace in the hole: grass-feeding does not mean the animals are eating grass out at pasture—they can be fed *indoors* with silage (fermented grass) or hay (dried grass). Furthermore, grass-fed labels regulate exclusively the diet the animals receive. For example, the 'UK Pasture for Life Scheme' requires that the livestock must be fed 100% on grass and that the breed of animals matches the available grassland. However, no requirements apply for handling, housing, transport, and slaughter (The Courtyard Dairy, 2017).

In countries like Brazil and Australia, the favourable climate allows a good part of the cattle to be raised or fed on pasture. But this does not apply for pigs—most of them live under extreme confinement.

Outdoor reared and outdoor bred pork

Outdoor reared and outdoor bred are unregulated labels used by some producers and retailers worldwide. UK's Waitrose supermarket chain claims all their 'Waitrose Essential' bacon, pork, ham, and sausage are sourced from outdoor bred pork.

Although these terms have no legal standing and, therefore, no law enforcement, they are described as: (a) outdoor reared pigs are reared in outdoor systems for approximately half their lives, while the sows remain outdoors their entire lives; and (b) outdoor bred pigs are born in outdoor systems but raised indoors after weaning, while the sows remain outdoors. Both labels permit tooth clipping and tail docking (Rivera, 2017).

Red Tractor pork

Red Tractor, the UK's largest food standards scheme, certifies 85% of pig farms in the UK (Rivera, 2017; Red Tractor, 2017). The Red Tractor labels for meat exceeds minimum animal welfare legislation, for instance banning surgical and chemical castration of pigs, and also include provenance criteria. Meat from the EU can also be labelled when meeting equivalent standards.

In Red Tractor's website, we find the following statement:

> [...] we have partnered with animal welfare experts and vets since 2000 to continuously strengthen our requirements on animal welfare. Together we work to ensure that our animals are healthy with the right living space, food and water. Our members include award-winning farmers who share our priorities to strive for the very best in animal health and well-being. (Red Tractor, 2017)

Nevertheless, farrowing crates, tooth clipping, nose ringing, and tail docking are permitted.

The space criteria follow the minimum requirements from England's 2007 Welfare of Farmed Animals Regulations. Although sow stalls are banned, the indoor pens that house sows and gilts (female piglets) can measure as little as 2.8 m lengthwise. Pregnant gilts and sows are allowed an area of 0.95 m^2 and 1.3 m^2, respectively, throughout their whole lives. After giving birth, they are further confined for up to five weeks in a farrowing crate. Finishing pigs (85–110 kg) are entitled to a minimum area of 0.65 m^2 per animal. Pens can have concrete-slatted floors, without bedding material (Viva!, 2020; Red Tractor, 2017). In other words, the animals are still confined and deprived of their natural behaviour.

The Red Tractor industry scheme has already accredited tens of thousands of UK pig farmers. While their meat is labelled, they were not required to make major changes to the way they raise the animals. Moreover, farmers are not prosecuted or even fined in case they fail to meet the scheme's standards (Viva!, 2020).

In 2015, a Red Tractor farm supplying pig meat to Morrisons supermarket was investigated by the 1994-founded charity Viva!. The investigation, published in the *Daily Mail*, revealed scenes of piglets living in extreme confinement in three-tiers-deep battery cages. Another Red Tractor farm featured farrowing crates, low hygiene standards, and even rotting piglets (Viva!, 2020).

Certified Humane Raised and Handled® pig meat

The Certified Humane Raised and Handled® label guarantees that pig meat and eggs meet strict animal welfare standards, according to the international certification program Humane Farm Animal Care. The non-profit organisation counts with a technical committee composed of scientists and veterinarians and is endorsed by the American Society for the Prevention of Cruelty to Animals. Diet, housing, bedding, handling, transport, and slaughter standards are periodically revised and translated into several languages. Accredited farms are found in Argentina, Australia, the Bahamas, Brazil, Canada, Chile, Colombia, Costa Rica, India, Jordan, Malaysia, Mexico, New Zealand, Peru, Singapore, the US, and Uruguay.

For example, sow stalls and farrowing crates are banned in pig farms. Total floor space and lying area must measure at least 3.5 m^2 and 1.5 m^2 for each sow (Certified Humane, 2020). Although these measures improve the quality of life of farmed animals, they are still regarded as a property from which to make a living. Imagine living in a comfortable cubicle where you are warm, well-fed, and have enough space to turn around. You have one pregnancy after another, and your children are taken away to become food. When your productivity declines, you are killed.

Can you see the difference between animal rights and animal welfare?

When the cure is worse than the disease

In this chapter, we have discussed some of the welfare labels related to red meat. You might have noticed it is easy to get confused about their practical meaning. The lack of standardisation among labels stems from the absence of strict legislation against animal exploitation in general—a reflection of our society's values. Animal welfare labels are mostly regulated by animal protection organisations and industry schemes, based on *voluntary* joining. The meat industry is probably concerned about losing a slice of the market were they not to conform to the incipient consumer's awareness.

In fact, when Red Tractor CEO David Clarke was interviewed by *Independent Ireland* on the advantages of outdoor bred pork, he declared: 'The advantages are some people are prepared to pay more money for it because they think it's better welfare' (Rivera, 2017).

The meat industry is gaining from our ambivalence. Do we consumers support animal abuse or not? If we care about animals, the last thing we should do is eat them.

Whether the welfare schemes are enforced or not, we must understand that animal suffering is *inseparable* from animal agriculture. Just like us, animals hold their lives in great esteem. They want to live and fulfil their purposes. They love and care for their children. Animals have physiological needs, emotional lives, and a rich intellect. Their body, milk, and eggs belong to themselves. And even if they were completely deprived of emotions and thoughts, they would still feel pain and discomfort. Objectifying sentient beings is unfair.

I understand people mean well when they pay more for meat carrying organic, free-range, and other labels. I value their effort, but we can and *must* do better. If our money is our vote (and it truly is), why pay for *lesser* suffering if we can simply say no to any amount of abuse?

Furthermore, people cannot control the source of their meat at all meals—or "higher welfare" meat is the only one they eat? Bear in mind that in the UK, where farmers brag about having the highest welfare standards in the world, 60% of nearly 400,000 sows are kept in farrowing crates until weaning (Viva!, 2016).

Interestingly, the meat industry argues against itself when organic and other labelled farms claim their product is more ethical (and generally more expensive). These farmers *recognise* the expression of natural behaviours by their animals is as important as having their physiological needs met. The meat industry further argues that higher animal welfare standards result in more nutritious meat. Well, if the meat industry itself admits that improved farming practices benefit the animals and their products, why do intensive, large-scale factory farms defend the animals are perfectly fine as long as they are safe and well-fed?

In addition, meat production is highly resource-intensive and severely detrimental to the environment, regardless of the production system. If we were to produce "ecologically friendly" meat within higher animal welfare standards, we would need even larger areas to provide adequate space for farmed animals and grow organic feed. This would entail further deforestation, enteric emissions, and resource use. When we choose the wrong path, the cure is worse than the disease.

In a meta-analysis published in *Environmental Research Letters*, US researchers Dr. Michael Clark and Dr. David Tilman found that grass-fed beef not only had higher land use requirements, but also 19% higher GHG emissions than grain-fed beef. The results are ascribed to the lower macronutrient density and digestibility of grass, as well as the lower feed-to-muscle conversion rates of grass-fed beef. As grass-fed animals raised outdoors grow slower, they are slaughtered 6 to 12 months older, giving rise to more enteric fermentation–driven emissions. The study included 742 agricultural systems and over 90 food products (Clark & Tilman, 2017).

To be noted is the fact that higher welfare husbandry increases production costs, which ranchers (especially small-scale ones) sometimes cannot afford.

Red meat is red with blood, slavery, and suffering. It is not sustainable, it is not healthy, and it is not (*and will never be*) ethical.

Grazing our forests

There might be no sea ice left in the Arctic by 2035, according to a *Nature Climate Change* study (Guarino et al., 2020). This result is in line with another scientific paper published in the *Geophysical Research Letters* that predicts the Arctic Ocean will be sea ice–free during summer before 2050 (Notz & SIMIP Community, 2020). This research shows how fast the climate crisis is escalating, despite our denial. As global rising temperatures melt the glaciers, the resulting melt ponds absorb more solar energy than ice due to the lower reflectivity of water. The outcome is further global warming, ice melting, and sea level rising.

The ice sheets of Antarctica have lost an astounding 2.4 ± 1.3 trillion tonnes of ice from 1992 to 2017 (Shepherd et al., 2018). The ice loss rate increased from approximately 66 billion metric tonnes/year in 2007 to 199 billion metric tonnes/year in 2017. By keeping this rate, sea levels can increase by more than 15 centimetres by 2100, considering the sole contribution of Antarctica's ice sheets (Shepherd et al., 2018). If all 27 million km^3 of ice contained in Antarctica's ice sheets were to melt, global sea level could rise by 58 m (Fretwel et al., 2013).

Sea ice melting and other climate events can kill people, destroy ecosystems and coastal areas, exterminate species, and trigger severe economic crisis, hunger, and refugee fronts. For instance, the 2003 heatwave that affected Europe resulted in crop losses valued at 13 billion euros and 70,000 deaths (Foer, 2019). In July 2021, record-breaking temperatures of 54.4°C and 49.6°C affected Death Valley (California) and British Columbia (Canada), respectively, resulting in fires, hospital visits, and even deaths (Clarke, 2021). Events like this will become ever more common.

Currently, the two primary sources of anthropogenic carbon emissions are the burning of fossil fuels and animal agriculture. Animal agriculture, however, poses other serious environmental and social threats beyond GHG emissions, as mentioned previously (e.g., water scarcity, deforestation, extinction of pollinisers, ocean acidification, environmental contamination, soil degradation, diseases, and destruction of marine life). In 2018, 50 hog lagoons flooded in North Carolina, US, due to Hurricane Florence. The overflooded water contaminated crops and mixed with rainwater, carrying dangerous manure-originated bacteria into the soil and underground water. The North Carolina Department of Agriculture and Consumer Services prohibited the use of the crops for animal and human consumption on the grounds of health hazards. Almost a million households were advised to not consume water from private wells (Davis, 2018). Notably, hog lagoons, as well as cattle slurry, are significant sources of atmospheric methane (Petersen et al., 2016). If governments are serious about the Paris Climate Accord, they must tackle animal agriculture.

Although 85% of human-made CO_2 emissions originate from the burning of fossil fuels, climate change is not limited to *current* CO_2 emissions or to CO_2 emissions alone—it is caused by *cumulative* GHG (mainly $CO_2 + CH_4 + NO_2$) and aerosol emissions. Land-use changes (primarily agriculture-driven deforestation) are responsible for the remaining 15% of CO_2 emissions (CDIAC, 2017).

As already mentioned, but worth stressing, animal agriculture is the leading source of methane emissions, contributing 37% to them (Gerber et al., 2013). Methane can trap 28 times more heat in the atmosphere than carbon dioxide over 100 years (Shindell et al., 2009). To make things worse, animal agriculture is the primary emitter of N_2O, which traps 296 times more heat than carbon dioxide over

100 years. Nitrous oxide is mainly emitted by manure, enteric fermentation, and chemical fertilisers, which are widely used in the cultivation of monoculture feed crops (Koneswaran & Nierenberg, 2008; Shindell et al., 2009). Finally, sulphate emissions, which *cool* the globe, result primarily from the burning of fossil fuels.

It is the combined effect of these main emissions over time that causes climate change. Rao (2021) integrated the annual CO_2 emissions from land-use changes and fossil fuel burning over 8,000 years, as land-use changes have been occurring ever since. CO_2 emissions from land-use changes (1,678 gigatonnes) surpassed those of fossil fuel sources (1,089 gigatonnes). Therefore, land-use changes are in reality the leading source of CO_2 emissions within this timeframe (Rao, 2021).

No wonder why several months of worldwide lockdown in 2020 contributed to a negligible cooling effect of 0.01°C by 2030, according to a study published in *Nature Climate Change* (Forster et al., 2020). The authors estimate that green energy policies and fossil fuel disinvestment could avert a global warming of mere 0.3°C by 2050.

We must also consider that the livestock sector occupies an area the size of Africa, which stores little carbon. A further area equivalent to South America is currently used to grow crops, mostly to feed animals that will become food (Anthropocene, 2013). By eliminating animal agriculture, we can further sequester carbon via regeneration of the soil and vegetation. The 37% share of the global land area used exclusively for grazing stores just 2% of the world's land carbon (Rao et al., 2015). This is what experts call Carbon Opportunity Cost (Searchinger et al., 2018).

Based on these assumptions, animal agriculture is estimated to be responsible for a staggering 87% of annual anthropogenic GHG emissions (Rao, 2021).

A projected 1,814 gigatonnes of CO_2 could be sequestered by transitioning the world to a plant-based economy, where animals are not used for food nor any other products (Rao, 2021). Such a restored economy would bring about important outcomes other than climate healing: habitat and biodiversity regeneration and lesser dietary-related diseases, resulting in lower mortality rates and reduced health costs.

Röös and Nylinder (2013) discuss how the estimation of carbon emission from livestock systems is highly uncertain, 'since most emissions arise from biological processes that are difficult to control and model, and there is high variability in management practices, climate conditions, and soil characteristics.' But again—even if carbon emissions attributable to animal agriculture are considered to be as "low" as 14.5% of global GHG emissions, as estimated by FAO, ignoring this is irresponsible and unintelligent, to say the least.

Unquestionably, we must reconsider our energy sources and work towards a "clean energy" economy. Nonetheless, it is much smarter and faster to tackle our diet first and then resolve the complex issue of fossil fuel burning (Rao, 2021). In fact, according to the analysis of Rao (2021), the global warming impact of our annual methane emissions is nearly three-fold higher than the impact of our annual carbon dioxide emissions. Not to dismiss is the fact that a shift away from oil, natural gas, and coal will bring technological challenges and environmental impacts in the short term. We will need to convert all our machinery so they can operate with alternative energy sources. Vehicles, industrial and household equipment, and building heating systems will have to be adapted to more renewable energy sources, such as solar, wind, tidal, and biofuels. Many will have to be completely replaced. This retrofit will require technology advancements, significant investment, and disruption of our lifestyle to some extent. This transformation cannot happen overnight. But dietary change can.

According to the IPCC, we have a carbon budget of 226.8 gigatonnes of carbon (counted as of 2013) if we want to keep average global warming below 2°C in relation to pre-industrial levels (IPCC, 2013). If our current 9.1 gigatonnes/year emission rate persists, our budget will be reached by 2038. Please note that this estimate is much more conservative than those of other authors, such as Rao (2021). Most importantly, even if we managed to limit global warming to 2°C, this scenario would be a *complete disaster*. For example, 2°C would be enough to raise sea level by half a meter, forcing millions of people to migrate. Big cities such as New York, Dhaka (Bangladesh), and Karachi (Pakistan) would become uninhabitable. Over half of all animal and plant species would go extinct, 20–40% of the Amazon rainforest could turn into savannah, and crop yields would decrease considerably (Foer, 2019).

Suppose we want to minimise the damage ascribed to climate change, rather than only ensuring our survival. In that case, we need to limit the global warming to 1.5°C—this means our budget decreases from 226.8 to 34.5 gigatonnes of GHG. Still, it would be a very difficult world: heat waves killing people and animals, unstable weather destroying crops, and species extinction at alarming levels. Yet, the meat and dairy industries project to *expand* their production. If their emission rates persist, the livestock sector alone will emit more than 27.2 gigatonnes of carbon by 2050 (IATP & GRAIN, 2018). Yes, you read it right—27.2 gigatonnes out of the disastrous 34.5 gigatonnes of CO_{2eq}. This means that if *all industries* were completely shut down, except for the meat and dairy sectors, we would still reach 1.5°C at some point by 2050.

Climate change is a *fixable problem*, and virtually everyone can be part of the solution. In an interview in November 2020 with Roger Hallam, one of the founders of Extinction Rebellion, climate expert Dr. Peter Carter said: 'We are in an emergency of our climate and an emergency of our oceans. [...] It's now the survival of our children. Not of our grandchildren anymore... This is now an existential threat.' He emphasised: 'The most effective, definitively effective, immediately effective, readily doable action that everybody in the world can do is go vegan. In theory, we could all do that. If we do that, emissions will drop immediately' (Extinction Rebellion, 2020). Dr. Carter, M.D., has served as IPCC reviewer in several reports and co-authored a book titled *Unprecedented Crime: Climate Science Denial and Game Changers for Survival* along with Elizabeth Woodworth. He is also founder of the Canadian Association of Physicians for the Environment and the Climate Emergency Institute.

According to Clark and Tilman's analysis cited in the previous section, ruminant meat (beef, goat, and lamb) has impacts 20–100 times higher than plant crops per kilocalorie of food produced, considering GHG emissions, land use, energy use, acidifying potential, and eutrophying potential. The impacts of pork, poultry, milk, eggs, and seafood production are 2–25 higher than for plant-based foods (Clark & Tilman, 2017).

Researchers Poore and Nemecek reached similar conclusions: 'Meat, aquaculture, eggs, and dairy use ~83% of the world's farmland and contribute 56–58% of food's different emissions, despite providing only 37% of our protein and 18% of our calories.' There is a wide variation in the environmental impact of the same food across producers, which represents a great mitigation opportunity. However, animal products proved to be the least sustainable foods in virtually all five environmental indicators (land use, different types of emissions, and water scarcity), but especially carbon emissions. The authors evaluated 40 products, which represent around 90% of the global protein and calorie consumption (Poore & Nemecek, 2018).

Measures to mitigate the impact of food production range widely from fertilising methods (e.g., organic versus synthetic) through pasture management to agronomic practices (e.g., monoculture versus polyculture, intercropping, catch crops, enhanced irrigation, among others). In the context of beef production, though, the authors show that the primary carbon emissions source is the animals' enteric fermentation. Figure 15 depicts how the carbon footprint of meat production is inherent to the animals' biology, taking as example beef farms with below median GHG emissions (Poore & Nemecek, 2018).

In fact, Clark and Tilman reported that 52% of GHG emissions in grain-fed beef systems result from enteric fermentation. This number rises to 61% in grass-fed beef (Clark & Tilman, 2017).

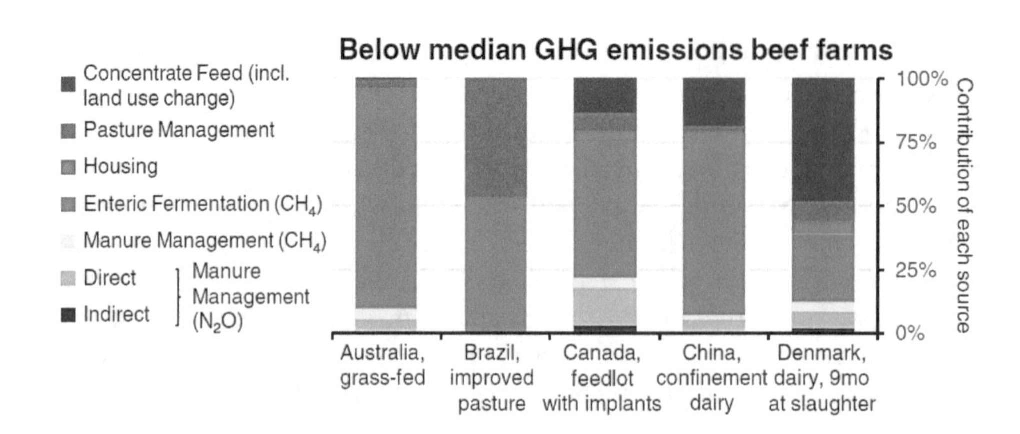

Figure 15. Contributions of emission sources for different beef producers. Source: Poore & Nemecek (2018).

And even though animal agriculture studies have been focusing on GHG reduction through pasture and manure management, production systems, and even methods to limit enteric fermentation, we are talking about curbing a *single* environmental impact. Others must be addressed too. See the example of forest slash-and-burn, widely used throughout much of the world for agricultural purposes. The vegetation is cut down with saws or using chains dragged by bulldozers and then burned. Typical examples of large-scale slash-and-burn are in the Amazon basin to clear land for animal agriculture and feed crops, and in Indonesia to replace forests with palm trees for oil production.

According to official data from Terra Brasilis, a deforestation monitoring tool developed by Brazil's National Institute for Space Research (Inpe), 10,851 km^2 of vegetation were destroyed in the Legal Amazon in 2020 (Terra Brasilis, 2021). Legal Amazon is a socio-geographic division in Brazil, covering approximately 5,200,000 km^2, that encompasses all 9 states in the Amazon basin. The Amazon basin comprises 40% of the world's remaining rainforest and 25% of terrestrial biodiversity (World Bank, 2019). Legal Amazon covers 3 different biomes: all of the Brazilian Amazon, 37% of Cerrado, and 40% of Pantanal (Terra Brasilis, 2021).

In the Amazon biome alone, 10,321 km^2 of forest were destroyed in 2020 (Terra Brasilis, 2021). By comparison, Jamaica and Lebanon have an approximate total area of 11,000 km^2 and 10,400 km^2, respectively.

Legal Amazon homes 59% of the Indigenous lands and around 355,000 Indigenous people (Terras Indígenas no Brasil, 2021a). Violent conflicts between native people, miners, loggers, and ranchers have intensified in the region over the last years. In fact, the risk of genocide of Indigenous populations in Brazil was mentioned in the UN Human Rights Council in 2021 (Terras Indígenas no Brasil, 2021b).

Brazil is one of the world's largest meat producers and exporters, behind only the US when it comes to beef production. Furthermore, Europe and China are major markets for livestock grains and cereals from Brazil. Approximately 80% of the soy cultivated in the Amazon basin is grown for animal feed uses (Global Forest Atlas, 2020).

As not even pasture-free and free-range animals are 100% fed on grass, the world's livestock industry is a major consumer of crops. Farmers from China, the UK, and the EU import large amounts of South American corn, soy, and wheat to feed their animals. Neighbouring Cerrado, a biodiversity-rich grassland savannah area that covers many Brazilian states, including Mato Grosso, is also being slashed to give place to soy plantations. Cerrado shelters 10,000 species of plants and 5% of all the living species on the planet (WWF, 2016). Around half of the UK soy imports come from Mato Grosso alone. Almost 60% of the 2,248 Amazon fires recorded in June 2020 happened in Mato Grosso (Morrison, 2020). Hence, the consumption of animal products across the globe is fuelling deforestation in the Amazon and Brazilian Cerrado.

In Europe and many other places, the livestock industry has propagated the importance of locally produced meat and dairy, in order to promote an image of sustainability. While choosing local food is important in reducing our carbon footprint and strengthening local economies, transportation comprises only a small fraction of the environmental impact of highly resource-intensive products (e.g., <1% for red meat) (Ritchie, 2020). Furthermore, the term 'local' does not mean much when the production chain relies on inputs from other parts of the world.

For instance, 'the volumes imported [by the UK from Mato Grosso alone] are so large that 93,000 hectares of soy plantations are needed—equivalent to more than half the size of Greater London' (Morrison, 2020). But before someone claims (erroneously) that tofu and other soy-derived products are responsible for this, let us remember that animal agriculture (including aquaculture) uses an astonishing 75% of the soy grown worldwide (WWF, 2016). The remainder is divided between non-human uses (e.g., biodiesel and pet food), vegetable oil, and human food.

> Europe's intensive livestock sector relies on soy, most of it imported from South America, to meet the demand for meat and dairy products. Demand for soy within the EU uses an area of 32 million acres in South America, out of a total of 114 million acres of soy production. This is equivalent to 90% of Germany's entire agricultural area. The main European importers of soy are countries with large industrial-scale pig and chicken production. (WWF, 2016)

In a globalised food economy, animal products consumption is closely tied to tropical forest destruction, no matter where on the planet the consumer is.

A comprehensive study published in the *Lancet Planet Health* journal (Springmann et al., 2018a) assessed the health, nutritional, and environmental implications of different diets. Nutrient levels, diet-related and weight-related disease mortality, as well as environmental impact were examined for more than 150 countries from all world regions. A growing body of evidence shows that plant-based diets benefit both the environment and human health (Poore & Nemecek, 2018; Ranganathan et al., 2016; Willett et al., 2019; Springmann et al., 2018b). But the differential of this study is the use of the following factors altogether: (a) several environmental indicators (other than GHG emissions); (b) different scenario designs (progressive reductions of animal-based foods or calories, flexitarian, pescatarian, vegetarian, and vegan balanced diets); and (c) countries with different income levels.

The replacement of animal foods with plant-based ones (from 25% to 100%) in high- and middle-income countries improved nutrient levels, reduced premature mortality up to 12%, lowered GHG emissions by 20–84%, reduced cropland use by 12–29%, nitrogen application by 22–38%, and phosphorus application by 25–35%, but increased freshwater use by 4–16% (Springmann et al., 2018a).

Although GHG emissions and nitrogen application were reduced in all regions, increases in cropland use, phosphorus application, and freshwater use occurred in low-income countries, reflecting differences in typical yields, agricultural practices, and technological level (Springmann et al., 2018a). Noteworthy, the high rates of undernutrition (in terms of both calories and nutrient intake) in these countries explain the increase in some environmental indicators when a balanced diet (be it plant-based or not) becomes the norm.

Notably, the effect of processed red meat (a proven carcinogen) and whole grains (source of fibres, phytochemicals, and other important health-promoting nutrients) was not computed in the study, which would further impact the health indicators—negatively for processed meat and positively for whole grains.

In the US, animal-based foods represent approximately 90% of the agricultural land use and 85% of the food production-related GHG emissions. In comparison with the average world diet, the US consumer ingests an additional 500 calories per day, of which 400 derive from animal products (Ranganathan et al., 2016). Countries where people rely mainly on meat, dairy, and eggs to meet their protein requirements are the ones who would benefit the most by replacing these products with plant-based protein sources (Springmann et al., 2018a).

Figure 16 shows that *all regions* in the globe exceed the recommended daily protein ingestion (roughly 0.8 g/kg body weight), with a *per capita* global average consumption of 68 g protein/day. Given the adverse health effects and the environmental burden ascribed to animal-based foods, there is no compelling reason why we should not make the shift to plant-based living (Ranganathan et al., 2016).

Protein Consumption Exceeds Average Estimated Daily Requirements in All the World's Regions, and is Highest in Developed Countries
g/capita/day, 2009

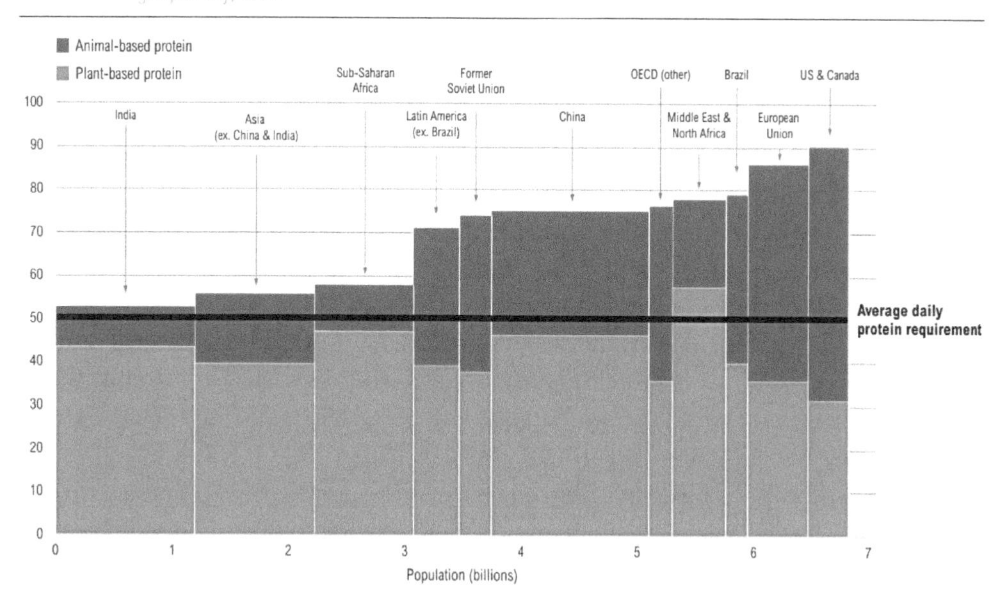

Figure 16. Protein consumption patterns around the world. Source: Ranganathan et al. (2016).

Another significant impact of animal agriculture not so widely discussed is manure overload and wastewater from farms and processing plants. Slurry and solid manure produced by large-scale animal farms pose a serious environmental and public health problem due to excessive phosphate, potassium, and nitrate levels in the soil and waterways, microbial contamination of waterbodies and crops, and GHG emissions from dung fermentation. This results in increased risks of enteric disease transmission, air pollution and related health conditions, algal blooms, habitat degradation, and further global warming. Research published in *Nature Sustainability* in 2018 estimated a faecal matter global production of 3.9 billion tonnes in 2014, of which 79% come from livestock animals. Faecal biomass is forecasted to reach 5 billion tonnes by 2030 (Berendes et al., 2018).

Animal waste has become a particularly severe problem in many countries, which must export it to be treated elsewhere. In Northern Ireland, only 1 out of 21 lakes is within the limits set by EU's water framework directive (Greene, 2021). A shocking 98% of conservation areas exceed critical nitrogen levels, with some by over 300%. According to advisory bodies, the country may need to export 35% of the animal waste produced to bring water and soil quality back to acceptable levels (Greene, 2021). In the Netherlands, 93% of cattle farms in 2020 produced excess animal manure (The Netherland's Environmental Data Compendium, 2021). In China, nitrogen and phosphorus inputs to rivers increased 2–25 times from 1970 through 2000, largely accompanying the growth of animal agriculture. Direct discharges of animal manure to rivers accounts for 20–95% of nutrients in the central and southern rivers and >66% in northern rivers (Strokal et al., 2016).

Wastewater from farms, slaughterhouses, butcheries, and meat processing plants is also a concerning issue. These residues exhibit high microbial counts and vast amounts of organic and inorganic matter, which entail high costs and resource use with water treatment. In addition, GHG emissions from slaughterhouse effluents are higher than for most other food products (Poore & Nemecek, 2018).

Animal agriculture is working *against* the Paris Agreement and the UN Sustainable Development Goals, which include, for example, climate change mitigation and sustainable food production. Despite the robust evidence on the unsustainability of meat production, producers are not legally required to disclose their carbon emissions and other environmental indicators, let alone put mitigation measures into place. IATP and GRAIN examined the commitment of the world's largest meat and dairy companies to measuring and publicly reporting their GHG emissions. Out of 35 companies, only 4 provided comprehensive and credible emissions estimates (IATP & GRAIN, 2018).

While 80–90% of the meat and dairy emissions are related to the supply chain (from feed crops cultivation to enteric fermentation methane), 29 companies underreported or simply disregarded the supply chain carbon footprint in their mitigation plans. In clear contradiction, the six remainder companies expect to reduce emissions in the supply chain *and* increase production at the same time (IATP & GRAIN, 2018).

Meat and dairy companies should include supply chain emissions even when they do not raise and slaughter the animals themselves. After all, their raw materials (animal body parts and their fluids) do not come at a zero environmental cost. Furthermore, they cannot shift the responsibility to their suppliers, whose profit margins are already minimal. When the carbon footprint of feed production and animal raising are excluded from the computations, the overall emissions of meat and dairy companies can be underrepresented by more than 80% (Gerber et al., 2013).

The emissions of the meat and dairy sectors of these 35 companies were estimated by IATP and GRAIN based on the open corporate data on production volumes and the FAO Global Livestock Environmental Assessment Model (GLEAM). The 5 biggest meat and dairy companies together emit 524 million tonnes (or 578 mt) of CO_{2eq} per year, more than ExxonMobil (523 million tonnes), Shell (461 million tonnes), and BP (406 million tonnes) (Figure 17).

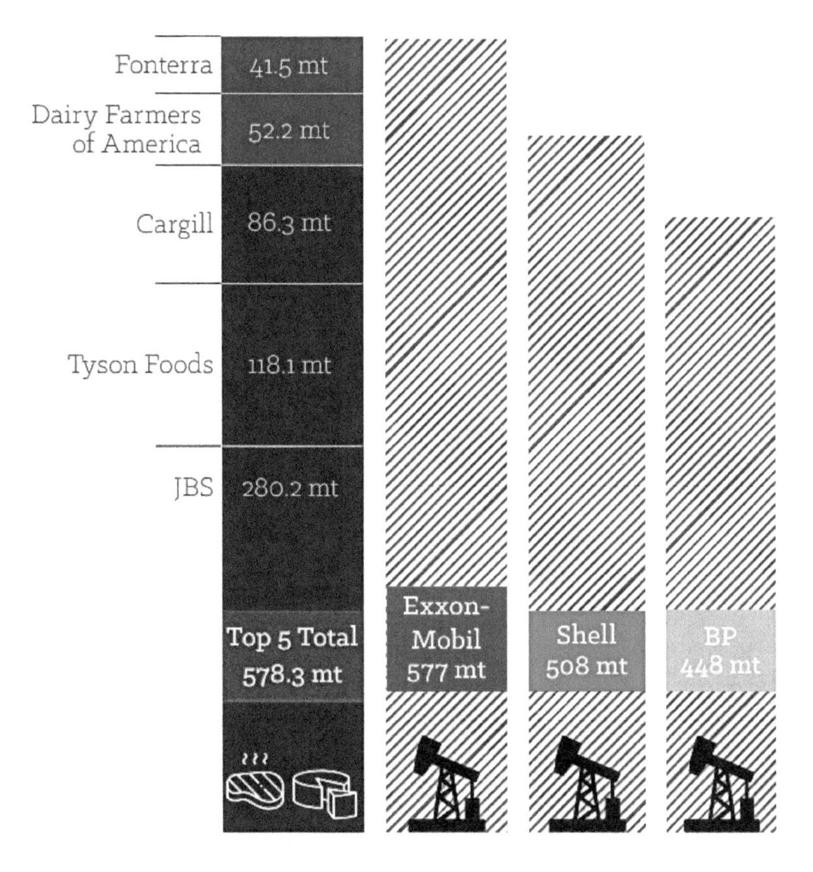

Figure 17. Annual carbon emissions of the top five meat and dairy industries compared to major oil companies. Note: 1 mt = 0.90718 million tonnes. Source: IATP & GRAIN (2018).

In addition, the GHG emissions of the 20 biggest dairy and meat industries together (933 mt or 846 million tonnes CO_{2eq} per year) surpass those of strong economies, such as Germany (818 million tonnes) and France (421 million tonnes) (Figure 18).

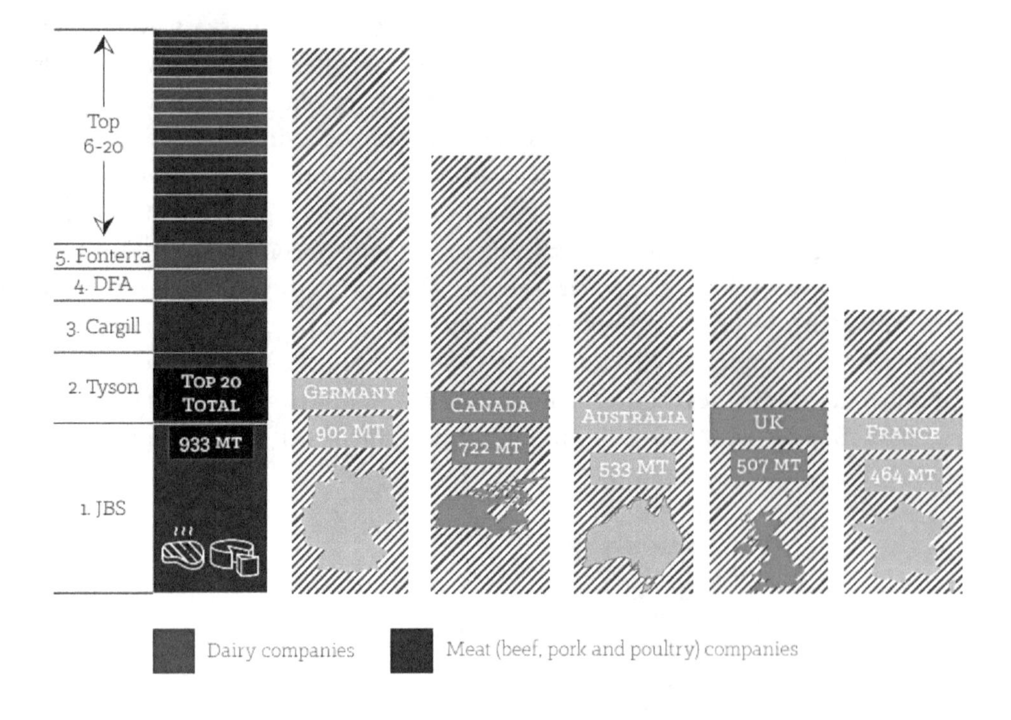

Figure 18. Annual carbon emissions of the top 20 meat and dairy industries compared to major world economies. Note: 1 mt = 0.90718 million tonnes. Source: IATP & GRAIN (2018).

While governments do not hold animal agriculture accountable for its carbon footprint and consumers remain in a state of apathy, we approach the point of no return. Tyson expects to increase its beef and poultry production by 3–4% annually. At the same time, US-based Tyson promised to reduce its carbon emissions by 30% from 2015 to 2030, without any robust plans on how to curb their on-farm emissions. Although Cargill admits their supply chain accounts for around 90% of their carbon footprint, supply chain emissions are not included in their GHG accounting (IATP & GRAIN, 2018).

A 2021 study carried out by researchers from the New York University (NYU) revealed how meat and dairy companies will exceed the carbon emission target of their headquarter countries. For instance, the EU's leading pork producer Danish Crown will consume 42% of Denmark's emissions targets set in the Paris Agreement by 2030. In turn, Fonterra in New Zealand and Nestlé in Switzerland will surpass 100% of these countries' total emissions budget in the same period (Lazarus et al., 2021).

Another long-term investigation, conducted by DeSmog and covered by *The Independent*, corroborates that animal agriculture is underreporting its emissions, primarily by narrowing the activities they include in the estimations (Christen, 2021). Furthermore, the meat industry (and the animal exploitation sector in general) counts on many tactics to distract the public from their environmental and social impacts. The strategies include (Christen, 2021):

- Downplaying the impact of animal farming on the climate and ecological crises;
- Claiming that meat is necessary to a healthy diet;
- Disguising the detrimental effect of meat consumption on human health;
- Casting doubt on the efficacy of plant-based diets to halt the climate emergency;
- Overstating the effect of animal agriculture innovations on the reduction of the industry's environmental impact;
- Claiming that vegan diets are elitist;
- Portraying itself as part of the solution to hunger and food insecurity while hiding the sector's massive resource use and environmental footprint.

Bottom line: legal regulations, efficient control, and strong sanctions will be crucial to curb GHG emissions from food production and avoid an environmental calamity.

A study published in science journal *One Earth* in 2020 discusses how the 60-billion-euro agriculture subsidies through the EU Common Agriculture Policy (CAP) could be used more efficiently (Scown et al., 2020). Although CAP's main objectives are to support sustainable food production and ensure farmers a viable income, current CAP spending is aggravating income inequality and environmental degradation. CAP's subsidies are being directed mainly to wealthier regions with the most polluting and biodiversity-impactful practices. More than EUR 24 billion out of the EUR 60 billion of 2015 CAP payments benefited regions where average farm incomes are above the EU median income. Intensive animal agriculture regions like the Netherlands, UK, Ireland, Belgium, and Denmark are receiving billions of euros every year despite having the greatest carbon footprint. Additionally, the benefited farms are not required to control or report their emissions.

Most importantly, animal agriculture is not regulated by free supply and demand. Federal subsidies, tax breaks, corporate grants, and insurance policies created a fail-proof food system where the livestock sector is protected regardless of reductions in market demand (Simon, 2013; Dorning, 2020).

US taxpayers are paying USD 38 *billion* in direct and indirect subsidies every year for the killing of animals and environmental destruction—even when they are vegan. Conversely, subsidies towards fruits and vegetables total USD 17 *million* (Simon, 2013). In April 2020, the US government approved a USD 15.5 *billion* coronavirus bailout for meat and dairy farmers (Dorning, 2020). No wonder why these resource-intensive foods can be cheaper than vegetables in some countries.

Shooting ourselves in the foot

Livestock contributes largely to the Agricultural Gross Domestic Product (AGDP) of many countries. In the US, for example, meat and poultry plants employed 29.3% of food and beverage manufacturing employees in 2018, versus 9.4% in the fruit and vegetable sector (USDA, 2020). These facts might explain the global disregard for the environmental agenda in food production and the labour rights of slaughterhouse staff amid the coronavirus pandemic.

Up to the 12th of June 2020, more than 24,000 COVID-19 cases had been linked to US meat plants. This might not seem a large number in relation to the two million people who tested positive in the country by that date; however, infection rates near large meat factories were more than twice the national rate. According to a June 2020 news piece, 287 out of the 1,000 workers tested positive for coronavirus in the JBS beef plant in the city of Hyrum, Utah in May that year (Samuel, 2020). The employees protested in front of the Hyrum plant for its closure, requesting the right to take paid leave regardless of whether they are sick. The employees (most of them migrants) had to keep showing up to work, putting their families at risk, as they could not afford to lose their jobs. Coincidentally or not, northern Utah saw an alarming rise in coronavirus cases, from 117 on the 28th of May to 837 two weeks later.

As a result of 1,500 infected employees in the Tönnies meat factory in Gütersloh, Germany, in June 2020, two districts had to be put under lockdown, Gütersloh and the neighbouring Warendorf (Janjevic, 2020). In August 2020, three counties in Ireland (Kildare, Laois, and Offaly) were put under local lockdown due to the high number of coronavirus cases in meat factories, affecting local businesses and people's lives. In the first week of the month, 86 positive COVID-19 cases were recorded at O' Brien Fine Foods, a meat factory in Timahoe, Laois (Keaveny, 2020). From the beginning of the pandemic until mid-August 2020, a total of 1,450 meat plant workers were tested positive for COVID-19 in Ireland, which has fewer than 5 million inhabitants (Molony, 2020a).

The pandemic has exposed the substandard work conditions in the Irish meat sector—around 90% of the workers are not entitled to sick pay and their workload in a year is nearly 5 weeks longer than in other EU countries (Donnelly, 2020a). Many of them are migrants who live in rough conditions (including room sharing) and carpool to work, which contributes to spreading the virus. According to Nora Labo, from the Cork Operative Butchers Society, a branch of the Independent Workers Union, 'On minimum wage with no overtime, [meat workers] had to work extremely long hours—with no time to go to the doctor or to learn English to communicate any problems.' She claims the production pace in meat factories has not slowed down during the pandemic: 'When some workers got ill, others had to do more work' (Molony, 2020a).

Labo also points out: 'Many workers we know are being housed by their employment agencies which, seeking to maximise profit from the accommodation they provide to their employees, crowd as many people as possible into each house they let' (Kiernan, 2020). She continues by saying that for years, a great number of Romanian workers were employed in meat plants in Munster without PPS number (individual reference number required for all dealings with Irish public service agencies, including Revenue), annual leave, illness benefit, or child allowance (Kiernan, 2020).

To make things worse, the Health and Safety Authority (HSA) was not being notified of COVID-19 positive cases in meat factories, according to Patricia King, leader of the Congress of Trade Unions (Molony, 2020a). This means the number of cases officially reported might well be underestimated. Moreover, only 39 of 149 meat plants in Ireland were inspected by HAS. In addition, meat plants received an advance notice ahead of HSA inspections in 30 out of 39 cases—in general, high levels of compliance were found, according to HAS Chief Executive Sharon McGuinness (Molony, 2020b). What a surprise.

At the beginning of the COVID-19 lockdown, the Irish Department of Agriculture had advised beef companies not to avail of Temporary Veterinary Inspectors over 70 years of age for safety reasons. Nonetheless, according to parliament member Patricia Ryan TD, 50–70 vets older than 70 had remained working after having signed waivers to not sue the employer if they contracted coronavirus (Donnelly, 2020b).

In São Miguel do Guaporé, in the Brazilian state of Rondônia, over 60% of all the COVID-19 infections in the municipality originated from a JBS plant (Brasil Notícia, 2020; Portal Rondônia, 2020). As a result, the Public Prosecutor's Office of Rondônia (MP-RO) and the Labour Public Prosecutor's Office filed a lawsuit requesting the facility's closure. The appeal was approved by the Labour Court, but the prosecution further requests the payment of BRL 20 million (around EUR 3.55 million) to the workers for moral damages.

You might be asking why slaughterhouses and meat factories are COVID-19 hot spots. First, because these places feature the perfect conditions for the development and spread of pathogens, such as extreme confinement and antibiotics overuse in the farms of origin, contamination of the meat with faecal and pulmonary microorganisms, and the presence of high amounts of blood and viscera. Second, meat plants are humid and are kept at low temperatures to prevent the meat from decomposing, which are perfect conditions for the virus to spread. Third, meat plant workers are positioned close together along processing lines, rendering social distancing impossible. Fourth, the proximity between people in toilets and canteens is unavoidable. And fifth, noise in the factories cause workers to shout to communicate, with the air-cooling systems dispersing contaminated saliva droplets throughout the plant.

Human and animal lives are being sacrificed because we are unable to make the most basic connections.

Other pandemics have been affecting the livestock sector, but have not received much media attention. For instance, a significant decline in pork meat production from 2018 to 2020 resulted from the outbreak of African swine fever (ASF) in China. The country lost 41% of the herd in the 12 months following October 2018

(Reuters, 2020). The ASF virus, which belongs to the Asfarviridae family, is a highly contagious haemorrhagic pathogen that is virtually 100% fatal to pigs but does not seem to affect humans. However, novel asfarvirus-like sequences found in human serum and sewage have the potential to transmit to humans (Loh et al., 2009). ASF is endemic in farms in Africa and has been sweeping Asian and European countries. Poland, for instance, has faced two outbreaks in early 2020 (Kość & Standaert, 2020). While small-scale pig farmers around the world are struggling to survive, record profits are recorded by the biggest producers, who can incur high costs to contain the virus spread within their herds (Standaert, 2020).

In his 2013 book *Meatonomics*, David R. Simon explains how economic, political, and marketing strategies generate artificially low prices and overconsumption of animal products in the US. While solid science confirms the detrimental effect of meat and dairy on human health, USDA dietary guidelines continue to include these products. Annually, the country spends USD 557 million to promote animal products versus a measly USD 51 million to promote fruits and vegetables. At the same time, diseases related to meat and dairy consumption cost the US more than USD 300 *billion* every year.

According to Simon, a 50% excise tax on meat and dairy could save 172,000 human lives every year (Simon, 2013). Meanwhile, corporate animal agribusinesses pay US lobbyists around USD 18 million annually to influence farm policy to their clients' advantage (Agriculture Fairness Alliance, 2020). Tyson Foods and WH Group/Smithfield Foods topped the 2019 meat clients list with lobbying investments of USD 1,334,159 and USD 1,460,000, respectively (Open Secrets, 2019).

Indeed, when the UN FAO published its first report on the GHG emissions of animal agriculture, in 2006, the organisation faced strong pressure from lobbyists. According to Dr. Samuel Jutzi, then director of the FAO's Animal Production and Health Division: 'You wouldn't believe how much we were attacked' (IATP & GRAIN, 2018). A new report was subsequently published in a collaboration between FAO and animal agriculture groups to "reassess" the calculations. The update lowered the result from 18% to 14.5% of global GHG. Note that FAO's 18% estimate already underrepresented the actual emissions from animal agriculture for not including the Carbon Opportunity Cost (IATP & GRAIN, 2018).

A sustainable industry incorporates the impact of their activities on natural resources, living beings, employees, and consumers without compromising future generations. But as consumers, we have a responsibility too. Each cent we spend tells industry and governments a message: *we support these ethics and want more of this product.*

Is your money creating a better world?

Investigations

Investigation 1 – Mexican slaughterhouses

This investigation was conducted in 21 Mexican slaughterhouses in 2016. Animal Equality's team recorded scenes of extreme cruelty and neglect, at odds with the Mexican Federal Animal Health Law. This investigation comprises small-scale abattoirs with outdated installations. Industrial slaughterhouses operate on a much larger scale, fast-paced processing, with modern and cleaner infrastructure. However, the slaughter process does not differ significantly between big and small abattoirs, which in essence involve stunning, killing, bleeding, eviscerating, and butchering the animals. There is no way to forcibly take someone's life without causing physical pain and emotional trauma.

In Figures 19a and 19b, pigs covered in mud are unloaded at the slaughterhouse. One of them was suspended by the hind leg while still alive and fully conscious.

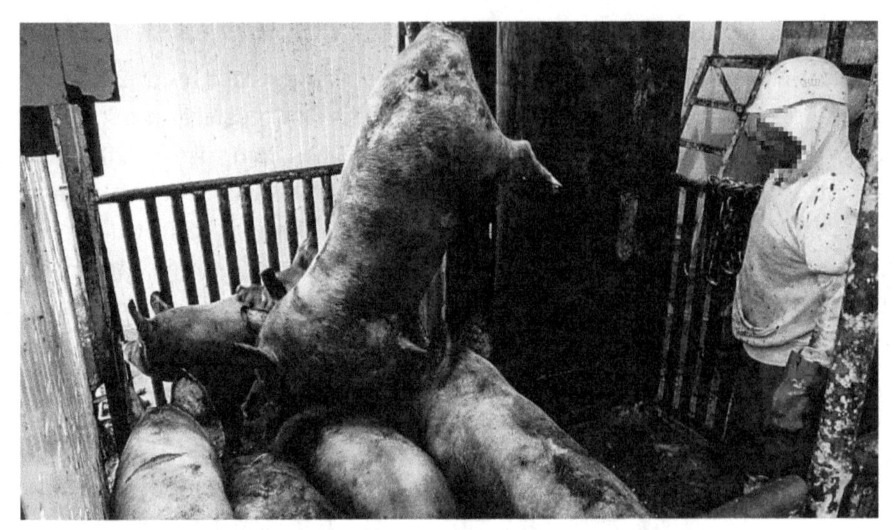

Figures 19a and 19b. Pig slaughter at a Mexican slaughterhouse in the state of San Luis Potosí.

Figures 20a to 20c show scared pigs walking towards slaughter while witnessing the killing of other animals. Some people claim the solution to the problem of eating meat is relying on small-scale, local production. This might solve a tiny slice of the problem (e.g., supporting small businesses), but it certainly does not address the ethical aspect of consuming animal flesh. Small- and large-scale meat plants have each their pros and cons, with animal suffering implicated in *either* case.

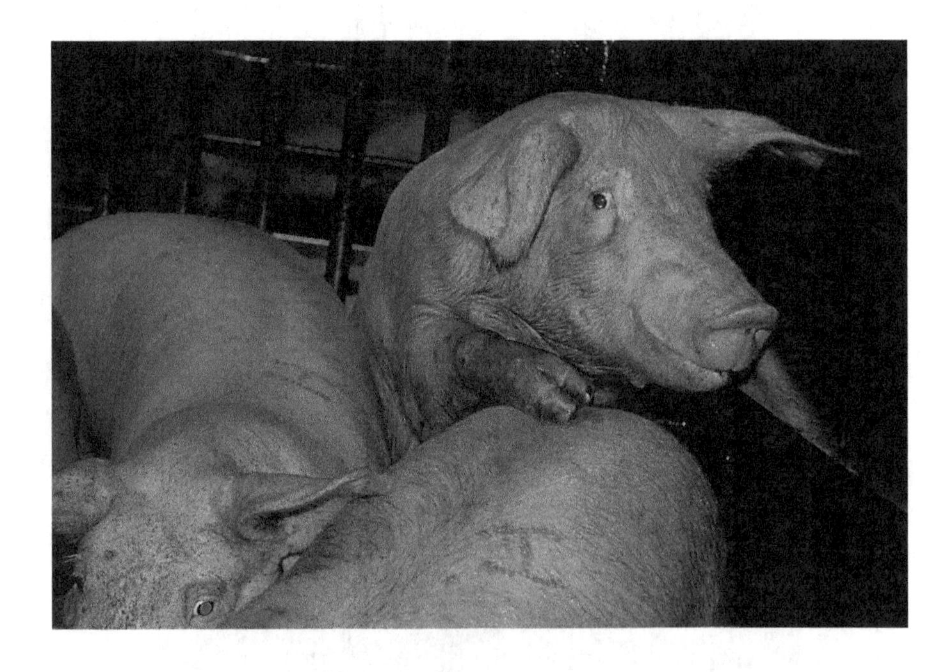

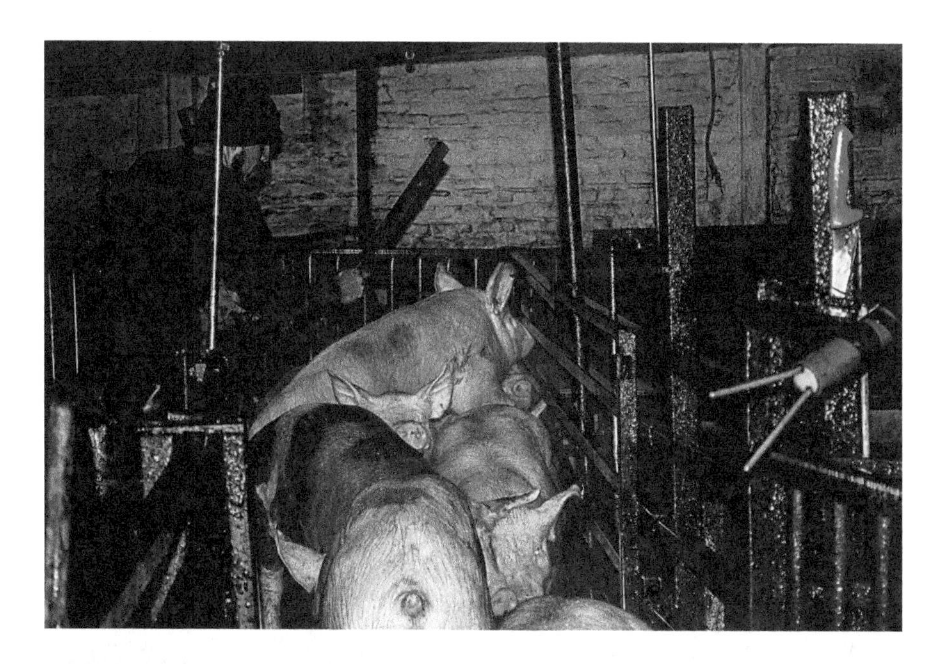

Figures 20a to 20c. Pig slaughter at a Mexican abattoir in the state of Aguascalientes.

In Figures 21a to 21c, we see a pig full of injuries over the side of his body. We also see workers stunning a pig in a sadistic way, with the sensors touching the animals' neck and eye. Another pig was restrained with a tight rope around his neck. The animal was stabbed while still conscious. Animal Equality's investigators revealed that *most* animals died in severe pain and agony, utterly aware of what was happening to and around them.

"Humane slaughtering" means the animals are stunned before being killed and exsanguinated. However, pre-slaughter procedures are severe sources of stress, fear, and risk of injury. In addition, stunning is not always successful, even in highly mechanised industrial-scale abattoirs. In smaller slaughterhouses, often called "artisanal" slaughterhouses, the human factor plays an even greater role in animal welfare levels during slaughter, with further and unnecessary suffering arising from inadequate training and deliberate cruelty.

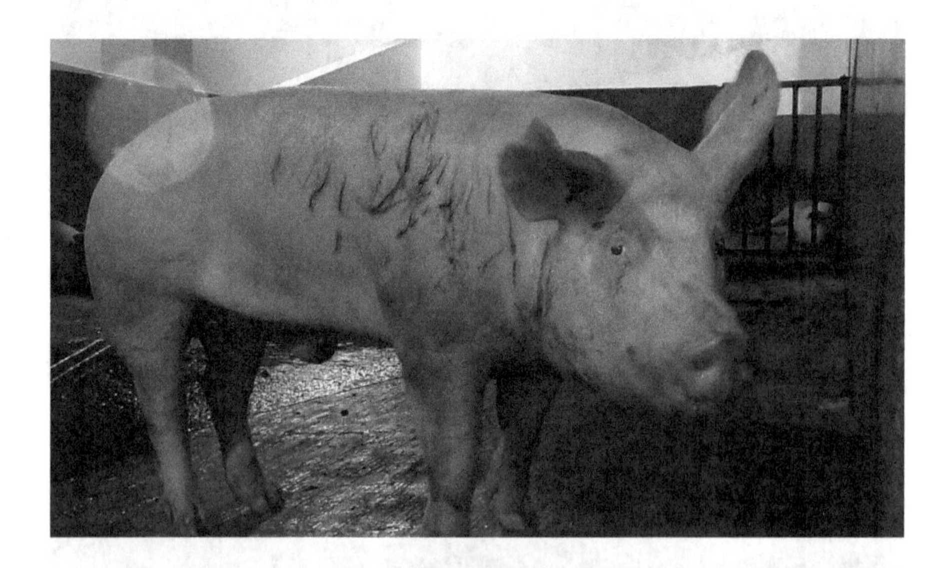

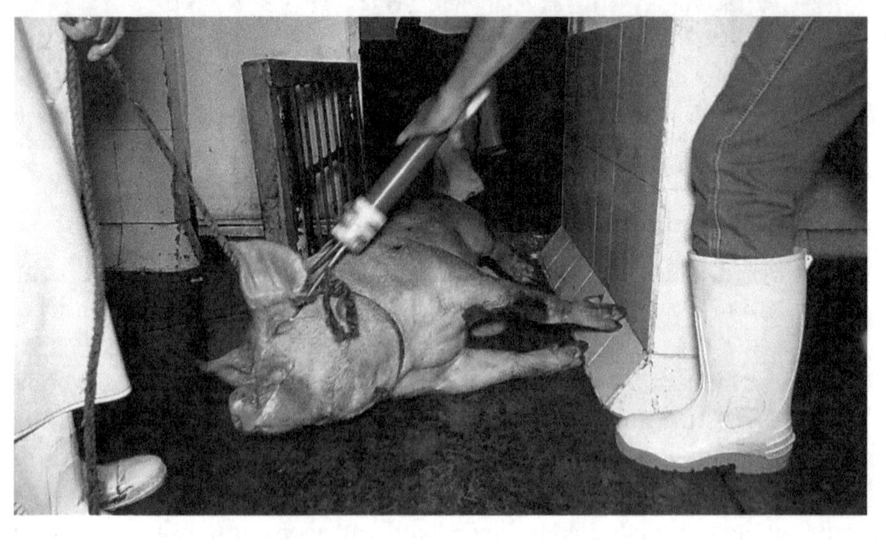

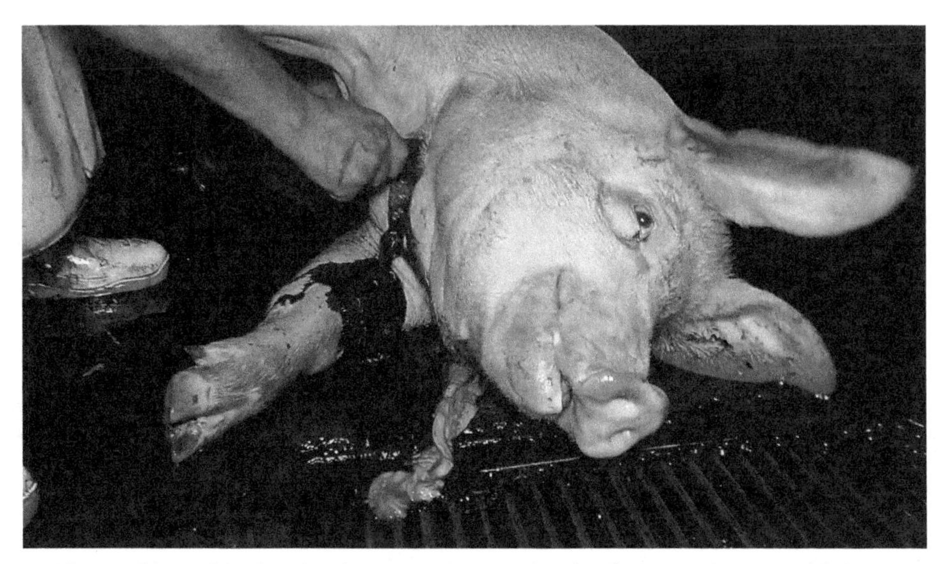

Figures 21a to 21c. Pig slaughter at a Mexican slaughterhouse in the state of Colima.

Figures 22a and 22b shows a worker chasing a terrified pig with an axe in his hand. The pig was killed without appropriate stunning. When faced with similar videos and pictures, some consumers claim their meat does not come from such places. How can they tell? Or do they buy their meat directly from the abattoir after having watched the slaughtering process with their own eyes? And is that the only animal flesh they consume? What about ready-to-eat products they buy at the supermarket? And what about the food they order on the go, in restaurants, caterers, etc.?

There is *no way* to control where meat comes from and how the animals were treated unless we raised and killed them ourselves. And once again, *using* sentient beings is morally unjustifiable, regardless of the treatment.

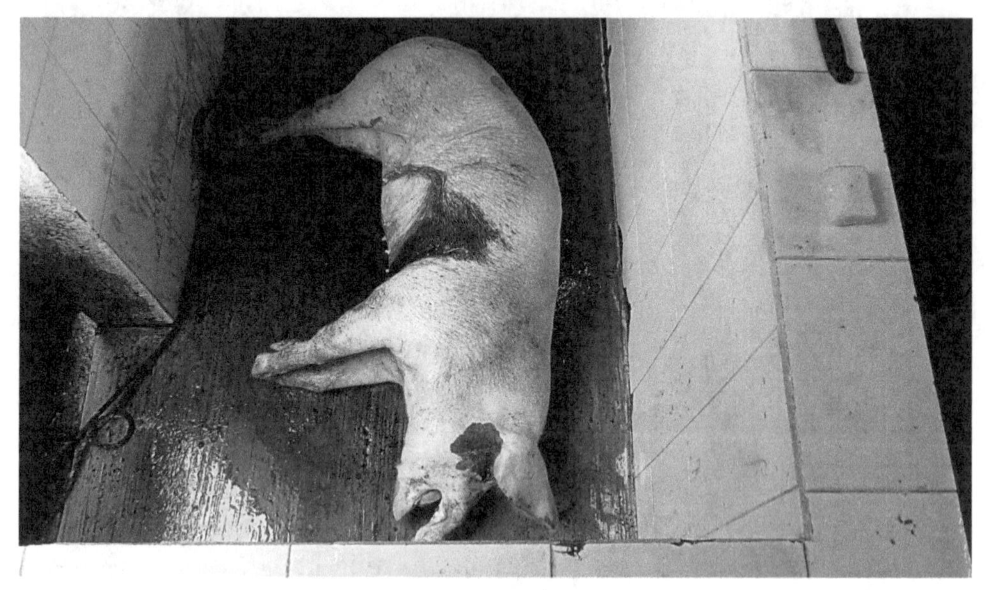

Figures 22a and 22b. Pig slaughter at a Mexican slaughterhouse in the state of Nayarit.

Figures 23a to 23c shows pigs being chased and hit with an axe in full sight of the others. Contemplate the emotional load people ingest when they eat pork, ham, and bacon.

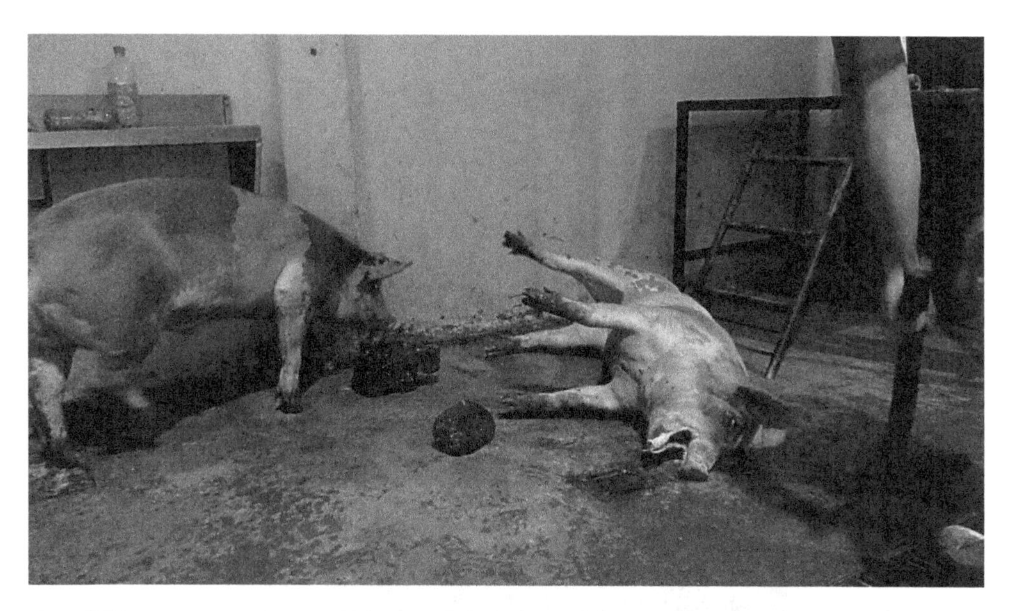

Figures 23a to 23c. Pig slaughter at a Mexican slaughterhouse in the state of Zacatecas.

Figures 24a to 24d shows once again the fear of pigs waiting for their turn to be slaughtered. Even when the animals are appropriately stunned, how humane can killing an innocent creature for sensory gratification be?

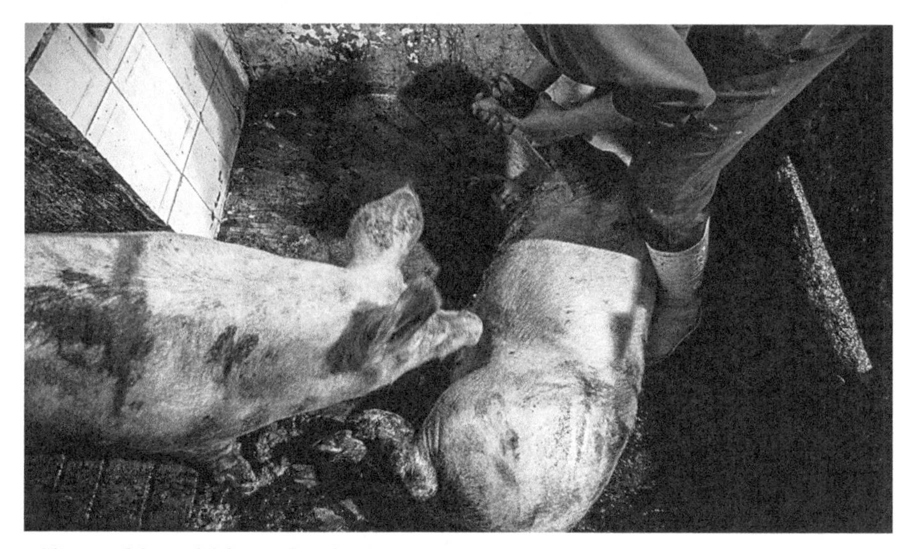

Figures 24a to 24d. Pig slaughter at a Mexican slaughterhouse in the state of Jalisco.

Animals sense death approaching from the smell of blood and viscera and the vocalisation of the others (Figures 25a to 25f). The trauma is worsened when they get to see other animals being killed in front of them.

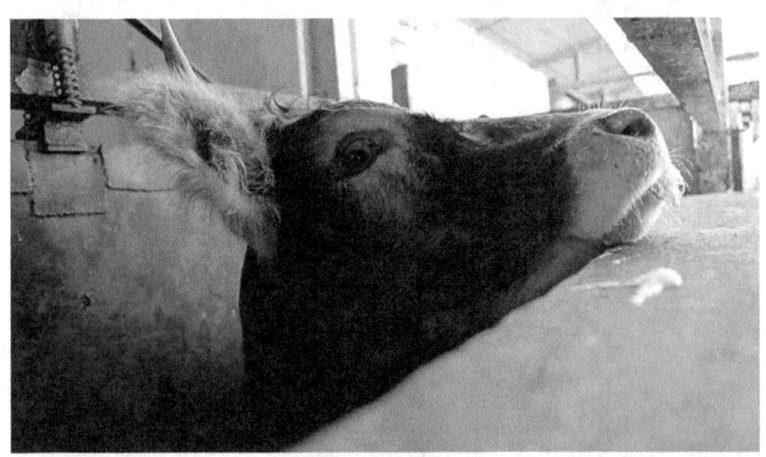

Figures 25a to 25f. Cattle and swine slaughter at a Mexican slaughterhouse in the state of Jalisco.

Low-yield dairy cows are slaughtered, and the milk in their udder is discarded along with the blood (Figures 26a and 26b). Their flesh is sold to meat processors as lower-value meat. Mince and burgers are common destinies for dairy-breed beef.

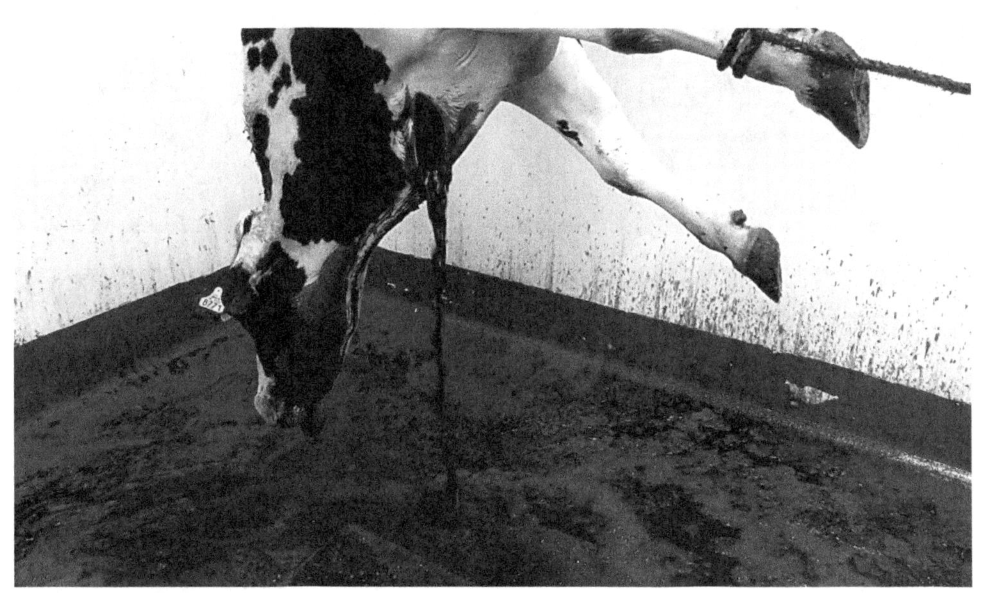

Figures 26a and 26b. Lactating cow slaughtered at a Mexican slaughterhouse in the state of Jalisco.

Deliberate animal torture is common in farms and slaughterhouses. Multiple cases of severe abuse are recorded and reported to local authorities by animal rights organisations every year. Unfortunately, they rarely end up in a prosecution.

Figures 27a to 27e shows the workers torturing pigs by poking, cutting, and stabbing them with a knife. One of them is suspended by the hind leg with a chain. Another pig is stabbed in the chest by a worker while the other man twists the animal's arm. There are generally no legal consequences for the employees. When cases such as these reach the media, the farm claims this is not standard practice and that the workers involved were dismissed. That simple. As for consumers, they console themselves by saying these are isolated cases (only they are not) and return to the usual apathy.

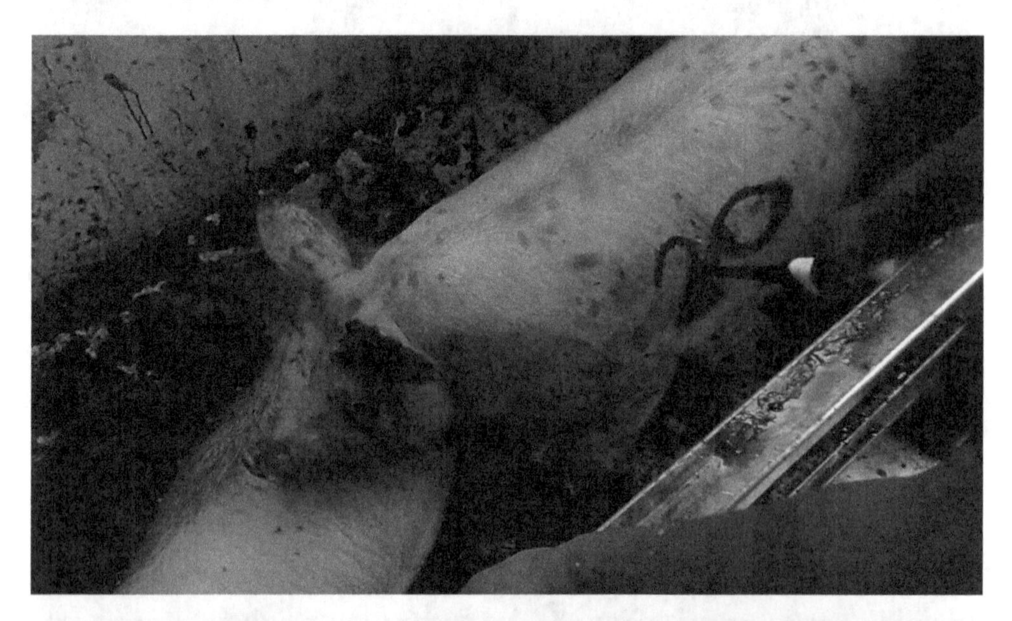

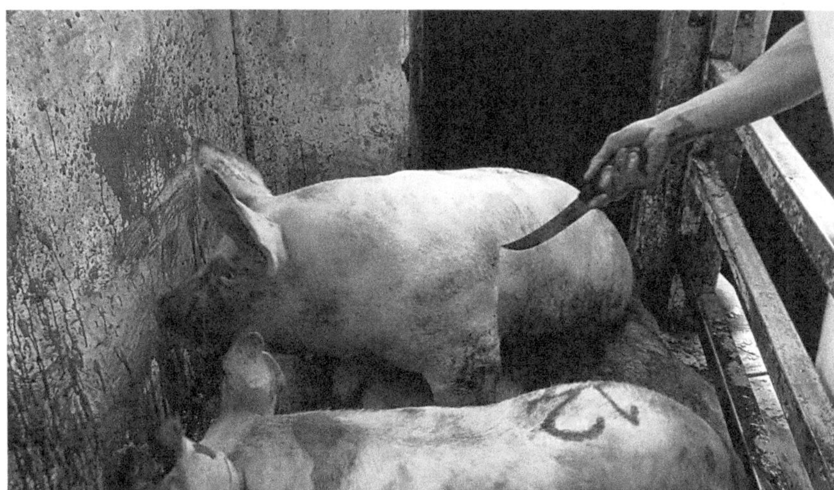

Figures 27a to 27e. Animal torture at a Mexican slaughterhouse in the state of Jalisco.

INVESTIGATION 2 – PIG FARM IN ITALY

In 2019, Animal Equality Italy published an investigation exposing shocking animal treatment in a pig farm in Lombardy (Animal Equality, 2019). The investigation was broadcasted nationwide on Tg2, the news programme of Italian state-owned television channel Rai 2. The farm was visited by the organisation's team along with Tg2 journalist and host Piergiorgio Giacovazzo, who witnessed thousands of pigs living under terrible conditions.

Animal Equality reported the farm to the authorities and started a petition addressed to the Minister of Agricultural Policies Gian Marco Centinaio and to the Minister of Health Giulia Grillo to close this farm and increase supervision in factory farms. The investigation was also published on social media. Nonetheless, the farm continued to operate.

As long as the demand for animal products continues, little will be done to release animals from a lifetime of suffering and pain. When we consume meat, dairy, fish, and eggs, we are voting for the exploitation to continue.

Figures 28a to 28c shows female pigs confined in sow stalls. Note that most of them had their tail docked, which was likely done without pain relief. In Figure 29, we see a tail found on the floor.

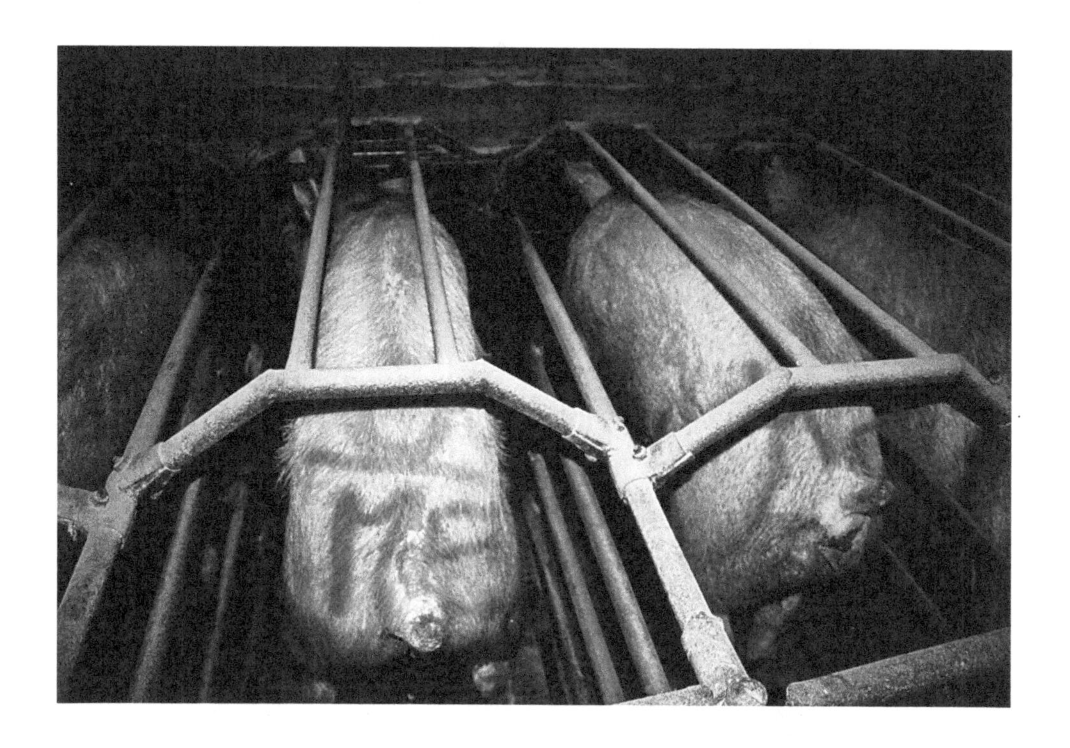

Figures 28a to 28c. Sow stalls in a pig farm in Northern Italy.

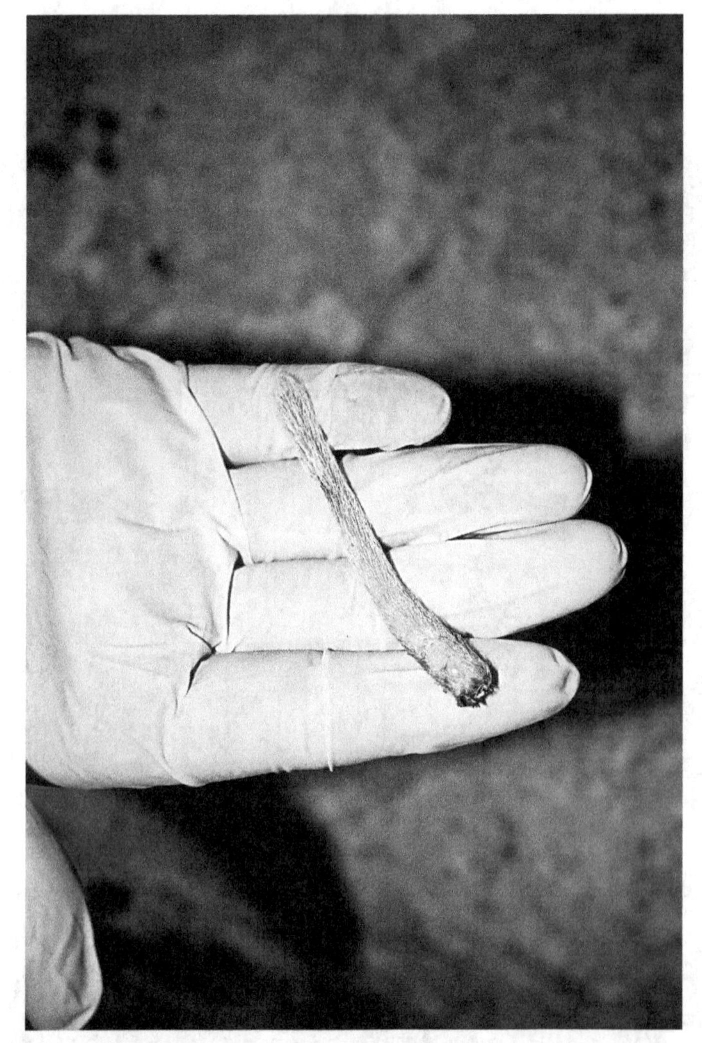

Figure 29. Pig tail found on the floor in a pig farm in Northern Italy.

Many pigs suffered from rectal prolapse (Figures 30a and 30b), a painful condition that is very common in commercial farms. Typical causes include: diarrhoea, especially associated with *Salmonella* spp., swine fever, and swine dysentery; constipation, most prevalent in sows close to farrowing; excessive straining during parturition; water shortage, causing the faeces to become harder; high doses of antibiotics that cause swelling of the rectum; mycotoxins that lead to rectal straining; rectal damage resulting from boars (non-castrated males) riding each other; intense coughing from respiratory diseases, causing the anus to prolapse; fast growth, especially in pigs on high-density diets; and cold, which forces the pigs to huddle and pile—if one of them coughs, the abdominal pressure will be relieved at the anus. The prolapse can be worsened due to trauma on feeders and pen divisions, especially in overcrowded pens. The other pigs can bite the prolapse, causing further injury.

Figures 30a and 30b. Pigs suffering from rectal prolapse in a pig farm in Northern Italy.

Severely injured, sick, and dead piglets were found at high numbers in crates (Figures 31a to 31c).

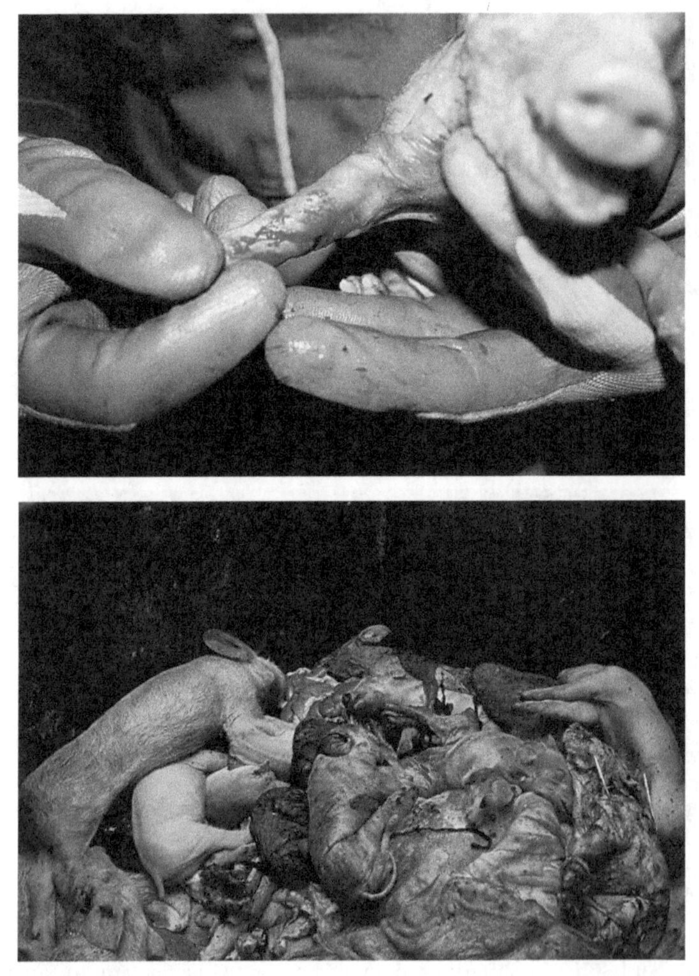

Figures 31a to 31c. Dead, sick, and injured piglets in a pig farm in Northern Italy.

Ear biting, a typical sign of stress, was found in many pigs (Figures 32a and 32b).

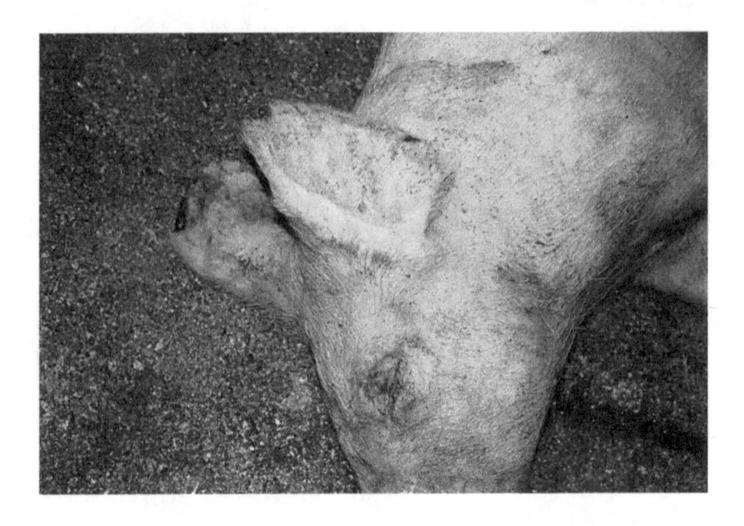

Figures 32a and 32b. Stress-related ear biting in a pig farm in Northern Italy.

Figures 33a to 33c shows animals suffering from abscesses and scrotal and umbilical hernias. The leading causes of hernia are genetic defects, bad administration of the umbilical cord of newborn piglets, and bacterial infections.

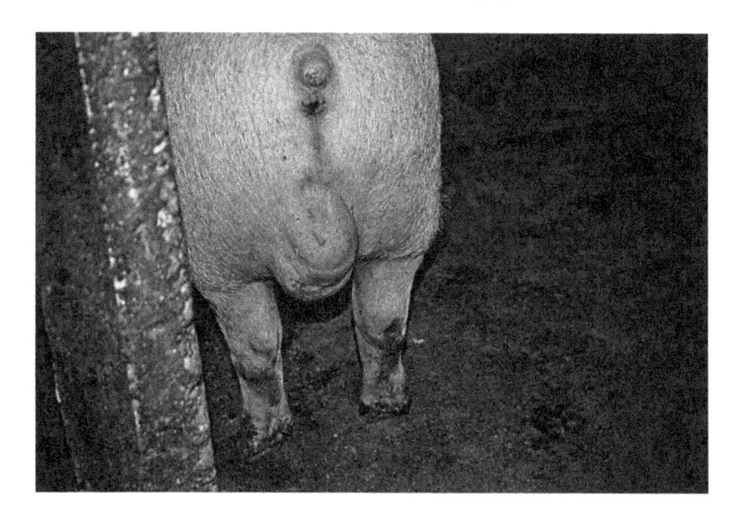

Figures 33a to 33c. Pigs suffering from abscesses and hernia in a pig farm in Northern Italy.

The cruelty witnessed by the investigators also included a worker who killed a pig by throwing it against the wall, pigs covered with wounds and bedsores in tiny cages, feeders full of faeces, animals covered in faeces and urine, and infestations of cockroaches, worms, and mice.

Out of the eight million pigs bred every year in Italian factory farms, roughly half are reared in Lombardy. More than half a million female pigs spend most of their lifetime in crates, deprived of their fundamental rights (Animal Equality, 2019).

The best way to halt this violence is to *stop fuelling it*. Your food choices are a vote. Choose wisely.

CHAPTER 3

POULTRY, MEAT, AND EGGS

Imagine you give an apple *and* a chick to your three-year-old child. Would your kid eat the fruit and pet the chick or do the opposite? I hope your baby chose to eat the fruit and leave the poor animal alone, but let us imagine they ripped the chick's head to eat their flesh. Would you find it natural? I think most people would worry about their kid's psychological health. Take a moment to reflect on that *with sincerity.*

The fact that young children know in their hearts that animals are not food shows that the current narrative—animals exist to serve our needs—is artificial, manipulated, and goes against our inner wisdom. Even though we do not believe it, human nature is one of compassion. Our first impulse when seeing someone drowning is to save them. Our immediate reaction when noticing someone is about to be run over is to shout or push the person away from the vehicle. Our natural response when watching a person beating an animal is to do our best to stop it. We do these things even though we do not have any relationship with the one who is in danger. Deep down, we know they are all sentient beings who want to live and have the right to do so—this is our natural ethics. Hatred, prejudice, and *insensitivity to others' pain* are taught, are passed from generation to generation, are absorbed from the society around us.

Even those who argue that humans were essentially hunters in the past will agree that we are not in the Palaeolithic anymore. The world evolved. Most of us have plentiful food choices at our disposal.

Now let us say you put your child in a room with several animals—chickens, pigs, dogs, and cats. Would your baby treat them differently? Unless taught otherwise, any child would probably like to play with them all. Distinctions between "companion animals" and "farm animals" are based on convenience—we as a society choose which animals to protect and which to abuse, hand-picking the most docile to be used as food.

Attitudes towards animals change slightly from one geographical region to another due to the local biodiversity and religious traditions. Therefore, a species that is deemed food in some countries would never be eaten in others; however, the base mindset remains unchanged: 'it is okay to abuse animals.'

Fortunately, our inner wisdom (and science too) tells us animals are sentient beings who feel pain, fear, anxiety, and a range of complex emotions, and as such, they have rights. Nonetheless, at any given moment, we exploit 26 billion chickens for their flesh and eggs. Not to mention all other birds who live a dreadful life so we can satisfy our taste buds—ducks, turkeys, pheasants, quails, and others.

As the last exercise, let us imagine your kid spent the afternoon playing with lovely yellow chicks. You prepared some chicken nuggets for dinner. Would you tell your child those nuggets are the flesh from those cuddly animals?

What if they explicitly asked what the nuggets are made of? Would you tell the truth?

The very fact that we must lie (or at least omit information) indicates we are doing something *wrong*. By lying to our children about something fundamental (the food they are putting inside their bodies), we are deceiving them. We are teaching them to disconnect from reality when it is convenient to do so. We are passing onto them a confusing code of ethics—we must not harm animals, but we can eat their flesh. The truth is that following this utilitarian, plastic ethics is much more complicated than simply doing the right thing. Living according to adjustable principles is confusing, counterproductive, and lacks integrity.

Several kinds of birds are used for their flesh, eggs, and liver secretions (a.k.a. foie gras). Nevertheless, I will focus on egg-laying hens and broiler chickens (chickens raised for meat) for practical reasons. Chickens are the most abused animals on Earth. They have a short reproductive cycle and have been manipulated over the decades to lay more eggs or grow at unprecedented rates to raise productivity and profit. In the 1970s, a standard chicken meat company would process 3,000 animals per hour. This number increased to 8,000 per hour in the next decade and 15,000 nowadays (Abramovay, 2020).

The global chicken population was 25.9 billion in 2019, up from 14.4 billion in 2000 (Shahbandeh, 2020). That means more than three chickens for every person in the world. The numbers include egg-laying hens and broiler chickens. In 2018, the chicken population was concentrated in 3 main countries: 6.4 billion in China, 2.2 billion in the US, and 1.5 billion in Brazil (Ritchie & Roser, 2019).

The US is the world's largest producer of chicken meat, having produced over 22.3 million tonnes in 2018. China and Brazil come at second and third, with a 2018 production of 20.1 million tonnes and 15.5 million tonnes, respectively. Collectively, European countries also produce large amounts, just below the US production (Ritchie & Roser, 2019). Brazil is the leading exporter of poultry meat in the world (NCC, 2019).

The global production of chicken meat grew more than 12-fold between 1961 and 2014 (Ritchie & Roser, 2019). From 2012 to 2020 only, it increased from 83.3 million to 100.6 million tonnes, with domestic consumption increasing steadily in the main producing countries (Shahbandeh, 2021; USDA, 2020).

China is the leading producer of eggs worldwide, with a 2019 production of an astonishing 661 billion eggs. In the same year, the US produced 113 billion eggs (Shahbandeh, 2020). If we consider the annual egg production *in weight*, China (31.7 million tonnes) is followed by the US (6.5 million tonnes), India (5.2 million tonnes), Brazil (2.8 million tonnes), and Russia (2.5 million tonnes), according to 2018 data (Ritchie & Roser, 2019).

Due to advances in poultry science, the annual productivity of US hens raised from 264 eggs per animal in 2000 to 296 in 2020 (United Egg Producers, 2021). In the wild, hens will lay up to *50 eggs a year*, depending on the strain. "Spent" hens (as the industry calls laying chickens whose egg production declined) are slaughtered, and their meat is used in processed human food and pet food. Over the 12–18 months they are entitled to live in the egg industry, hens suffer from weak bones and muscles, skin injuries, and other health conditions. Therefore, their meat is deemed less valuable than that of broiler chickens. The injuries are "masked" when their flesh is minced, cooked, and added to soups, sausages, hamburgers, and other processed food.

Just like pig and cattle abattoirs, chicken slaughterhouses are coronavirus hotspots. With fewer workers due to COVID-19 infections, chicken abattoirs are having their routine disrupted. In the UK alone, over 400,000 birds were culled between January and August of 2020. The usual culling method in the UK is degassing—in other words, suffocation. Generally, the chickens are housed in specialised containers, which are filled with CO_2. Such containers are brought to the farm by depopulation companies (Kevany, 2020a). 'Depopulation' is a politically correct term for the killing of animals no longer deemed profitable. The slaughter of chickens is hardly carried out on-farm; rather, it is usually done in slaughterhouses that have the capacity to kill up to two million birds per week (Kevany, 2020a).

During the coronavirus pandemic, billions of farmed animals are being euthanised worldwide due to staff shortage and unstable market demand. After all, restaurants, cafeterias, and hotels are not operating at full capacity, if at all. Moreover, a significant number of pigs and chickens are being culled due to diseases such as swine fever, bird flu, and brucellosis. For safety reasons, their flesh cannot enter the food chain. Furthermore, millions of spent dairy cows and egg-laying hens are *routinely* slaughtered every year for second-value meat. Now think with me:

- Animal welfare is in the best interest of animal agriculture whenever it can get more value from the animals as it entails costs to the producers;
- *even* when we are talking about "high-value" animals (young egg-laying hens and dairy cows, ready-to-slaughter pigs, broiler chickens, and cattle), these creatures are submitted to painful practices;
- animal welfare during handling and transport is crucial to the meat industry because stressed and injured animals mean lower-value flesh (e.g., bruised chicken breast and DFD and PSE red meat). Still, cases of neglect, malpractice, abuse, and traffic accidents are typical;
- deliberate cruelty by farm, transport, and abattoir workers is commonplace.

Considering all of this, how do you think "spent," sick, and unwanted animals sent for slaughter are treated?

As coronavirus cases among US abattoir workers have reduced the nation's slaughtering capacity, millions of farmed animals were culled on-farm in the summer of 2020 (Kevany, 2020b). Up to May of that year, an estimated ten million hens had been killed by water-based foam smothering (Kevany, 2020b). This method relies on the generation of an atmosphere depleted of oxygen by using gas-filled foams (similar to firefighting foams) inside the poultry house. The filling gas can be an inert gas like nitrogen, which induces unconsciousness and death by anoxia (EFSA, 2019a).

Although foam suffocation is approved by the USDA Animal and Plant Health Inspection Service (APHIS), EFSA advises against it on ethical grounds. 'A blanket of high-density foam is created and spread to cover all the birds. The dense foam blocks the airways resulting in death by suffocation. In general, death due to drowning in fluids or suffocation by occlusion of the airways is not accepted as a humane method for killing animals, including poultry' (EFSA, 2019a; Benson et al., 2007).

Despite EFSA's recommendations, culling by foam suffocation is carried out in Europe. Livetec Systems, for example, is a UK-based company specialised in depopulating farmed animals. Their Nitrogen Foam Delivery System (NFDS) is one of the technologies offered for killing poultry (Livetec Systems, 2021). On Livetec's website, we find a range of on-farm killing solutions, all compliant with the UK's Animal and Plant Health Agency (APHA), ranging from NFDS to containerised gassing units. But what if these "solutions" were meant for cats or dogs? Or *humans*? Would we be browsing their website with casualness?

If your answer is 'no,' there is no reason why we should be suffocating farmed animals.

In April 2020, a couple of Minnesota contract egg farmers had their entire 61,000 bird flock culled due to the lower demand during the coronavirus pandemic. The decision was made by Daybreak Foods, a company that owned the flock and paid to feed them, and supplied eggs to Cargill. The laying hens were degassed with CO_2 a day after the notice. When asked about the birds, here is what one of the farmers told a reporter from the *Star Tribune*: 'Don't sugarcoat it. It is what it is. It's painless for the birds. I don't have a thing against that, but [...] when they euthanized the birds, that was our paycheck euthanized' (Belz, 2020).

Common killing methods for birds are grouped into four main categories: (1) electrical; (2) modified atmosphere; (3) mechanical; and (4) lethal injection. Electrical methods comprise water baths (especially in large-scale killing), head-only electrocution, and head-to-body electrocution. Euthanasia by modified atmosphere can be carried out by whole-house gassing, whole-house gassing with gas-filled foam, gas mixture in containers, and low atmospheric pressure. Main mechanical killing techniques include captive bolt, percussive blow to the head, cervical dislocation, neck cutting, and *maceration*. The latter is applied to day-old male chicks (EFSA, 2019a). Yes, you read it right: chicks are ground alive in the egg industry.

Now let us have a look at standard practices in the production of chicken meat and eggs.

Egg production

There are three main egg production systems: (1) extensive, in which the hens are raised outdoors and there is little nutritional, productive, and sanitary control, generally limited to family consumption; (2) semi-extensive, in which the birds are raised in semi-confinement, with access to outdoor areas for certain periods, or inside pens the entire time, a mixed diet of feed and vegetables, and nutritional and sanitary control; and (3) intensive, in which hens of specific strains live under extreme confinement, with a strictly controlled environment, diet, and supplementation.

Eggs produced in semi-extensive systems can be called cage-free or free-range, depending mainly on the possibility of access to external areas. Cage-free hens may live indoors at all times, while free-range chickens must be allowed to roam outdoors for at least a period of the day. Egg production can also take place within the organic system, which involves outdoor access, the absence of allopathic medication and chemical fertilisers, prohibition of certain painful practices, and an organic diet (Certified Humane Brasil, 2017). We will detail these differences in the following sections.

The *conventional* egg production system is based on cage-housing. Pens filled with wire cages piled across several floors house hundreds of thousands of birds, each cage holding up to ten hens (Greene & Cowen, 2014). This is what is called *battery cages*. The animals occupy an area the size of an A4 paper sheet during their entire adult lives. They cannot roam, forage for food, nest, dust bathe, or feel the sunlight on their little bodies. They cannot even *spread their wings*. Controlled diet, performance enhancers, and artificial lighting are applied to maximise egg production.

Battery cages are the primary housing system for laying hens worldwide. Any improvements (e.g., enriched cages, absence of cages, or outdoor access) require investment from the farmers and larger areas. Therefore, alternative systems entail a higher end price for the eggs, which only a few people are prepared to pay for.

In China, the leading egg producer in the world, only 10% of the 529 billion eggs produced yearly come from cage-free hens. In light of the growing awareness of consumers, the country is looking at switching to cage-free egg production systems to remain competitive in the international market (The Poultry Site, 2019).

In the US, 71% of hens (231.7 million) were raised under the conventional system until early 2021, with the animals crammed into cages their entire productive lives. Only 6.8% of egg-laying hens were organically raised and another 22.5% were raised as non-organic cage-free. Although the cageless numbers are low, there has been progress over the last decade. The percentage of cage-free egg production increased from 4% in 2010 to 29% in March 2021 (United Egg Producers, 2021).

In Brazil, intensive production is also the main system. Most of the egg farms operate under this model because the confinement enables to raise a larger number of birds in the same space, in addition to standardising the herd and facilitating disease control and husbandry practices.

Battery cages were banned in the EU in 2012, after a 13-year phaseout, through the EU Council Directive 1999/74/EC. Nevertheless, the EU Directive allows *enriched cages* to be used. This means that most EU producers still use cages, only larger and provided with enrichments, such as perches and litter (European Commission, 1999).

According to 2020 data from the European Commission, there are approximately 372 million laying hens in the EU. Of those, 48.0% live in enriched cages, 33.9% are housed in barns without cages, 11.9% are free-range, and 6.2% are organically raised (European Commission, 2021).

Note that the old battery cages in EU member states provided a minimum area of 550 cm^2 per animal. Currently, enriched cages (also called furnished cages or enriched colony cages) must provide a minimum area of 750 cm^2 per hen, comprising 600 cm^2 of usable space, a nest, forage material, claw shortening devices, and at least 15 cm of perch per animal (European Commission, 1999). In the US, the situation is even worse: the United Egg Producers recommend a space allowance of 432–555 cm^2 per hen (United Egg Producers, 2017).

Nonetheless, a study has shown that a hen needs 540–1,006 cm^2 to turn, 653–1,118 cm^2 to stretch her wings, and 540–1,005 cm^2 to ground scratch (Dawkins & Hardie, 1989). As you can see, we still have a long way to go in terms of animal rights. Giving enslaved animals tiny glimpses of "freedom" amidst an entire life of exploitation is far from recognising their inherent rights.

Battery cages (whether enriched or not) restrict movement and typical behaviours, leading to physical and psychological problems in laying hens. Young hens who have not started to lay are called pullets. Pullets are transferred to cages at around 16 weeks of age, where they stay until the end of the laying phase, at approximately 72 weeks of age.

Encaged hens suffer from sore and even atrophied limbs, foot deformities, and overgrown claws. The claws can get stuck in the cage flooring, causing severe injuries. Stress-induced feather pecking, vent pecking, and cannibalism add to their suffering. Exposed skin from injurious pecking is a source of inflammation and infections, other than being uncomfortable. The lack of exercise combined with the high calcium demand for egg-laying can lead to osteoporosis, regardless of the housing system. Osteoporosis and consequent bone fractures are especially prominent in caged hens (Webster, 2004).

The most common fractures are in the legs and keel bone. Bone breaks can happen at any stage throughout their hatch-to-slaughter lifetime and rarely go treated. Injured hens are normally slaughtered, as the treatment plus labour costs typically surpass their commercial value. Many fresh and healed bone fractures are detected only at slaughter. Just like in humans, bone fractures result in both acute and chronic pain in hens (Webster, 2004).

Switzerland outlawed battery cages in 1992 due to the intense distress they cause hens (Häne et al., 2000). Norway and Bhutan did the same in 2012, and other countries are following the lead (Rise, 2018; Humane Society International, 2012).

But what about cage-free and organic eggs? Are they any better in terms of animal welfare? Let us understand the differences between egg production systems before we discuss other relevant welfare issues in laying hens.

Cage-free, free-range, organic, and other labels

It is easy to get confused by the many labels when choosing eggs on a supermarket shelf. The popular belief is that cage-free and free-range eggs come from happy hens roaming large meadows, eating a wide variety of vegetation and insects, and peacefully interacting with each other. But is this the reality? Spoiler alert: unfortunately, not.

First, cage-free is a term that refers exclusively to the absence of cages. That is, all egg production systems other than conventional cage and furnished cage are referred to as cage-free, even if the hens *never* get to see daylight. Also, the housing type, diet, and husbandry practices vary considerably for different cage-free systems. For instance, barn housing, aviary housing, and organic production are all within the cage-free category, despite the significant differences between them.

Second, 'free-range' and 'cage-free' are often used interchangeably, which is incorrect. Even though free-range hens are cage-free, cage-free chickens can never get to see the sunlight, sharing indoor spaces with thousands of other birds. In turn, free-range hens *must be allowed outdoors*—although the duration and quality of the outdoor environment vary.

Noteworthy, the quality of life of cageless hens changes considerably between farms and production systems (e.g., cage-free x free-range x organic). Legislation and non-governmental certifiers impose rules on the diet, housing, grazing area, drug use, and management practices for each system. But even in the best-case scenario, the birds are still regarded as property.

And third, while free-range hens must be given access to outdoor areas, they may spend little time outside the shed compared to the amount of time they are kept indoors. Furthermore, indoor stocking densities can be remarkably high—up to 300,000 hens per shed. That is, free-range egg production can be *large-scale.*

The standards behind the 'free-range' label vary from country to country and are sometimes unregulated or uninspected. In Brazil, for example, free-range egg production must follow normative ABNT NBR 16437:2016, of the Brazilian Association of Technical Standards (ABNT). The guidelines contemplate the maximum stocking density inside (7 birds/m^2) and outside (0.5 birds/m^2) the sheds, the quality of the outdoor space (which must contain vegetation and sheltered areas), and the time spent outdoors (at least 6 hours per day). Synthetic pigments that alter the colour of the yolk, anticoccidials, and performance enhancers are forbidden (Russo, 2019; ABNT, 2019).

In the EU, similar requirements apply, with a major exception: free-range hens can spend up to *four months* with no open-air access (European Commission, 2017). The 16-week period aims to protect egg producers from financial losses due to influenza outbreaks, which require keeping hens confined as a protective measure. To be noted is that, the 16-week derogation came into force in late 2017, but a 12-week no outdoor access period was already allowed since 2008 (European Commission, 2017). In other words, eggs from hens confined for prolonged periods are still sold as free-range in Europe.

In the US, the free-range egg guidelines are exceptionally vague: there are no government specifications over the quality of the external environment and the amount of time spent outdoors. According to the USDA's website: 'For [free-range eggs], we verify they are produced by hens that are not only housed in a way that allows for unlimited access to food and water and provides the freedom to roam within the area like cage-free hens but also gives the hens continuous access to the outdoors during their laying cycle' (USDA, 2017). To put it another way, a small passage through the shed wall to an outdoor concrete patio can be considered legal. Given that the sheds are crammed with birds and that natural hierarchies develop between hens, birds that are too far away from the passage or are less dominant may rarely get to leave the shed (United Poultry Concerns, 2019).

Private schemes such as Certified Humane set their own guidelines, in general ensuring higher welfare levels than national or regional legislation (Certified Humane, 2017).

To sum up, free-range hens are not necessarily raised in backyards or small farms, and long-term confinement can be legal. As private organisations, governmental agencies, and industry groups adhere to differing standards of what 'free-range' eggs mean, it is difficult to assure the hens had a "decent" life.

To be noted is that free-range, barn, and aviary housing are all non-cage egg production systems where the hens roam inside a shed equipped with perches, nest boxes, and automatic feeders and drinkers for all or part of their lives. The barn and aviary systems differ in that aviaries are provided with multilevel structures, which improve the welfare of hens (Ahammed et al., 2014). Among these three systems, only the free-range allows outdoor access. Stocking density, feeding, and other husbandry practices are generally controlled. For instance, a maximum stocking density of 9 hens/m^2 of usable area is allowed in EU barn-laid egg production, according to the EU Welfare of Laying Hens Directive (Egg Info, 2017).

Rice husk, wood shavings, or straw are commonly used on the floor in barns and aviaries. Synthetic lawn can also be used in aviaries (Ahammed et al., 2014). Aviary is a multi-tier shed equipped with ramps, perches, and nest boxes where hens can move freely and disperse over several levels. Differently from barns, where eggs and manure can be collected by the staff, egg and manure collection is generally automated in aviaries (Ahammed et al., 2014). Aviaries can house large flocks and, thus, be used in large-scale production while allowing the hens to engage in some of their natural behaviours.

Organic eggs can be certified by national, international, and private schemes. In the EU and the UK, the housing conditions for organic hens follow the EU Organic Regulations. Egg-laying hens must be fed an organic diet and be free-range; that is, they are either reared outdoors or usually in barns with outdoor access. A maximum flock size of 3,000 birds and a maximum stocking density of 6 hens/m^2 of usable area are mandatory. Cages are banned in organic egg production (Egg Info, 2020a).

In order to promote a better environment for the animals to engage in their natural behaviours, enrichment must be provided in indoor areas (e.g., nest boxes, perches, and litter). Litter must cover a third of the ground surface for scratching and dust bathing. Growth promoters, antibiotics, feed containing GMO grains, synthetic fertilisers, pesticides, herbicides, and animal-origin ingredients are banned (Egg Info, 2020a). Forced moulting and beak trimming are strictly prohibited in organic egg production.

In Brazil, organic egg production is regulated by Law 10,831 of 23/12/2003 and Directives IN 46/2011 and IN 17/2014 from the Ministry of Agriculture, Livestock and Food Supply (MAPA). Technical Normative ABNT NBR 16437 also applies to organic eggs, although it does not have the force of law. The Brazilian standards for space allowance, husbandry practices, and diet are similar to the EU's.

Organic accreditation can also be made by Certified Humane (present in many countries around the globe). Certified Humane eggs must meet additional requirements, other than the ones set by legislation, such as at least 15 centimetres of perch for each hen and specific labelling guidelines (Certified Humane, 2020).

Non-governmental accreditation schemes are available worldwide. In Brazil, the UK, the US, and many other countries, organic, cage-free, and free-range egg farms can be certified by Certified Humane. The conditions usually surpass those set by national regulations (Certified Humane, 2020). UK organic and non-organic egg farms can be certified by the British Lion Scheme. This scheme represents more than 90% of the UK production and is managed by the British Egg Industry Council (BEIC). Organic eggs carrying the British Lion Quality label must be produced under additional standards other than those set by the EU Organic Regulations (it is not clear yet how Brexit will affect the standards). These include a maximum stocking density of 2,000 birds per hectare and the provision of popholes for outdoor access with minimum height and width, which must be open at least 8 hours a day (Egg Info, 2020a; Egg Info, 2020b).

Important to highlight is that the current global demand for eggs cannot be supplied exclusively by free-range, organic, and cage-free production due to land requirements.

Sherwin et al. (2010) assessed several welfare indicators in hens housed under four systems: conventional cage, furnished cage, free-range, and barn. The research was carried out on 26 flocks of varying sizes across the UK, reflecting real-life commercial conditions. The hen samples were visited 3 times for 47–53 weeks since the beginning of their laying phase (16–20 weeks of age). An alarming 56% of the hens had experienced a bone break during their lifetime, and 47% of all hens suffered a keel fracture during the laying phase.

If we replicate this finding to the population of egg-laying hens in the UK back when the research was conducted (33 million hens in 2010), we conclude that over 18 million birds had at least 1 bone fracture during their productive lives. Also, conventionally caged hens exhibited five times more recent keel bone fractures compared to hens housed in the other systems. The authors associated this result with the higher incidence of osteoporosis in conventional cage systems (Sherwin et al., 2010).

Casey-Trott et al. (2017) evaluated the effect of age and housing system on the prevalence of keel bone damage in laying hens. Four flock replicates of 588 pullets were reared in either conventional cages or aviary system. When laying age was reached, the hens were housed in either conventional cages, 30-bird furnished cages, or 60-bird furnished cages. Keel bone fracture assessments were done at 30, 50, and 70 weeks of age. Hens reared in conventional cages when pullets had a higher mean percentage of fractures (60%) than those reared in the aviary system (42%). This finding was explained by the positive effect of exercise during pullet age, which can result in a lower prevalence of keel fractures throughout the laying period. However, the prevalence of keel fractures in adult hens did not vary across the housing systems. Moreover, fracture prevalence increased with age in all housing systems or cage types.

Rufener and Makagon (2020) reviewed the scientific literature on keel bone fracture in egg-laying hens published between 1989 and 2019. The review reports on the relationship between keel bone fracture prevalence and several factors, such as strain, age, and housing system. Brown hens were slightly more prone to keel bone fracture than white hens, irrespective of their age. Keel bone fracture incidence increased significantly between 25 and 35 weeks of age, the latter being considered the lay peak. This increase is associated with the high daily calcium requirements in egg production. In fact, when hens are fed calcium-deficient diets, their skeleton starts to supply the calcium demand for eggshell formation.

Moreover, the prevalence of keel bone fracture was lowest in conventional cages, intermediate in multilayer aviaries, and highest in furnished cages and barns (also called single-tier system) (Rufener & Makagon, 2020). These results can be ascribed to the risks of falls associated with perches and multiple tiers (Sandilands et al., 2009; Stratmann et al., 2015).

The mixed results on the occurrence of keel bone fracture in caged versus uncaged hens indicate that:

- Bone fractures are dependent on additional factors (e.g., diet and genetics), which can impact bone strength. When comparing results from hens of different ages or belonging to different strains, we might not be able to isolate the effect of the housing system;

- the presence of perches increases the incidence of falls, both in cages and cage-free systems, which can result in keel fracture;
- flock management and husbandry practices play a crucial role in bone break prevention, regardless of the housing system. Hence, the provision of quality training to farmworkers and third parties (e.g., transport and slaughterhouse personnel) that include animal welfare and ethics may reduce the prevalence of keel bone fracture and other injuries in laying hens;
- the method used to detect bone fracture varies across studies (e.g., palpation, dissection, and radiography). Some methods exhibit higher accuracy than others, leading to overestimation or underestimation of fracture occurrence;
- welfare risks exist in all housing systems, although at different levels, since factory farming inherently deprives hens of their freedom and fundamental rights.

Sherwin et al. (2010) evaluated important animal welfare indicators other than keel bone fracture across different housing systems (conventional cage, furnished cage, free-range, and barn): gentle feather pecks given (pecks/hen/min), feather damage score, hens with feather damage (%), hens using perches (%), body weight (kg), vent-pecked hens (%), hens with old keel fractures (%), hens with recent keel fractures (%), abnormal egg calcification (%), eggs with bloodstains (%), weight of hens found dead (%), and faecal corticosterone (ng/g dry matter). Plasma and faecal corticosterone concentrations are indicative of fear and stress levels, which result in increased activity of adrenal glands. The researchers assessed the individual condition of all hens from the 26 flocks in the study sample in the beginning, middle, and end of the laying period. An additional post-mortem assessment was carried out with 150 hens from each flock. All three visits comprised a health check, collection of fresh faecal samples, behavioural observation (e.g., injurious behaviours and frustration vocalisations), and assessment of fear.

Post-mortem evaluations included skin damage, beak deformities, vent damage (not from pecking), evidence of vent pecking, foot condition, parasites, plumage damage, plumage soiling, keel protrusion, keel deformation, and fractured keels. The farmers were also asked to provide weekly information on calcification spots and bloodstains on eggshells. Eggshells displaying calcium carbonate deposits can be a sign of delayed egg-laying due to stress, while vent prolapses and ruptures can indicate excessively large eggs were laid. Potential causes of mortality for the hens found dead each week were also provided by the farmers (Sherwin et al., 2010). The study's main results were:

- Free-range hens exhibited the highest rates of vent pecking and gentle feather pecking given. However, feather damage score and proportion of the flock with feather damage were the lowest;

- free-range hens exhibited the worst skin damage and keel protrusion;
- barn hens had the highest feather damage score and the greatest percentage of hens with feather damage;
- barn hens had the most significant faecal corticosterone concentrations, almost twice the lowest concentration found in free-range hens;
- barn hens had the highest prevalence of poor plumage condition, old fractures, emaciation, and abnormal egg calcification;
- furnished-caged hens used the perches the most and had the lowest faecal corticosterone;
- conventional-caged hens sustained more fractures at depopulation;
- the lowest prevalence of problems occurred in hens in furnished cages;
- 6.8% of hens placed at the start of lay were found dead or were culled;
- all four housing systems showed an increase in mortality rate with age.

The authors concluded that despite the differences in animal welfare between housing systems, stress levels were concerning in *all of them*. The high rates of emaciation, plumage loss, and fractures across all farms indicated low welfare levels in industrial egg farming (Sherwin et al., 2010).

Worthy of note, we have been discussing only a few animal welfare problems facing the egg industry, with a focus on differences between housing systems. Let us now have a look at other relevant issues pertinent to all production systems.

Typical welfare issues in commercial laying hens

As previously discussed, bone fractures and osteoporosis (sometimes referred to as 'cage layer fatigue') is widespread in commercial layers. Calcium is depleted from the hens' bones due to the high egg production, especially when fed diets deficient in calcium, phosphorus, or vitamin D (Olgun & Aygun, 2016). Furthermore, the absorption of dietary calcium and phosphorus is dependent on vitamin D_3 (Leeson, 2015). Hens can get vitamin D_3 from the feed or by direct exposure to sunlight, which is typically limited or nonexistent in industrial production. The reduced mobility, especially in cage systems, further contributes to bone fragility (Rodriguez-Navarro et al., 2018). Common signs of osteoporosis are weakness, inability to stand, thin-shelled eggs, and paralysis. Avian osteoporosis causes pain and increases the incidence of fractures, leading to higher mortality rates and economic losses (Webster, 2004; Olgun & Aygun, 2016).

Osteoporosis in hens is not an isolated problem, but rather a widespread welfare issue in egg production. In the UK, osteoporosis-related deaths in caged layers are estimated to account for 30% of total mortality (Farm Animal Welfare Council, 2010).

Worth mentioning, a hen that has suffered from osteoporosis through part or all of her productive life has experienced pain and frustration *even* if the problem did not result in her death. Unfortunately, data on the level of the physical and emotional pain of egg-laying hens due to avian osteoporosis is unavailable. But would that make any difference for farmers and consumers? Would they stop producing and consuming animal products?

We would not need objective measurements of pain levels if we simply respected other sentient beings.

Interestingly enough, we *do* have a wealth of data supporting the extreme pain and distress farmed animals face. But indisputable evidence cannot touch hearts that have *chosen* to be immune to empathy and kindness. And this may explain the misery in which humanity is living now. *We create the world we live in.*

Beak trimming (erroneously called debeaking) is a common practice worldwide in all egg production systems other than organic. The purpose of beak trimming is to reduce stress-related injurious pecking and cannibalism due to the intense confinement, both leading to economic burden for the farmers. Although birds naturally peck each other as a social behaviour (which is called pecking order), the rate and aggressiveness of pecking are aggravated by stress and frustration. Injurious pecking and cannibalism can result in a mortality of up to 30% of a flock (Poultry Hub, 2020a).

The beak is commonly trimmed using a hot blade, a cold blade (e.g., scissors), a machine with electrically heated blades, or an infrared trimming machine. Trimming by infrared is currently preferred because it does not leave an open wound, while the other methods can cause bleeding from the beak and physical injury during handling (Glatz & Bourke, 2006). Beak trimming causes both acute and chronic pain and can also impair the animals' ability to eat and drink (Sherwin et al., 2010).

Birds have their beaks trimmed at various ages, depending on the farm's practices. Most commonly, beak trimming is carried out soon after hatching, up to ten days old, but can also occur later. Whenever beak regrowth occurs, adult birds can have their beaks retrimmed (Glatz & Bourke, 2006). Several countries have outlawed beak trimming on the grounds of animal welfare, such as Norway (1974), Finland (1986), Sweden (1988), Denmark (2013), Germany (2017), and the Netherlands (2019).

Some strategies exist as an alternative to beak trimming, such as genetic selection, light control, use of devices to restrict vision, environmental enrichment, nutrition, and beak abrasives. Let us have a look at some details of each approach (Poultry Hub, 2020a):

- Genetic selection: strain selection and molecular technology are employed to reduce the birds' mortality rate and propensity to pecking behaviour. With so much genetic manipulation, we cannot really say animal foods are natural products, can we?;
- Light control: hens are housed under low light levels so they cannot see each other well. This seems to reduce feather pecking and cannibalism, though at the expense of eye abnormalities and reduced egg-laying capacity;

- Devices to restrict vision: spectacles are fitted to the nares of the hens, preventing them from looking ahead. This approach is not only bizarre, but also causes eye irritation, eye infections, and abnormal behaviours;
- Environmental enrichment: "toys" are fitted to cages to distract the hens and reduce aggressiveness. In cageless systems, adequate perch space and quality litter in barns and the existence of trees and hedges in open areas help the birds feel safer and more relaxed;
- Nutrition: diets rich in insoluble fibre (e.g., rice hull, oat hull, and millrun) minimise cannibalism, while balanced nutrition reduces pecking mortality;
- Beak abrasives: abrasive materials are added to the birds' feed to trim their beaks and reduce pecking-related injuries.

Forced moulting is another critical ethical issue in egg production. Forced moulting consists of inducing the process of moult so that "spent" hens have a second egg-laying cycle. In nature, chickens lose their feathers every year around autumn, and hens lay very few or no eggs for some weeks to a few months. This is *natural* moulting. Over this period, the temporary respite will ensure the nutrients the hens ingest are redirected to feather regrowth. After moulting, the hens undergo a new period of enhanced egg-laying since their reproductive tract had the chance to regenerate. Post-moulting egg productivity and egg quality (e.g., egg mass, eggshell thickness, and eggshell:yolk:albumenratios) improve compared to the end of the previous laying cycle.

Hens are forced-moulted by submitting their bodies to intense stress, usually by combining dietary restriction (partial or total) with light manipulation. Forced moult is done in substitution of killing spent hens (at around 12 months of age) and replacing them with pullets. Moulted hens can have their reproductive cycle extended by more than 30 weeks (Mitrović et al., 2016). Whether forced moulting is more financially viable than the depopulation and replacement of hens depends on several factors: stock size and homogeneity, market demand, current egg price, mortality rate during moulting, and flock replacement costs (HSUS, undated; Molino et al., 2009).

Egg producers force-moult hens by either starving them or feeding them with nutrient-deficient feed (e.g., with low or zero dosages of such crucial nutrients as calcium, phosphorus, sodium, and amino acids). Alternatively, a low-density diet may be provided (e.g., ground corn, soybean hulls, and alfalfa meal) (Molino et al., 2009; Patwardhan & King, 2011). Moult regimes include withholding feed for up to 28 days, but usually 7–14 days, sometimes in combination with water deprivation for 1–2 days. Alternatively, reduced feed rations are provided to the birds (15–60 g/day/hen).

In addition, "daylight" hours are reduced (HSUS, undated; Molino et al., 2009). As the vast majority of hens are kept indoors at all times, reducing daylight hours means manipulating light intensities to mimic shorter autumn/winter days. When total feed deprivation is replaced by the provision of a strategically poor diet, the hens do not necessarily suffer from hunger. However, their bodies detect the undernutrition and translate it into a faster moult. During moult, the hens show clear signs of distress, such as weakness, aggressiveness, stereotyped pacing, and immunological depression (HSUS, undated; Silva et al.; 2017).

Hens are force-moulted until 10–35% of their body weight is lost (Andreatti Filho et al., 2019; HSUS, undated). In Australia, it is permitted by law to starve hens for up to 24 hours (including feed *and* water). There is no time limit for the provision of a restricted diet of 40–60 g/day/hen (AWC, 2002). Even though forced moulting by *fasting* is prohibited in the EU through the Council Directive 98/58/EC, forced moulting by *dietary restriction* is practised across Europe (Flock & Anderson, 2016). Induced moult is widely observed in many parts of the world, including in major producer and consumer countries, such as the US, Brazil, EU, Australia, Mexico, and Turkey (Bracke et al., 2019; Teixeira et al., 2014; HSUS, undated; AWC, 2002).

Induced moulting is such a common practice in the egg industry that thousands of scientific papers are available on the topic, written by researchers from all over the globe. A quick search on the Science Direct database (www.sciencedirect.com) in October 2020 by the keyword 'forced moult' resulted in 3,593 papers. Most of them discuss the impact of different moult regimes on post-moult performance, egg quality, bird health, flock mortality, and other variables, with a view on *increasing efficiency* and profit in the egg industry. A slim fraction of these studies (<8%) focuses on animal welfare, with most of them addressing the relationship between animal welfare and financial outcomes (e.g., suffering increases mortality rates). None of them mentions the obvious and most efficient way of preventing hens from suffering: *living vegan*.

Mortality rates during the moult can be as high as 25% depending on the moult regime, genetic strain, housing type, and flock management (Garcia et al., 2001). This explains the abundance of studies on optimum moulting methods.

In well-managed systems and under lower stress levels, around 7–12% of layers die during their reproductive age (~20–72 weeks) of different causes (FAO, 2003; Fulton, 2017). The world's chicken population does not decline due to these high mortality rates because the birds are reproduced to keep a constant or growing stock.

Fulton (2017) assessed the 15 most common "natural" causes of death in 16 commercial egg farms in Michigan, US. The researcher visited the farms once a month for 38 months (from June 2011 to July 2014) and recorded results on bird health, body conditioning, skeletal integrity, and potential causes of daily mortality. To determine the actual cause of death, a sample from each flock underwent necropsy every month.

The most common causes of mortality, from the most to the least prevalent, were: egg yolk peritonitis, hypocalcaemia, gout, self-induced moult, salpingitis (inflammation of the oviduct), bird caught by spur, intussusception or volvulus (twisted intestine), cannibalism, tracheal plug, septicaemia, fatty liver syndrome, hepatitis, persecution, and prolapsed vent. The study detected additional morbidity causes, such as hyperthermia during summer, trauma, coccidiosis, ovarian neoplasia, egg binding, urolithiasis, peritonitis (not induced by egg yolk), leg fracture, wing fracture, tumour (other than ovarian origin), exsanguination, and cardiomyopathy (Fulton, 2017). Is an omelette worth all this pain?

With increasing urbanisation, consumers grew to believe that all foods are available any time of the year—and at a constant price. Nevertheless, food production and distribution are naturally subjected to seasonal changes, climate conditions, and geographical constraints. Advancements in food technology, agronomic sciences, and veterinary medicine increased and widened the supply of foodstuffs, but at a high cost to the environment and to farmed animals. Hens, for example, are deceived by the artificial lighting into laying eggs for extended periods. Besides, their productivity increased from some 50 eggs a year to more than 300 eggs a year over the last century due to selective breeding and other husbandry practices. The same happens to cows and all other females whose reproductive cycle is exploited.

Cows do not give milk the whole year since they do *not* get pregnant soon after birth in the wild. They were not designed to fulfil the whims of human beings. We are not the centre of the Universe (seriously, we are not). Chickens lay eggs periodically, just as women ovulate every month. The purpose of eggs in nature is to give life to offspring. There is balance and wisdom in Nature—and we are disrupting them.

The egg-laying cycle varies according to the hen's age and strain, and even from individual to individual. Are we women all alike? Then why would females of other species be identical? On a farm, not all hens are the same age or lay eggs with the exact same characteristics, let alone at a specific productivity rate. Therefore, induced moulting helps farmers ensure a constant egg supply throughout the year.

The appalling disregard for animals in our society reflects on the scientific terminology, as in 'spent hens,' for instance. A recent scientific paper published in the high impact factor journal *Agriculture and Natural Resources* reads: 'Induced [moulting] of hens, when appropriately done, can provide a way of *recycling* old hens by improving the rate of egg production, albumen quality and eggshell quality during the postmolt period' (Gongruttananun & Saengkudrua, 2016, emphasis added). We are investing resources into *recycling* animals. How sad.

This kind of speciesist language reminds me of how we have named animals according to the human "need" they fulfil. Eating animals (or their ovulation, liver fat, and breast secretions) is not a necessity but rather a choice, a preference, a whim. Look at the term 'farm animals.' Throughout this reading, you might have noticed I prefer to use the term 'farm*ed* animals' instead. That is because farms are a human invention, and some animals happened to be chosen by us to be farm*ed*. The terms 'dairy cattle,' 'meat chicken,' and 'layer hen' are typical signs of the value (or devalue) we attribute to animals.

Imagine the Earth has been populated by aliens, who decided that Asian people are remarkable for their tender meat, Caucasian people have delicious eyeballs, Hispanic and Native American people are great for carrying loads, and Black females produce tasty breastmilk. We have therefore been renamed according to our *utility* (just like we have categorised animals): 'meat humans' for Asians, 'eyeballers' for Caucasians, 'beasts of burden' for Hispanics and Native Americans, and 'dairy humans' for Black people. Do you notice the speciesism? Can you see how we have reduced animals to machines?

Let us now go back to other animal welfare issues in the egg industry. Egg yolk peritonitis (EYP) is common among both commercial and backyard-laying hens, with a higher incidence in commercial flocks. Despite the popular belief, the yolk forms *before* the eggshell. When the yolk goes into the peritoneal cavity, instead of going down the oviduct (where the egg is formed), the cavity serves as a perfect medium for the growth of pathogenic bacteria.

E. coli is the most common bacteria isolated from hens affected with EYP. According to Srinivasan et al. (2013), 'egg peritonitis might be caused by either the translocation of intestinal *E. coli* into the peritoneal cavity or by the movement of cloacal *E. coli* into the oviduct followed by the ascension of these bacteria up the oviduct, through the infundibulum, and into the peritoneal cavity.' The *E. coli* infection can infiltrate through the main body cavity (called coelom), causing widespread peritonitis. EYP can lead to death due to septicaemia in a few days, if undetected.

As birds are prey in the wild, they tend to hide when they are feeling unwell to avoid being caught by predators or bullied by dominant birds within the flock. This trait makes it difficult to detect EYP at an early stage when an antibiotic treatment could still save the hen. Not surprisingly, signs of EYP—abdominal swelling, lack of interest in the surroundings, poor appetite, and eyes partially or fully closed—go rarely noticed in factory farms housing up to hundreds of thousands of birds. Also, how to detect that a caged hen is quieter than usual if she is constrained in such a way that she cannot be herself? In the 38-month-long investigation conducted by Fulton et al. (2017) with 71,487,540 hens, mortality rates due to EYP were 26.6%.

You might be asking the reason for the occurrence of ectopic yolk within the coelom in the first place. Well, EYP is primarily caused by respiratory diseases (e.g., avian infectious bronchitis), which damages the reproductive tract of birds (Ennaji et al., 2020). But other reasons exist—for example, vent trauma by pecking and physical trauma leading to egg breaking inside the body. All of the above factors are potentially found in commercial flocks, where: (a) the transport of young pullets from the hatcheries poses substantial injury risks; (b) the birds are subjected to high stress levels due to the confinement, leading to injurious pecking; (c) respiratory diseases develop in response to high ammonia concentrations, high dust levels, and the spread of pathogenic bacteria and viruses in the poultry house; and (d) *E. coli* infections (colibacillosis) are one of the five most common diseases in layers (Gingerich, 2016). Further conditions for the spread of viral and bacterial infections leading to EYP are found in poorly managed systems, such as those with inadequate sanitation, unbalanced nutrition, insufficient ventilation, and water contamination.

By the way, infectious bronchitis in chickens is caused by the *coronavirus* IBV, short for avian infectious bronchitis virus (Ennaji et al., 2020). Yes, IBV and SARS-CoV-2 are viruses from the same family (i.e., the coronaviruses). We seem to have mastered the art of creating pandemics through animal agriculture.

Importantly, any pathogen present in a hen's ovary or oviduct is prone to be present inside her eggs. *E. coli*, for example, is a pathogen involved in serious foodborne outbreaks, especially the serotype *E. coli* O157:H7. Every year, *E. coli* O157:H7 is responsible for an estimated 73,480 illnesses in the US alone, resulting in 2,168 hospitalisations and 61 deaths (Rangel et al., 2005). In the EU, *E. coli* infections are the third leading zoonotic-origin disease in humans. The number of reported cases raised from 5,901 in 2013 to 8,161 in 2018 (EFSA, 2019b). In humans, *E. coli* causes gastrointestinal infections, with the typical symptoms being fever, nausea, and vomit. The infection can develop into a haemolytic-uraemic syndrome, especially in small children and elderly people (Rangel et al., 2005).

The natural habitat of *E. coli* and other dangerous pathogens (e.g., *Salmonella* spp., *Listeria* spp., and *Campylobacter* spp.) is the intestines of humans and animals. Not a surprise, thus, that raw milk, dairy products, meat, and eggs are the top carriers of these pathogens (FDA, 2019a). The contamination of the soil and water bodies by faeces can spread the pathogens to vegetable crops, fish, and seafood. Therefore, whenever plant-based foods (e.g., ready-to-eat, washed salad) are involved in foodborne outbreaks by gastrointestinal pathogens, it is because the products have been contaminated with faecal matter from humans or animals carrying the bacteria (FDA, 2019a). Plants are *not* natural hosts of *E. coli*, *Salmonella* spp., and *Listeria* spp., but nonhuman and human animals are.

The consumption of eggs (especially raw or undercooked) is the *number one* cause of salmonellosis in humans. Common symptoms of *Salmonella* spp. infections are abdominal cramps, diarrhoea, and fever, but also lethargy, headaches, and even blood in the urine and stool in more severe cases (FDA, 2019b). In the US, *Salmonella* spp. causes an estimated 1.35 million illnesses, 26,500 hospitalisations, and 430 deaths per year (CDC, 2019). In Europe, a total of 91,857 cases of human salmonellosis were registered in 2018. Over 30% of the foodborne outbreaks reported in EU member states were caused by *Salmonella* spp. That year, salmonellosis cases explicitly linked to the consumption of eggs caused 1,801 illnesses, 341 hospitalisations, and 2 deaths in the EU (EFSA, 2019b).

Please note these numbers refer to *reported and confirmed* cases. The incidence of foodborne diseases is commonly underestimated because people hardly report the case if the symptoms are mild. Salmonellosis is the second main gastrointestinal infection in humans, after campylobacteriosis (246,571 reported cases in 2018)

(EFSA, 2019b). That leads me to another important point: the primary source of campylobacteriosis is the consumption of poultry meat.

Eggs can also be responsible for listeriosis, a potentially deadly disease caused by *L. monocytogenes*. The bacteria are typically found in fish and fishery products, meat and derivatives, cheese, raw vegetables, and eggs. Common symptoms include fever, vomiting, severe headache, and mental confusion. *L. monocytogenes* infections are especially dangerous for the elderly, pregnant women and immunosuppressed patients. Listeriosis can lead to meningitis and septicaemia, and cause miscarriages and stillbirths (Rivoal et al., 2013).

In the EU, the incidence of human listeriosis has been rising over the past decade, with 2,549 cases recorded in 2018. Although the prevalence of listeriosis is much lower than that of salmonellosis and *E. coli* infections, the proportion of hospitalisation cases and mortality rates are substantially higher. In EU countries, 97% of the 2018 infections by *L. monocytogenes* resulted in hospitalisation, and the mortality rate was 15.6%. Fatality rises in people aged 64 years and over (16.2%), and further in patients older than 84 years (17.9%) (EFSA, 2019b).

This leads me to another relevant topic: food safety. Food producers and retailers are responsible for the microbiological, chemical, and physical safety of their products. This means they must ensure the food does not pose any health risks for consumers, provided they follow the cooking and storage instructions. Microbiological risks concern viruses and bacteria that are hazardous to human health, either because they cause diseases or produce toxins. In contrast, chemical and physical risks refer to chemical compounds (e.g., mycotoxins, residues of pesticides, antibiotics, and heavy metals), insects, and objects found in the product (e.g., bones, hair, wood chips, plastic, glass, or metal fragments). Also, allergens (e.g., milk, egg, soy, wheat or gluten, and shellfish) must be indicated on the label.

Animal products are the top sources of microbiological-related and chemical-related foodborne diseases. Classic microbiological examples are *Salmonella* spp. in eggs, poultry meat, and red meat, *Campylobacter* spp. in poultry meat and milk, *Listeria* spp. in dairy products and ready-to-eat meat, and *C. perfringens* in beef and poultry meat. According to EFSA, most foodborne outbreaks in 2018 were linked to animal products, in the following order: (1) eggs and derivatives; (2) meat and derivatives; (3) fish and fishery products; and (4) milk and dairy products. That year, 20.2% of all strong-evidence food outbreaks were caused by the consumption of eggs and derivatives (EFSA, 2019b).

In April 2018, 207 million eggs produced by Rose Acre Farms were recalled in the US due to *Salmonella* spp. contamination. Rose Acre Farms has 17 facilities spread across 8 states. After 23 people from 9 states got ill, officials from the USDA ran inspections and attributed the cases to the company's facility in North Carolina. The plant produces 2.3 million eggs per day from 3 million hens and distributes the product to restaurants and retail stores in many states under several brands. This was the largest egg recall since 2010, when eggs produced by Iowa farms caused a major outbreak that affected 1,500 people (Phillips, 2018).

In 2019–2020, several US states and government agencies FDA and Centers for Disease Control and Prevention (CDC) issued food safety alerts on the consumption of eggs after *L. monocytogenes* infections were linked to hard-boiled eggs produced by Almark Foods of Gainesville, Georgia. Eight listeriosis cases were reported in five states, resulting in five hospitalisations and one death. The eggs were recalled from stores, and people were advised not to use already purchased eggs. The outbreak was considered resolved by March 2020 (CDC, 2020a). Following the initial recall for hard-boiled eggs, Almark Foods expanded it for other 88 products containing eggs produced at the Georgia plant with best-if-used-by dates through March 2020. By the time the recall was announced, the contaminated eggs had been already distributed nationwide under dozens of brands (Geske, 2019).

In March and April 2019, several egg recalls were issued across Australia over potential *Salmonella* spp. contamination. The recalls involved brands such as Steve's Fresh Farm Eggs, Southern Highland Organic Eggs, and Ash and Sons Eggs's Blue Mountains, among others. The New South Wales Health Agency urged buyers not to consume the eggs (Ruiz, 2019).

Research carried out in China revealed that 0.5% of the 33,288 commercial eggs tested were contaminated with *Salmonella* spp. (Li et al., 2020). Another study conducted in five egg processing plants in France over a year found that 8.5% of raw egg products were contaminated with *L. monocytogenes* (Rivoal et al., 2013). Chemaly et al. (2008) tested 200 egg-laying flocks in France from caged and cage-free (floor) systems. *L. monocytogenes* was detected in 15.5% of the flocks, with higher percentages in cage-free hens. Another study from France found a 30.9% prevalence of *L. monocytogenes* in caged layer hens (Aury et al., 2011). Is it wise to risk our lives for a food preference?

Pathogenic viruses and bacteria that affect birds are also a concern for farmers because they can reduce egg yield in laying hens and sometimes kill entire flocks of layers and broilers. Newcastle Disease and Avian Influenza (H5N1), for example, are viral diseases commonly found in chicken farms around the world. In chickens, these diseases manifest as respiratory, circulatory, and even nervous symptoms. Mortality rates among birds are high in both diseases. Humans infected with the Newcastle Disease virus or the H5N1 virus usually exhibit flu-like mild symptoms. However, pneumonia and acute respiratory distress syndrome can develop (Bertran et al., 2017; NHS, 2018). What the egg industry does not tell you is that both viruses can be found *inside eggs*. That is one of the crucial reasons why proper cooking is recommended.

Another welfare issue in egg-laying hens is egg binding (or dystocia), which is sometimes confused with EYP. Hens that have difficulties passing an egg within an average amount of time are considered egg-bound. The problem can be caused by improper function of the oviduct or by malformed or overly large eggs that obstruct the oviduct. Egg binding causes discomfort and pain and frequently leads to death. If the egg breaks inside the hen's body, eggshell fragments can perforate the abdominal cavity. Moreover, the presence of the ectopic yolk leads to inflammation and ultimately septicaemia, as in EYP. An egg lodged in the oviduct for a prolonged time may also compress the kidneys and blood vessels, inducing excretion and circulatory abnormalities, renal dysfunction, and ultimately shock (Poultry DVM, 2020a; Saunders, 2013).

Important underlying causes of egg binding are chronic egg laying, lack of exercise, unbalanced diets, hypocalcaemia (low blood calcium level), older age, neoplasia, obesity, stressful events, mycotoxins from the feed, and premature egg production. Breeds manipulated to exhibit a high egg production (e.g., Leghorns and New Hampshire Red) are highly prone to egg binding. Egg binding is more prevalent among commercial layers than in backyard hens due to the unnaturally high productivity, confinement, and low exposure to sunlight, which depletes their blood calcium levels.

Egg-bound hens can display a distended abdomen, a swollen vent, difficulty breathing, and may look weak or clearly straining to pass the egg (Poultry DVM, 2020a; Muthulakshmi et al., 2015). Possible treatments involve supportive care to help the hen pass the egg, management measures (such as placing the hen in a quiet, temperature-controlled environment, away from other birds), surgery, and hormone therapy (to lower her reproductive activity) (Poultry DVM, 2020a; Saunders, 2013).

In large commercial flocks, especially in cage systems, egg-bound hens go easily unnoticed and end up dying a painful death. Also, the goal of farmers is to derive income from their flocks, thereby treating egg-bound hens, especially older ones, is generally seen as a waste of resources. Besides, a hormone treatment to suppress or speed down the birds' laying activity goes entirely against the objective of laying hen farming—to optimise productivity.

Laying hens are also highly susceptible to Fatty Liver Haemorrhagic Syndrome (FLHS), which is characterised by large amounts of fat in the liver, leading to hepatic rupture and sudden death by haemorrhage (Squires & Leeson, 1988). FLHS is caused by a combination of nutritional, hormonal, genetic, and environmental factors, with a remarkably high incidence in overweight layer hens, especially those raised in cage systems (Shini et al., 2019). Surplus dietary energy is the primary cause of fat deposit in the liver. Caged hens are especially affected because they do not burn off the excess calories. According to a survey in commercial farms in Australia, 74% of necropsied birds from caged systems were diagnosed with FLHS. The percentage dropped to 0–5% in free-range hens. Out of 651 birds necropsied over 3 months, around 40% died due to FLHS (Shini et al., 2019).

Unbalanced diets lacking in antioxidants, such as vitamin E and selenium, contribute to the onset of tissue rancidity, increasing the chances of FLHS to develop (Maurice et al., 1979). Mycotoxins from contaminated cereals can also induce lipid build-up in the liver, rendering it more prone to FLHS (Bryden et al., 1979). High levels of oestrogen also contribute to the onset of FLHS by disturbing the lipid metabolism (hypercholesterolaemia and hypertriglyceridaemia) (Shini et al., 2020).

Chronic laying increases oestrogen levels, which also explains the high incidence of FLHS in commercial flocks. Heat stress also aggravates FLHS (Akiba et al., 1983). Hens suffering from FLHS show few or no symptoms. Prevalence estimates in egg farms are thereby understated, as necropsies are not conducted regularly unless in the case of a disease outbreak.

The provision of a diet rich in lipotropic nutrients (which accelerate fat burning) is a classical strategy to promote healthy liver function and to lower fat levels in the liver of farmed birds. Therefore, nutrients such as vitamin B_{12}, methionine, chlorine, carnitine, biotin, L-tryptophan, selenium, and others are regularly added to the birds' feed or water. This leads me to the false claims promoted by anti-vegans that plant-based foods lack essential nutrients. As the myth of protein deficiency in vegans no longer convinces people (as it is *not* backed up by science), the new bait is 'B_{12} deficiency.' However, omnivores are supplemented indirectly through the consumption of animal products.

Natural or synthetic nutrients, antibiotics and other drugs, pesticides, fertilisers, and GMOs ingested by the animals accumulate in their tissues, organs, and bloodstream. These substances make their way into our bodies through the consumption of animal flesh and secretions. Since hatching, chickens are supplemented with vitamin B_{12} (and many other nutrients) at specific amounts according to the type of animal and life stage. Figure 34 by DSM, one of the world's leading suppliers of nutrients to the feed, food, and pharmaceutical industries, displays the nutritional requirements for poultry. Yet animal agriculture insists that meat, dairy, and eggs are *natural* sources of essential nutrients. Are they?

POULTRY[1]

Category/Phase	Duration	Vit. A	Vit. D_3[2]	25OHD3 (Hy•D)[2]	Vit. E[3]	Vit. K_3 (Menadione)	Vit. B_1	Vit. B_2	Vit. B_6	Vit. B_{12}[7]	Niacin	D-Pantothenic acid	Folic acid	Biotin	Vit. C[8][9]	Choline
		I.U.	I.U.	mg	mg	mg	mg	mg	mg	mg	mg	mg	mg	mg	mg	mg
Broilers																
Starters	1 - 10 days	11000-15000	3000-5000	0.069	150-300[4]	3 - 4	3 - 4	8 - 10	4 - 6	0.02 - 0.04	60 - 80	15 - 20	2 - 2.5	0.2 - 0.4	100-200	400-700
Growers	11 - 24 days	10000-12500	3000-5000	0.069	50-100[5]	3 - 4	2 - 3	7 - 9	4 - 6	0.02 - 0.03	60 - 80	12 - 18	2 - 2.5	0.2 - 0.3	100-200	400-700
Finishers	25 days - market	10000-12500	3000-5000	0.069	50-100[5][6]	3 - 4	2 - 3	6 - 8	4 - 6	0.02 - 0.03	50 - 80	10 - 15	2 - 2.5	0.2 - 0.3	100-200	400-600
Broiler breeders																
Starters & growers	0 - 18 wks	10000-12000	3000-4000	0.069	80-100[4]	3 - 5	2 - 3	6 - 8	3 - 5	0.02 - 0.03	30 - 60	13 - 15	1.5 - 2	0.2 - 0.4	100-150	350-700
Layers (& male breeders)	19 wks - end	12000-15000	3000-5000	0.069[10]	100-150[5]	5 - 7	3 - 3.5	12 - 16	4 - 6	0.03 - 0.04	50 - 60	15 - 25	2 - 4	0.25 - 0.4	100-150	350-700
Hen & duck layers																
Starters (pullets)	0 -10 wks	12000-13000	3000-4000	0.069	50-100[4]	3 - 3.5	2 - 2.5	6 - 7	4.5 - 5.5	0.025 - 0.03	50 - 60	15 - 17	1 - 1.5	0.15 - 0.2	100-150	200-400
Rearing (pullets)	10 wks - 2% lay	10000-12000	2000-3000	0.069	30-35	3 - 3.5	2 - 2.5	5 - 6	3 - 5	0.02 - 0.025	30 - 60	12 - 15	1 - 1.5	0.1 - 0.15	100-150	200-400
Layers	Laying phase	8000-12000	3000-4000	0.069	20-30[5]	2.5 - 3	2.5 - 3	5 - 7	3.5 - 5	0.015 - 0.025	30 - 50	8 - 12	1 - 1.5	0.1 - 0.15	100-200	300-500
Layer breeders																
Growers & layers (& male breeders)	0 wks - end	10000-15000	3000-4500	0.069[10]	50-100[8]	2 - 5	2.5 - 3.5	10 - 12	5 - 6	0.02 - 0.04	45 - 60	15 - 20	2 - 3	0.25 - 0.4	150-200	300-500
Ducks and geese		12000-15000	3000-5000	0.069	40-80	3 - 5	2 - 3	7 - 9	5 - 7	0.02 - 0.04	60 - 80	10 - 15	1 - 2	0.1 - 0.15	100-200	300-500
Partridges, quails and pheasants		12000-13500	3000-4000	0.069	50-80	2 - 4	2 - 4	5 - 7	4 - 6	0.03 - 0.05	50 - 80	15 - 25	1.5 - 2	0.15 - 0.25	100-200	400-600
Ostriches and emus		12000-16000	3000-4000	0.069	40-60	2 - 4	3 - 5	10 - 20	6 - 8	0.05 - 0.1	80 - 100	12 - 20	2 - 4	0.2 - 0.35	200-250	600-800

[1] Added per kg air-dry feed. [2] Local legal limits of Vitamin D_3 activity need to be observed. [3] When dietary fat is higher than 3% then add 5 mg/kg feed for each 1% dietary fat. [4] Higher level for optimum immune function. [5] Under heat stress conditions increase level up to 200 mg/kg. [6] For optimum meat quality increase level up to 200 mg/kg. [7] Use upper level as reference for animal protein free diets and when cobalt is supplemented at very low levels or removed. [8] Recommended under heat stress condition and to enhance reproductive performance in breeders. [9] Use ROVIMIX® STAY-C® (ascorbyl-monophosphate) for reducing losses during processing. [10] Improve hatchability by using MAXICHICK®, the combination of Hy•D® (25-OH-D_3) with 60 ppm of CAROPHYLL® Red (Canthaxanthin). The use of MAXICHICK® is covered by a DSM Nutritional Products international patent application WO 2010 057811 nationalized in US and Europe amongst others. MAXICHICK® is a trademark registered in Europe and the US amongst others.

Figure 34. Recommended supplementation for poultry per kg of feed. Source: DSM (2011).

In humans and nonhuman animals, vitamin B_{12} is essential in the production of red blood cells and to the maintenance of mental function (Poultry DVM, 2020b). Farmed animals must be supplemented with vitamin B_{12} as they do not produce it in significant amounts. Vitamin B_{12} or cobalamin is produced *primarily by microorganisms* present in the soil and to a lesser extent in the intestine of human and non-human animals. The ingestion of producing microorganisms (through the pasture, soil, or vegetables with soil residues) enables the microbial synthesis of B_{12} in the digestive system of humans and animals by fermentation. Differently from humans, though, ruminants (such as cows, goats, and sheep) are polygastric. This means they have a multi-compartment stomach that permits a longer digestion. Longer digestion implies an extended fermentation, which explains the fact that the tissues of certain animals can store exceptionally high amounts of $B_{12.}$ To be noted is that dietary cobalt is necessary for the synthesis of vitamin B_{12} (Blezinger, 2008).

But before someone goes saying there is no difference in *producing* an essential vitamin or *being a host* for microorganisms that produce it, let us analyse four essential points:

(1) Obviously, not all animals are ruminants. Pigs, chickens, and fishes, which make up the majority of animals used as food, are not ruminants;
(2) in modern livestock agriculture, only a slim fraction of farmed animals relies on foraging or are even grass-fed. Therefore, farmed animals ingest modest amounts of B_{12}-secreting microorganisms in comparison to their wild counterparts;
(3) in animal agriculture, ruminants are traditionally supplemented with cobalt to prevent the onset of vitamin B_{12} deficiency. This is especially valid for weaners, who are generally deprived of drinking their mother's milk. Soil type, pasture species, and climate conditions may reduce the availability of cobalt in pastures. Hence, not even free-range animals have their cobalt/B_{12} requirements guaranteed. According to veterinarian Dr. Anna Erickson, from the Australian Department of Primary Industries and Regional Development: 'colostrum provides vitamin B_{12}, but milk has low levels' (Erickson, 2019);
(4) although the benefits of sanitisation obviously outweigh the cons, the consumption of treated water and washed food reduces our counts of B_{12}-secreting microorganisms to negligible levels.

In summary, farmed animals receive B_{12} and other key nutrients in their feed and water regularly, which means non-vegans are indirectly supplemented with this vitamin. Even cattle, who are ruminants, may need supplementation (Blezinger, 2008).

In chickens, vitamin B_{12} deficiency manifests as nervous system dysfunction (for instance, leg weakness), but also as perosis (bone deformation), gastrointestinal conditions, poor feathering, and lowered egg hatchability. Active laying hens must be supplemented with 0.015–0.025 mg of vitamin B_{12} per kg of feed (Poultry DVM, 2020b).

But as you might have imagined, human nutrition (let alone animal welfare) is not the primary reason for diet supplementation in farmed animals—performance is. Studies show that B_{12} deficiency in hens results in reduced egg size. On the other hand, adequate B_{12} supplementation improves laying performance and hepatic health levels. As a "bonus," vitamin B_{12} concentration in the egg yolk responds promptly to supplementation levels (El-Katcha et al., 2019), which serves both as a diet adequacy indicator and as a marketing strategy.

'Eat more eggs, they are good for you'

In recent years, a dangerous misconception has been growing around the fact that dietary cholesterol is harmless to our health. This is partly a reflection of conflicting research on the relationship between cholesterol from food intake and plasma cholesterol levels in the last decade (Barnard et al., 2019). Such controversy benefits animal agriculture, since eggs, red meat, poultry, dairy, shrimp, and shellfish are cholesterol-rich products. In fact, cholesterol is found *exclusively* in animal products (Bellows & Moore, 2012). One large egg, for example, contains 275 mg of cholesterol, which nears or exceeds the recommended daily cholesterol intake—300 mg/day for healthy individuals and 200 mg/day for people at risk of heart disease (Spence et al., 2010; Bellows & Moore, 2012; USDA, 2015a). According to the WHO, ischaemic heart disease and stroke are the world's leading causes of death, which killed over 15 million people in 2016 (WHO, 2018).

But how come decades of sound science about the consumption of cholesterol-rich foods and health risks (Weggemans et al., 2001; Spence et al., 2010; Rueda & Khosla, 2013; van der Made et al., 2014; Berger et al., 2015; Rouhani et al., 2018; Vincent et al., 2019) have been dismissed so easily? There are three main underlying reasons: (1) the influence of animal agriculture lobbyists on public health policies is growing (Agriculture Fairness Alliance, 2020; Open Secrets, 2020); (2) research funded by the egg industry on the effects of dietary cholesterol and plasma cholesterol has spiked since 2010 (Barnard et al., 2019); and (3) people want science and public policies to confirm their version of reality.

We sometimes convince ourselves that if we believe something is real, that thing will become true. We have all been there. It is like when we regret having our hair cut so short and try to find articles stating that trimming your hair makes it grow faster. We may *choose* to dismiss every article saying otherwise because it makes us feel better. When it comes to food, it is crystal clear that animal products are unsustainable and unhealthy. But for those who cling to their habits and preferences even in important matters, it is easier to disregard inconvenient truths. Ironically, these people keep tilting towards science support and science denial according to the direction in which the scientific evidence points. As if believing breathing asbestos (or eating animals) is healthy would benefit you.

Before we discuss industry-funded research, let us understand the role of cholesterol on human health.

Cholesterol, a lipid naturally produced by our organism, is involved in such essential functions as the synthesis of steroid hormones and vitamin D, as well as fat metabolism. The human body requires a certain amount of cholesterol for proper functioning. Dietary cholesterol can increase our blood cholesterol levels to such an extent as to clog our arteries, increasing the risk of heart attack and stroke (Bellows & Moore, 2012). Typical risk factors associated with high cholesterol levels include a poor diet (high in animal products and low in fruits, veggies, whole grains, vegetable oils, and nuts), lack of exercise, excess weight, smoking, and genetics (Bellows & Moore, 2012; USDA, 2015a).

The transport of cholesterol from foods to our bloodstream is carried out by lipoproteins called low-density lipoprotein (LDL) and high-density lipoprotein (HDL). LDL and HDL are also called "bad cholesterol" and "good cholesterol," respectively. LDL build-up in our arteries reduces the blood flow, causing cardiovascular conditions. Conversely, HDL transports LDL from the blood to the liver for subsequent digestion and removal from our body. That is why HDL/LDL ratios are as important as total cholesterol (HDL + LDL) (Bellows & Moore, 2012; USDA, 2015a).

The consumption of saturated fat and trans fat raises LDL ("bad cholesterol") levels. Besides, trans fats lower HDL ("good cholesterol") concentrations. *Both* saturated fats and trans fats are found in abundance in animal products. Although plant-based foods like coconut oil and palm oil are rich sources of saturated fat, the vast majority of vegetable products contains *little to zero* saturated fat. The opposite holds for animal products—whole milk, butter, eggs, cheese, meat (especially organ meats such as liver), and others are all high in saturated fats (Bellows & Moore, 2012; USDA, 2015a).

While *artificial* trans fats (also called hydrogenated fats) are found in highly processed foods (e.g., snacks and cookies), *natural* trans fats occur in products from ruminants, such as dairy and meat (FDA, 2018). There are no trans fats or cholesterol in plants. Synthetic and natural trans fats have the same impact on human health—they lower HDL and increase LDL. Despite the same chemical structure and effects on human health, trans fats obtained *synthetically* are currently banned in many countries, including the US, while *naturally* occurring trans fats (abundant in animal products) are not (FDA, 2018).

On the other hand, unsaturated fats (both mono and polyunsaturated) increase the HDL concentration in the blood. Unsaturated fats are typically found in vegetable oils (e.g., olive, sunflower, soy, canola, and safflower), nuts, avocados, seeds, and fatty fish (e.g., tuna and salmon). Foods rich in fibre (such as whole grains, fruit, and legumes) also lower LDL levels. Bear in mind that, polyunsaturated fats are *essential* fats, meaning they are required for good health but are not produced by our body. Therefore, polyunsaturated fats (abundant in algae and plants) *must* be ingested (Bellows & Moore, 2012; USDA, 2015a).

Although decades of robust science have shown that diet has a crucial effect on human health—for instance, that "good" fats are typically associated with plant-based foods, while "bad" fats are mainly related to animal products—health agencies continue to include animal foods in their dietary guidelines. A US study with 4,920 patients with atherosclerotic cardiovascular disease aged 40–85 years and tracked for 17 years revealed that the intake of drugs to lower cholesterol increased from 37% to 69% between 1999 and 2016 (Vega & Grundy, 2019). Well, if diet-related diseases (e.g., heart attack, type 2 diabetes, and others) entail societal health costs, why are governments not making proper use of the strong body of evidence on the relationship between diet and health?

Please note that in this section, we are discussing *one single* aspect of diets—fats and cholesterol. Many other serious health reasons exist for us to shift away from animal products. Who is profiting from denial? The safe bet is animal agriculture, of course, but perhaps also the pharmaceutical industry and insurance companies? No matter what the answer is, the average citizen is not gaining at all.

A systematic review of the impact of egg consumption on blood cholesterol levels conducted by US physicists revealed that the percentage of industry-funded studies increased from 0% in the 1950s to 60% in the 2010–2019 period (Barnard et al., 2019). The review comprised 211 relevant articles searched on scientific databases PubMed and Cochrane Central Register of Controlled Trials in early 2019, using the words 'egg' and 'cholesterol' together. The papers were filtered to include only intervention studies that permitted the assessment of particular effects of eggs on blood cholesterol levels, leaving 153 studies. These were split between industry-funded (59) and non-industry-funded studies (94). Studies were categorised as industry-funded whenever financial support from egg or poultry industries or companies promoting egg-derived nutritional products *were reported by the authors.*

The researchers concluded that industry-funded studies were more prone to disregard non-significant increases in plasma cholesterol concentrations than non-industry–funded studies. Also, industry-funded studies were more likely to draw conclusions that were inconsistent with the results. Furthermore, many studies (either industry-funded or not) did not report the significance of the results. By 'significant' and 'non-significant' data, the review refers to the level of statistical significance (95% or higher, denoted by $p<0.05$). The main results were:

a. None of the studies reported significant cholesterol *decreases* from the ingestion of eggs;
b. *Non-significant* cholesterol *decreases* were reported by 6% of non-industry–funded studies and 8% of industry-funded studies;

c. Cholesterol *increases* were reported by 93% of non-industry–funded studies (of which 51% statistically significant, 21% not significant, and 21% significance not reported);
d. Cholesterol *increases* were reported by only 51% of industry-funded studies (of which 34% statistically significant, 39% not significant, and 14% significance not reported);
e. Conclusions that were inconsistent with the results were reported by 49% of industry-funded studies and 13% of non-industry–funded studies.

It is relevant to state that the systematic review did not intend to detect whether data fabrication occurred, but whether the conclusions were supported by the studies' results, taking statistical significance into account. It became clear that industry-funded studies were biased into concluding that eggs do not raise cholesterol levels, despite the data indicating otherwise.

The systematic review by Barnard et al. (2019) also cites another review by Griffin and Lichtenstein (2013) on the association between dietary cholesterol and plasma lipoprotein profiles. The 2013 analysis assessed 112 articles searched on the MEDLINE database using combinations of the words 'cholesterol,' 'dietary,' 'eggs,' and 'dietary cholesterol,' published within the 2003–2013 period. Only 12 of the 112 articles met the eligibility criteria and were thus included in the review. The authors concluded that the impact of food cholesterol on plasma lipids concentrations is modest and limited to population subgroups. More specifically, while the authors state that restrictions in cholesterol-rich foods are justified for people at risk of vascular disease, they cast doubt on the validity of these restrictions to healthy individuals due to interindividual variability. Interestingly, 10 out of the 12 articles evaluated were funded by egg industry programs. Furthermore, decades of evidence (before 2003) on dietary cholesterol and blood cholesterol were disregarded by using the one-decade inclusion criterium.

Worth mentioning, one of the authors served on the US Dietary Guidelines Advisory Committee in 2005 (Barnard et al., 2019). The other author is currently affiliated with Pfizer, the multinational pharmaceutical corporation. At the time the review was published, both authors were affiliated with the Jean Mayer USDA Human Nutrition Research Center on Aging at Tufts University (based in Boston), which is supported by the USDA and has "[...] made significant contributions to U.S. and international nutritional and physical activity recommendations, public policy, and clinical healthcare" (Tufts, 2020).

An article published in 2018 found that cholesterol from eggs is not well absorbed and, thereby, does not affect plasma total cholesterol levels (Kim & Campbell, 2018). In the 'funding' and 'conflicts of interest' sections, the authors declared the research was funded by the Egg Nutrition Center of the American Egg Board, among other funders. They also stated: "The authors declare no conflict of interest. Representatives from the American Egg Board–Egg Nutrition Center were not involved in the design implementation, analysis, or interpretation of data from this investigator-initiated study."

The American Egg Board's primary goal is to promote egg consumption in the country. Their website reads 'The American Egg Board (AEB) was created by an Act of Congress in 1976 at the request of America's egg farmers, who desired to pool resources for national category-level egg *marketing*,' and: 'Home to The Incredible Egg and Egg Nutrition Center, AEB is dedicated to *increasing demand* for all U.S. eggs and egg products. For more than 40 years, America's egg farmers have supported this mission by funding the AEB. The AEB is 100 percent farmer-funded, and those funds directly support the *research*, *education* and *promotion* necessary to market eggs.' They also state: 'America's egg farmers earned a return of USD 9.04 for every marketing dollar invested in the AEB' (AEB, 2020, emphasis added).

If cholesterol from eggs were indeed not harmful to human health, why would industry and scientists engage in research on the reduction of cholesterol levels in eggs? Scientific articles on strategies to reduce cholesterol and improve the fatty acid profile of eggs are profuse. These focus on varying strategies, from hen diet optimisation (Batkowska et al., 2021; Sun et al., 2018; Mattioli et al., 2016) to post-lay processing—for instance, ultrasonic-assisted enzymatic degradation of cholesterol (Sun et al., 2011).

Instead of focusing on the real issue (eggs are rich in cholesterol, which are terrible for our health), the egg industry insists on the fact that eggs are great sources of "high-quality protein." Yes, I agree eggs are rich in protein. But this isolated claim is as valid as the one advising people to drink water-based wall paint to hydrate their bodies. While wall paint can contain a large amount of water in its composition, it is also made of many harmful chemicals.

The marketing strategy of focusing on the unique benefits of a particular nutrient is largely used by animal agriculture. The "nutrient-rich" tactic is what made us associate milk with calcium, beef with iron, and eggs with lean proteins, while plant-based foods can be equally or more abundant than these products. In the words of US physician John McDougall:

> Focusing on the abundance of an individual nutrient accomplishes an even more insidious marketing goal; it diverts the consumer's, and oftentimes the professional dietitian's, attention away from the harmful impact on the human body of consuming all kinds of animal foods. In my 42 years [in 2010] of providing medical care, I have never seen a patient sickened by eating potatoes, sweet potatoes, corn, rice, beans, fruits, and/or vegetables (unspoiled and uncontaminated). However, during my everyday practice I have witnessed (just like every other practicing medical doctor has) a wide diversity of diseases, including heart attacks, strokes, type-2 diabetes, arthritis, osteoporosis, and cancer, from eating fresh killed and/or collected, as well as processed and/or preserved, animal-derived foods. (McDougall, 2010)

Who are the animal industry's narratives benefiting? Perhaps we could look to the word of the American Egg Board: 'We are honored to serve America's egg farmers.' They are certainly not serving vegans and not even consumers of animal products, much less the animals and the planet. They are serving those who profit from animal exploitation. Think about that.

Grated chick: an egg industry speciality

One of the most controversial aspects of egg production is the culling of day-old chicks. Globally, approximately *seven billion* male chicks are killed every year. Just like male calves, cockerels (male chicks) are by-products of an industry that exploits the reproductive cycle of females. Chick culling is standard practice across all egg production systems, *including organic*. In the chicken meat industry, both male and female birds have commercial value. In contrast, the low fattening performance of layer chickens renders them inappropriate for meat production. Therefore, male layer chicks are simply discarded.

By the way, there are no layer and broiler strains in Nature—chickens have been genetically manipulated since the last century to serve specific purposes.

While male layers are unwanted in the egg industry, their female counterparts will replace their mothers in the egg production chain. Stock replacement is the reason why egg-laying hens are fertilised (I hate to refer to animals as "stock," but that is what we reduce them to when we consume animal products). Only a small fraction of male layers is kept for breeding purposes. Culling methods include maceration using a high-speed grinder, asphyxiation by CO_2, and cervical dislocation—all of them carried out without anaesthetics. In many hatcheries, the chicks are simply thrown in plastic bags, where they suffocate to death. Please check the undercover investigations carried out by Animal Equality (www.animalequality.org) and see for yourself.

Maceration is the primary method in the US, where an estimated 260 million cockerels are culled per year (AVMA, 2020; HSUS, undated). Asphyxiation is the exclusive method used in the UK, where 40 million male chicks are killed annually (Saul, 2015). Asphyxiation is also the primary method in Germany and France, which together cull nearly 100 million male chicks every year (Deutsche Welle, 2020; Ohier, 2020). RSPCA approves the gassing and maceration of chicks, poults, and ducklings by trained personnel (RSPCA, 2017b).

One of RSPCA's goals is to improve the quality of life of farmed animals worldwide. If higher animal welfare–certified eggs involve the grinding or suffocation of baby roosters, can you imagine what the standards are in uncertified producers?

More important than improving the welfare standards in animal agriculture is ending animal exploitation whatsoever. As long as the animals are valued for their utility, violence and oppression will continue to form the basis of our society.

The negative image of chick culling among consumers menaces the egg industry. Therefore, alternatives are being evaluated that identify the sex of unborn chickens.

Considering that around 50% of layer chicks are male, in-shell sexing entails lesser costs to the farmers with housing and feeding animals that are commercially valueless.

In Ovo, a Dutch spinoff from Leiden University, developed a screening machine that sexes chicks within 1 second with 95% accuracy (In Ovo, 2020). According to studies, the embryo does not feel pain until the seventh day from incubation due to the early stage of the nervous system development (Krautwald-Junghanns et al., 2018). An egg typically takes 21 days to hatch after incubation. In Ovo has been working with industrial partners towards upscaling and integration of the equipment with current hatchery machinery (In Ovo, 2020).

France is planning to ban chick culling by maceration by the end of 2021, but the other culling methods will remain legal (Le Monde, 2020). Other countries, such as Germany and Spain, are working towards a ban on male chick culling in general (Deutsche Welle, 2020; Animal Equality, 2020a). Technologies such as the one developed by In Ovo can help the transition towards a chick culling–free industry. The spinoff has been granted EUR 2.5 million from the European Innovation Council (EIC) to speed up development (In Ovo, 2020). Nevertheless, it is unlikely that small hatcheries will be able to incur the costs of such technology.

Currently, the industry relies on the manual sexing of day-old chicks. Professional chicken sexers are expected to identify the sex of 800 to 1,200 chicks per hour with a 97–98% accuracy. Even though in-shell sexing can potentially save billions of day-old chicks around the world, we should not disregard all the other lives being taken in the egg industry—for example, spent hens. Moreover, ending animal use in general should be our main concern. We should not be turning to technology where a simple consciousness examen suffices.

Broiler chickens

Over the last century, chickens have been genetically manipulated to produce strains with specific purposes: meat production versus egg production. Chickens raised for meat production are called broiler chickens. Broiler chickens grow faster than laying hens in order to produce meat in a shorter period. Strains selected for high conversion to meat have their egg-laying capacity considerably reduced.

Slaughter age varies depending on genetics, production system, sex, diet, and final weight desired. But in general, broilers are slaughtered at 4–7 weeks of age, with organic and free-range birds reaching slaughter weight later, at 12–16 weeks (Bessei, 2006). Both males and females are used in poultry meat production.

As we have seen, spent hens are killed for second-value meat when their egg-laying performance declines. However, that is what layer flesh is: *lower graded meat.* The bulk of poultry meat in the food chain comes from broiler breeds.

Broilers are usually reared under intensive conditions in sheds of varying capacities, some as large as 20 m x 180 m. Even though there are alternative rearing systems, such as organic and free-range, the vast majority of chicken meat comes from farms that confine the birds in bare sheds their entire lives. Notably, only very small farms breed, raise, and slaughter the birds, and even subsistence farms depend on chickens and inputs produced on an industrial scale.

Most commonly, chicks are born in hatcheries, which incubate the eggs fertilised on broiler breeder farms. Large hatcheries incubate 30,000–100,000 eggs per day (de Carvalho et al., 2013). Broiler breeder farms raise parent stock, female and male birds that produce the fertilised eggs sent to hatcheries. In turn, parent birds are bought as day-old from primary breeding farms that breed pedigree chickens or purebreds. Birds born from purebreds are called great-grandparents. The mating of great-grandparents results in grandparents. From the mating of grandparents, hybrids called parent stock are born. The chicks from parent birds will become commercial broilers (Jiang & Groen, 1999).

The primary breeding sectors rely on genomics research to select the best genetic traits, depending on the purpose intended for the birds (eggs or meat). The pedigree birds' progeny reaches grandparent and great-grandparent generations. A typical broiler-chicken farm buys chicks from hatcheries and fattens them up to slaughter weight.

In summary, the production of chicken meat is an *integrated industrial process*, so even small producers are interconnected to a highly technological and large-scale system. The image we have in mind when we picture chicken farms is just the *last stage* of broiler production before slaughter.

Modern commercial broilers differ significantly from chickens from a century ago. They can reach two kilograms in a little over one month due to selective breeding, optimal feeding, disease control, and husbandry practices. Within 80 years, from 1925 to 2005, the time required to reach 1.5 kg dropped from 120 to 30 days (Bessei, 2006). The unnatural growth rates render broiler chickens particularly susceptible to skeletal deformities and rickets. Metabolic and infectious diseases, such as coccidiosis and dysbacteriosis, are typical. Respiratory conditions also arise in response to the high ammonia and dust levels in the barns. Particles from faeces, litter, feed, and feathers spread into the environment, irritating the animals' respiratory tract and carrying pathogenic microorganisms (Almuhanna et al., 2011).

The litter on the floor of poultry houses accumulates manure, which is rich in uric acid (80%), ammonia (10%), and urea (10%). Excreted uric acid and urea convert into ammonia by enzymes and microorganisms naturally present in the excrements. The main factors influencing the conversion rate of urea and uric acid into ammonia are air temperature, pH, and humidity. Ammonia is a toxic and odorant gas, which causes discomfort and injury to birds and humans. Therefore, a vast amount of research has focused on ways to control the ammonia levels inside barns, including managing the litter type, stocking density, housing type, ventilation rate, manure handling, bird activity, and even the chickens' diet (Naseem & King, 2018; Almuhanna et al., 2011).

The litter is normally replaced only once, after the birds leave for the slaughterhouse. This means the ammonia and dust concentrations in the barn and the moisture levels in the litter increase gradually over the weeks, impacting the welfare and health of the birds.

The lowest ammonia detection limit in humans and birds is 5 ppm, while concentrations as of 300 ppm pose immediate life risks. Exposure to concentrations as low as 35 ppm must be limited to a maximum period of 10–15 minutes for health reasons (Alltech, 2021). Researchers report that the exposure of birds to 20 ppm for long periods results in damage to the respiratory tract and debilitated immune system. In fact, brief exposure to high ammonia levels can be as damaging to the birds' organism as more prolonged exposure to moderate concentrations (Alltech, 2021). In terms of welfare, poultry show aversion towards environments with ammonia levels over 10 ppm (Jones et al., 2005).

A study was conducted with 4 flocks of female broilers housed in barns with 4 different ammonia concentrations (4, 11, 20, and 37 ppm) over 16 days. The effect of light intensity (10 lx or 100 lx) associated with 2 different day periods was also evaluated. The birds avoided the 2 highest concentrations, but spent the same amount of time in the sheds with 4 ppm and 11 ppm during bright periods. However, the broilers preferred the 4 ppm barns during dim periods (Jones et al., 2005).

In RSPCA-certified farms, ammonia and dust levels must be assessed daily, and ammonia concentrations must not surpass 20 ppm (RSPCA, 2017a). Nonetheless, ammonia levels in broiler houses commonly exceed 25 ppm, reaching up to 80 ppm, especially during winter, when artificial ventilation and heating are required (Yi et al., 2016). A summary of the harmful effects of ammonia inhalation by animal feed global company Alltech is shown in Table 2.

Table 2. Harmful effects of ammonia concentration in poultry and humans

Concentration (ppm)	Effects
5	Lowest detectable level
6	Irritation of the eyes and the respiratory tract
11	Reduced animal performance
25	Maximum exposure level allowed for a period of one hour
35	Maximum exposure level allowed for ten minutes
40	Headache, nausea, and loss of appetite in humans
50	Severe reduction in performance and animal health; higher pneumonia rates
100	Sneezing, salivation, and irritation of mucus membranes in animals
≥ 300	An immediate threat to human life

Source: Alltech (2021).

Globally, more than 70% of broilers are raised under industrial conditions. Intensive chicken farms house the vast majority of chickens in the EU, the US, the UK, Brazil, and China (CIWF, 2020). High stocking densities combined with insufficient ventilation frequently result in heat stress, as well as difficulty breathing, trachea irritation, air sac inflammation, eye damage, blisters on the feet, and burns on the chest and legs (called hock burns). High stocking densities also contribute to the proliferation of many viral and bacterial diseases, such as bird flu and necrotic enteritis (Tsiouris et al., 2015).

Broilers are reared cageless, but this does not mean they are free to roam outdoors, ground peck, or engage in other natural behaviours. Although there are different housing systems, most broilers are packed inside an overcrowded shed with limited or no sunlight their entire lives. The EU legislation, for instance, permits a stocking density of approximately 19 birds/m^2 (depending on their weight at slaughter). In practice, this means less than an A4 sheet of paper per bird (CIWF, 2020).

Broilers are still very young birds at the time of slaughter since the average chicken has a lifespan of 5–10 years and can live up to 15–20 years, depending on the breed (HSUS, undated). Farmed animals' lives are literally worth a meal.

During their short lives in these barren sheds, broilers have little space for exercise, which decreases as the birds grow. Intensively farmed chickens can spend much of their time lying down due to painful leg disorders, as their body is too heavy for their legs. Even though drinkers and feeders are available throughout the shed, lame birds have difficulty reaching them and may die from hunger or dehydration.

Cardiac problems are commonly seen among broilers, especially in males and fast-growing strains. Asymptomatic birds can die suddenly due to heart arrhythmia after a short wing-flapping convulsion. Other birds simply flip over and die on their backs. This is called Sudden Death Syndrome or Flip-over Disease. The most likely trigger of heart arrhythmia is stress. The only strategy known thus far to avoid Sudden Death Syndrome is to slow down the growth rate of broilers, especially during the first three weeks from hatching. The syndrome affects 0.5–4% of a flock (Sander, 2019a). This might seem like a low prevalence, but the numbers are sizeable if we consider that over 50 billion chickens are reared as a food source every year in the world. Millions of chickens die from a heart attack every year in the UK alone (CIWF, 2020).

Diet and light are manipulated to ensure a determined feed intake and proper feed conversion ratio (FCR). FCRs vary geographically and even from barn to barn in a single farm, with a national average of 1.8 kg of feed per kg of body weight gain in major producing countries, such as Brazil and the US (da Silva et al., 2019; NCC, 2020). In 2012, New Zealand's poultry company Tegel announced to have reached the world's best broiler conversion ratio, with an average of 1.5 (Thornton, 2012). In livestock operations, animals are seen and treated like machines; thereby, the FCR is an important performance indicator. The lower the FCR, the more efficient the animals are at converting feed into meat.

Worth stressing, animals are *considerably less efficient* in transforming nutrients into food compared to plants because much of the conversion efficiency is lost through the animals' metabolism and non-edible parts (bones, blood, internal organs, and others). Some might claim there is a use for virtually all non-edible parts of animals. However, this is also true for plants, algae, and fungi. The same way industry has created a use for animal by-products, it can develop applications for all parts of plants, algae, and fungi. It is past time industry changes its focus.

FCRs have been reduced primarily through advancements in poultry genetics and animal nutrition. Nutrition strategies include the use of exogenous enzymes that lower antinutritional factors in the feed, allowing for more efficient nutrient use (Aderibigbe et al., 2020). In the US, FCR in chicken meat production dropped from 4.7 in 1925 to 1.8 in 2019, whilst mortality decreased from 18% to 5% in the same period. Thereby, the market age decreased from 112 to 47 days (NCC, 2020).

The intensive selection of broiler chickens targeting specific traits (e.g., increased appetite and lower FCR) reduces sexual activity and fertility (De Jong & Guémené, 2011). Therefore, broilers (both males and females) used for breeding are kept on very restricted diets (a quarter to a half the normal ration) and must cope with chronic hunger in addition to the usual frustration and stress inherent to animal farming (De Jong & Guémené, 2011; De Jong et al., 2003; Savory et al., 1993).

The many life-threatening problems associated with the unnatural growth rates and the confinement do not manifest fully in broilers, as they are killed well before adulthood. However, broiler-breeders are slaughtered later (since the primary purpose they serve is stock maintenance) and thereby suffer from severe skeletal and cardiovascular conditions for longer. A UK study showed that 20% of chickens allowed to eat *ad libitum* had to be killed or simply died as early as 11–20 weeks of age due to severe lameness or heart failure (Savory et al., 1993).

Broiler-breeders can have their reproductive performance manipulated by the administration of hormone treatments—for example, gonadotropin-releasing hormone agonists (GnRHa) (Hezarjaribi et al., 2016). Hens are prone to several kinds of injuries related to overmating—for instance, due to males' sharp claws and spur bud (De Jong & Guémené, 2011). To prevent injuries to the hens during mating and to the farmers, breeder-broilers can be declawed, despurred, and have their beaks trimmed without anaesthesia (De Jong & Guémené, 2011).

Virus and bacteria spread go rampant in animal farms due to overpopulation, manipulated diet, confinement, selective breeding, and all practices aiming at greater production in the shortest time. Some common examples of microbial-origin diseases in broiler chickens are Newcastle Disease Virus (NDV), Avian Influenza (H5N1), Gumboro Disease, Marek's Disease, Fowl Cholera, necrotic enteritis, and salmonellosis. Chickens are also susceptible to parasites, such as protozoa, intestinal worms, mites, ticks, fleas, and lice. However, the (over)use of antibiotics and vaccination in poultry farms contribute to the relatively low mortality rates (Tiseo et al., 2020).

The global average use of antimicrobials in animal agriculture is estimated at 45 mg/kg for cattle, 148 mg/kg for chicken, and 172 mg/kg for pigs, considering 1 kilogram of meat produced. The global consumption of antimicrobials is forecasted to increase by 67% between 2010 and 2030, reaching nearly 95,000 tonnes (Tiseo et al., 2020).

Newcastle Disease Virus (NDV), which we briefly discussed in a previous section, is a highly pathogenic viral disease affecting birds that is transmissible to humans. Infected birds may exhibit gasping, coughing, mucus discharge from the nose, reduced appetite, nervous symptoms like twisted heads and paralysis, cyanosis of comb and wattle, swelling of tissues around eyes and neck, and diarrhoea. Most birds are asymptomatic in the first days of contraction. Mortality rates range from 10% to 90%, depending on the virus strain (Bertran et al., 2017).

Humans can contract NDV from contact with infected birds and their excretions and body fluids, but also with contaminated surfaces (e.g., equipment and feeders), shoes, and clothing. Infected people exhibit symptoms resembling flu, as well as mild conjunctivitis. NDV is found throughout the world, with a particular prevalence in developing countries (Bertran et al., 2017). An outbreak persisted in California poultry farms from 2018 to 2020 (CDFA, 2020).

Avian Influenza (H5N1), which we also mentioned previously in this chapter, is a severe and highly infectious viral disease in birds. Avian Influenza, also called bird flu, spreads primarily through contact between sick and healthy birds. Indirect spread through contaminated feeders and drinkers can also occur. Infected birds can present symptoms like those of Newcastle Disease Virus: cyanosis of the wattles, combs, and legs, diarrhoea, nasal discharge, coughing, sneezing, lack of coordination, reduced appetite, and ruffled feathers. People can contract H5N1 through contact with infected birds, their mouth, eyes, and nasal discharges, as well as their manure. Equipment, shoes, and clothing can also carry the virus (Bertran et al., 2017). Contamination can occur during husbandry procedures, transport, slaughter, and even while *preparing and cooking infected poultry meat* (NHS, 2018).

H5N1 symptoms in people include aching muscles, headache, high temperature and a cough, but also sickness, diarrhoea, chest pain, stomach pain, conjunctivitis, neurologic changes, and bleeding from the nose and gums. The symptoms can develop into pneumonia (NHS, 2018). In humans, H5N1 has a 60% mortality rate (WHO, 2011).

Gumboro Disease, or Infectious Bursal Disease (IBD), is a highly contagious viral disease of young chickens and turkeys. Clinical signs include a sudden drop of appetite, prostration, and watery diarrhoea. Despite the relatively low mortality rates (up to 20%), IBD is economically relevant because weight gain is delayed in infected broilers. Besides, vaccines are not as effective in chickens that had previous IBD, which predisposes them to other diseases. The disease spreads from chickens infected with the IBD virus to healthy birds through physical contact, feed, and water (Poultry Hub, 2020b).

Marek's Disease is also a highly contagious virus infection in chickens, turkeys, quails, and other birds. The virus affects the central nervous system and causes the formation of tumours in muscles, organs, and epithelial tissues. Symptoms include weight loss, grey or misshapen iris, roughened skin around feather follicles, vision impairment, purple comb due to inadequate oxygenation, and paralysis of legs, neck, and wings. The disease spreads mainly through dead skin cells from sick birds, but the virus can also be carried on clothing and shoes. Mortality can reach 100% (The Poultry Site, 2018).

As in Gumboro Disease, birds affected with Marek's Disease are immunosuppressed and, thus, vulnerable to other diseases. The virus survives at ambient temperature for up to 65 weeks and is resistant to some disinfectants. Marek's Disease is widespread all over the globe, with the virus continually mutating. Vaccination of day-old chicks is the primary preventive measure (The Poultry Site, 2018).

Fowl Cholera, or pasteurellosis, is a common bacterial disease in chickens, pheasants, turkeys, pigeons, and sparrows. The disease is primarily contracted through the intake of water and feed contaminated with the bacteria *Pasteurella multocida*. The bacteria can spread into the environment via shoes, clothing, equipment, and rodents. Even though some birds are asymptomatic when they die, infected birds generally exhibit loss of appetite, diarrhoea, coughing, ruffled feathers, lethargy, swollen joints, wattles, combs, and difficulty walking. Nasal and eye discharge can also occur. Sanitation, rodent control, and vaccination are considered the best measures to prevent the disease. As antibiotics will not eliminate the bacteria from the flock, eradication is only guaranteed with depopulation followed by disinfection. Antibiotics will only reduce mortality rates. High stocking densities and concurrent infections are predisposing factors for Fowl Cholera (Sander, 2019b).

C. perfringens are bacteria that naturally colonise the gastrointestinal tract of healthy birds. However, strains that produce the toxin NetB causes necrotic enteritis, an economically important disease due to the high spread and mortality rates within layer and broiler flocks alike. The bacteria and the toxin degrade the intestinal mucosa of chickens, resulting in tissue necrosis and, ultimately, death. Clinical symptoms include lack of appetite, diarrhoea, and depression. The subclinical disease manifests as increased FCR and slower weight gain (La Mora et al., 2020).

C. perfringens is one of the major three causes of foodborne diseases in humans, with the vast majority of cases linked to the consumption of poultry meat, beef, and pork. Depending on the strain, the bacteria's toxins can cause mild to life-threatening conditions. The most usual symptoms are abdominal pain and watery diarrhoea, with no vomiting or fever (La Mora et al., 2020). In the US alone, at least one million people get sick from *C. perfringens* type A food poisoning every year (McClane, 2014).

The most common zoonosis in humans is campylobacteriosis, with the consumption of poultry meat being the leading cause of infection by *Campylobacter* spp. In 2018, 70% of the reported foodborne diseases in EU member states referred to campylobacteriosis. Over 30% of the *Campylobacter* spp. infections that year involved hospitalisation (EFSA, 2019b). The approximately 250,000 annual cases recorded are probably underestimated, as people with mild symptoms rarely report the disease. EFSA estimates that the real number of cases approaches *nine billion* every year. According to the agency, campylobacteriosis costs the EU some EUR 2.4 billion a year (EFSA, 2019b).

In chickens, *Campylobacter* spp. infections can result in declined laying performance, intestinal inflammation, diarrhoea, and higher susceptibility to secondary diseases. The species *C. jejuni* can cause enteritis in newly hatched chicks, ultimately leading to death. However, as the bacteria are generally non-pathogenic in birds, infestations go frequently undetected, which poses serious food safety risks (Lee, 2019). All because we insist on eating animal flesh.

Salmonelloses are infections caused by different *Salmonella* spp. strains. Some of them have adapted specifically to poultry, such as *S. pullorum* and *S. gallinarum*. These strains pose few health risks to humans but cause, respectively, Pullorum disease and fowl typhoid in chickens. Paratyphoid species of *Salmonella* spp., however, are nonhost-adapted and can transmit to virtually all animals, both humans and nonhumans.

Pullorum disease, for example, is an acute systemic disease caused by *S. pullorum* that affects chickens younger than three weeks old. Sick birds appear weak and drowsy, show little appetite, have difficulty standing, huddle near heat sources, and have whitish diarrhoea. The faecal matter sticks to the vent, which can cause a fatal blockage if not cleaned. Defecating becomes painful for the chicks, who rapidly grow weaker and die within a couple of days after the first symptoms appear (Yeakel, 2019).

S. pullorum can be transmitted from laying hens to their chicks via eggs, through contact with infected birds, or by indirect contact with infected surfaces. Once infected, broilers and layers will become carriers of the pathogen indefinitely. Therefore, antibiotic treatment is not recommended, but rather the removal of the infected birds from the flock (Yeakel, 2019).

Paratyphoid *Salmonella* bacteria are *Salmonella* spp. serovars linked to human foodborne diseases, including *S. enteritidis* and *S. typhimurium*. Although treating poultry with antibiotics may reduce morbidity and mortality, the birds usually become longstanding carriers of *Salmonella* spp. Even asymptomatic birds are carriers of the bacteria, which can then be ingested through the consumption of eggs and poultry meat. As we have discussed in a previous section, salmonellosis in humans causes mild to severe gastrointestinal conditions and fever, being one of the top foodborne illnesses in the globe (EFSA, 2019b). Vaccination of the animals and good hygiene practices continue to be the best measures against *Salmonella* spp. Nevertheless, the asymptomatic infection in chickens and the high zoonotic potential render the eradication of *S. enteriditis* and *S. typhimurium* a significant concern in the food industry (Hernández, 2014).

Parasites like fleas and mites can also carry pathogenic viruses or bacteria and, therefore, infect birds and their products. In addition, pathogenic protozoa (single-cell organisms) and helminths (worms) can colonise the gastrointestinal system of chickens, causing severe diseases.

Coccidiosis, for example, is a disease caused by the protozoan parasite *Eimeria* spp., popularly known as coccidia. Coccidia eggs (or oocysts) can be ingested through contaminated feed and water or insects and worms. Even when the oocysts are in a latent state, they can sporulate in the bird's gut. The oocyst damages the intestinal tract of chickens, causing diarrhoea (sometimes bloody) and loss of appetite. Sick birds will eat less and have difficulty absorbing nutrients, which will hinder their growth rates. Coccidiosis is treated with antiprotozoals and antibiotics to prevent the onset of secondary microbial-origin diseases, such as necrotic enteritis. Humid litter and overcrowding are predisposing factors for coccidiosis (Tsiouris et al., 2015).

Worth mentioning, birds with coccidiosis eat less and have difficulty absorbing nutrients, which affects their conversion rates. Therefore, anticoccidials are used prophylactically in the intensive production of eggs and chicken meat to stimulate growth and improve performance.

But why am I discoursing on diseases that affect chicken? First, because they result in animal suffering. Second, because animal products are important carriers of pathogens that infect humans. And third, because farmed animals are treated with vaccines, antibiotics, antiparasitics, and varying medications, which create dangerous superbugs.

Contrary to the prevailing belief, antimicrobial residues *per se* in animal products are not the main health risk, but rather antimicrobial *resistance*.

To be noted is that antibiotics and other synthetic drugs are not permitted in organic poultry farms but are allowed in all other schemes: conventional indoor housing, indoor with higher welfare, and free-range. Let us see the details of those systems.

Free-range, organic, and other labels

Free-range and organic chickens have theoretically a better quality of life than those conventionally raised—although the abuse inherent to animal agriculture persists. Both free-range and organic broilers have outdoor access and opportunities for expressing their natural behaviours. These chickens are fed grains but can also eat pasture, raw vegetables, and bugs, rather than primarily corn- and soybean-based feed as in intensive-confinement systems. Their psychological and physical health is considerably better compared to housed birds, because they grow slower, exercise more, are given a varying diet, and get sunlight and fresh air.

The 'higher-welfare indoor system,' also called 'indoor enriched system,' is used in some farms, although it is not regulated by legislation. Private schemes like RSPCA set rules for the system. The purpose of the 'higher-welfare indoor system' is to give indoor-raised chickens more space and a richer indoor environment where they can forage and perch. Enrichment can include natural light, perches, and straw bales. Broilers can be from slower-growing breeds, which mean they will live one to two weeks longer than conventionally raised birds. Higher-welfare indoor chicken farms house 14–16 birds/m^2, rather than the stock density of 19 birds/m^2 set in the EU legislation for conventional broiler farms (CIWF, 2020; RSPCA UK, 2020).

In 2019, Tesco UK launched a higher-welfare indoor chicken product range. Tesco vowed to commit to welfare standards higher than those set by the government and the Red Tractor scheme (Ryan, 2019). This pledge came in handy to strengthen Tesco's own brands after some public scandals involving fictional farm brands and serious irregularities in one of its chicken meat suppliers.

In 2016, Tesco had launched seven meat, veggie, and fruit farm brands. The meat farm brands were Willow Farms for chicken, Woodside Farms for pork, and Boswell Farms for beef. However, these meat farms do not really exist, which triggered an angry reaction from the farming industry. The industry claimed the brands misled consumers into believing the products came from real farms, which give a perception of quality products (Mann, 2016). The *legitimate* Boswell Farm's owner, Linda Dillon, felt especially affected by the retailer's marketing strategy, as the farm (where she had been living for 21 years) offers cottage holidays and yoga and Pilates retreats. An association with cattle rearing and slaughter, she said, would harm their reputation (Mann, 2016).

Former Tesco's CEO Dave Lewis said the Willow Farms might be a made-up farm, but "the product truth is right," referring to the product's quality and safety (Armitage, 2017). Ironically, the chicken meat sold under Tesco's Willow Farms brand was found to include unsold meat from Lidl repacked and relabelled at a 2 Sisters poultry plant in West Bromwich, England. 2 Sisters Food Group, one of the UK's biggest chicken meat suppliers, has an annual turnover greater than GBP 3 billion (Wallop, 2017). An *ITV News/The Guardian* undercover investigation filmed the meat "repurposing" procedure. The footage also showed another breaching of food safety rules, such as frontline workers falsifying slaughter dates and using chickens that were seemingly picked up from the floor (Armitage, 2017; Wallop, 2017; Wood, 2017).

To make things worse, Tesco's Willow Farm chickens come from several suppliers across the UK. This reminds me of the argument I used before on the low probability of assuring the provenance of a product, especially when we eat out. Many people who claim to care about animals like to defend meat consumption by saying they only eat free-range, organic, or higher welfare meat. As if the animals had not had their lives taken for a meal in any of the systems... But anyway, do these consumers eat exclusively at home? Do they raise and slaughter their own animals? If the answer to one or both of these questions was 'no,' there is no way to guarantee the animals they are eating had a "good" life.

Despite the undeniable findings of the investigation on the 2 Sisters factory, spokespeople for Tesco and Lidl declared their commitment to rigorous food hygiene standards and pledged to conduct their *own* inspections on the processing plant (Wallop, 2017). It makes me wonder whose interests the new investigations would be serving.

Following the undercover investigation scandal, Tesco's CEO declared the hygiene issues were limited to the West Bromwich factory, which had been temporarily closed. He added that neither the UK Food Standards Agency (FSA) nor Tesco's own investigation found any issues in that plant (Wood, 2017). And life goes on.

Let us now talk about free-range chicken meat production. Free-range broilers must be provided with access to an outdoor range for at least half of their lifetime. The minimum outdoor space allowance, according to the EU legislation, for example, is one square meter per bird. The chickens can be housed in sheds at night or even during the day for protection against predators and inclement weather. Free-range broilers are slaughtered at eight weeks of age or older (CIWF, 2020).

Free-range certification is given by federal, state, local, or private accreditation schemes in line with national or international legislation and codes of practice on animal welfare, biosecurity, and food safety. Private certifiers enforce the minimum requirements established by legislation and often set additional standards to ensure higher levels of animal welfare. Each country or region has its own regulations for the production of free-range poultry meat. In Europe, for instance, the term 'free-range' is defined by EU Regulation 543/2008, while Regulation (EC) 1234/2007 sets the marketing standards for poultry meat. In Brazil, the criteria for free-range chicken rearing, slaughter, processing, and identification are set by the normative ABNT NBR 16389:2015 (FSAI, 2018; ABNT, 2015).

Some examples of private accreditation schemes are Red Tractor in the UK, Free Range Egg and Poultry Australia (FREPA) in Australia, Certifica Minas Frango Caipira in Brazil, and RSPCA in Australia and the UK.

Important to note is that, not all RSPCA- and Red Tractor-labelled chicken meat comes from free-range birds, since these schemes regulate the welfare of both indoor-raised and free-range chickens (Rivera, 2017).

Inspection of free-range farms is made by the accreditation company, government officials, or both. Nonetheless, non-compliance and false labelling are not atypical, either by farmers not meeting the totality of free-range farming requirements or by deliberately selling conventional poultry meat as free-range meat.

Worth noting, vague legislation and lax oversight compromise animal welfare and food safety. In the US, for example, the USDA free-range poultry regulations do not specify the size or quality of the outside range, nor the time duration the chickens spend outdoors in a day. This means that US free-range broilers can still spend most of their time confined indoors or have outside access only to a barren patch of land. The definition of free-range chicken meat on USDA's website is limited to: 'Producers must demonstrate to the Agency that the poultry has been allowed access to the outside' (USDA, 2015b).

In addition, inspection is sometimes not carried out in small poultry farms, meaning the consumers buying free-range chicken meat must rely on the farmer's word. According to Terry E. Poole, Professor Emeritus of the University of Maryland Extension, 'Small farm poultry enterprises can process up to 20,000 birds on the farm without having USDA inspection. This works well for direct marketing birds to the public, however if producers decide to sell their birds to restaurants, food stores, or caterers, it is recommended that they get USDA inspected' (Poole, 2015).

The last main broiler production system is organic. Organic broilers have a larger outdoor space allowance than free-range chickens: at least 4 m^2 per bird, during a minimum of one-third of their lifetime. Organic broilers also grow slower than their intensively reared counterparts and are slaughtered at 12 weeks of age or older. Often, more traditional breeds are used compared to intensive-raised broilers (CIWF, 2020).

Organic birds are typically smaller, lighter, and less well-feathered than free-range birds, mostly due to their breed. As organic birds are less robust, they might be kept indoors for the first 35 days of age (nearly a third of their lives) as a protection against adverse weather conditions (RSPCA, 2017a).

The major difference between organic and other systems is the ban on the use of GMOs, animal by-products, feed additives, antibiotics, growth promoters, hormones, and other synthetic drugs. The birds must be fed 100% organic feed, including grains and forages. In Europe, organic poultry broilers are reared according to the conditions laid down in EU Regulation 889/2008. In Brazil, organic chicken production must meet the standards set by Law 10.831 of 23/12/2003. In the US, organic standards are defined by the USDA National Organic Program. Nonetheless, the USDA regulation does not specify the time duration and physical area requirements for outdoor access. Therefore, the term "organic" ends up having little to do with higher animal welfare in US chicken farms. With a growing organic food market valued at USD 43 billion, USDA's vague standards benefit poultry farmers at the expense of commodified animals (Curry, 2017).

Organic certification can also be granted by private schemes, just as in free-range poultry meat. Some examples are the Soil Association in the UK, IBD in Brazil, and RSPCA in Australia and the UK. Each scheme sets its own standards for outdoor area quality, duration of outdoor access, stocking density on the range and in the poultry house, the existence of environmental enrichment in indoor areas, and other issues, generally exceeding minimal legislation requirements. Some accreditation schemes, such as Brazilian IBD, guarantee equivalence of standards in relation to export markets (IBD, 2019).

The certification agency or control authority must be informed whenever the animals receive veterinary drugs (e.g., antibiotics, vaccinations, parasite treatments, etc.) (Soil Association, 2020). A chicken ceases to be organic the moment it is given a synthetic drug. Phytotherapeutic and homoeopathic products are allowed, but these are little, if at all, effective against certain conditions—for example, viral and bacterial diseases. Stricter hygiene practices, the administration of prebiotics and probiotics, and lower stocking densities decrease the prevalence of certain diseases in organic farms, and thereby the need to use synthetic drugs. Nonetheless, access to pasture predisposes the animals to other diseases less seen in intensively confined chickens, such as coccidiosis and further parasitic and helminthic infestations (Kijlstra & Eijck, 2006).

Obviously, the fact that an animal is raised as organic does not mean it will not eventually fall ill. If treatment with natural drugs is not effective, animals reared in organic farms must be culled (Soil Association, 2020). However, farming animals means deriving profit from them. This raises the question of whether sick animals are denied treatment, so they keep producing "food." Treating sick animals so they can live a life beyond their commercial purpose is not contemplated.

And to whom the responsibility of rescuing some of these animals is transferred? To vegans, of course. Many sanctuaries worldwide rescue exhausted, sick, abused, and injured animals or take in animals that escaped from farms and slaughterhouses—using their own financial resources. Sanctuaries like Animal Place (US), Ahimsa – Vale da Rainha (Brazil), Eden Farmed Animal Sanctuary (Ireland), and numerous others across the globe endeavour to provide a happy life to animals that have only known abuse and slavery. The deplorable condition in which animals raised in organic and free-range systems reach sanctuaries shows the reality of livestock farming.

Speaking of sanctuaries, these are completely different from zoos and safaris. The animals are not bred, traded, put on the spot, or used to entertain people. Visit the website of the sanctuaries just mentioned and learn about the stories of the animals living there.

Industrial production became the bad guy

For some reason, we tend to believe that the *industrial* production of milk, eggs, and meat is the problem. As if small livestock farms do not rely on animal exploitation too. People associate high stocking densities and intense confinement typical of large-scale industrial farming with animal welfare problems (which is true), but fail to consider that space allowance is far from being the only welfare concern in animal agriculture. We have seen a wide range of causes of animal suffering other than confinement and animal per area rates.

Consumers also commonly believe (mistakenly) that deforestation, soil degradation, water pollution, carbon emissions, species extinction, and high resource use are limited to industrial, intensive farming. Nevertheless, we have seen in the previous chapters that free-range and organic farming can have a higher environmental impact than intensive farming due to the greater land requirements and lower conversion ratios.

Moreover, local, organic, and free-range farms are not able to supply the amounts of animal products we consume worldwide. Free-range raising of all animals used today for food would require many planet Earths. We simply do not have the required area.

Besides, local farms also buy inputs from distant markets. As grain prices largely impact the production and sale prices of meat, eggs, fish, and milk, farms will typically purchase feed ingredients from the most competitive suppliers. In pig-finishing farms, for example, animal feeding accounts for 80% of the production cost. In other words, ranchers buy inputs from suppliers that offer the best price.

Brazil and Indonesia are examples of major producers of animal feed ingredients that have increased production and reduced prices in response to relaxed enforcement of environmental laws. In Chapter 2, we have discussed how deforestation rates in the Amazon rainforest and Brazilian Cerrado are causally linked to animal product consumption in other parts of the world.

Another common misconception is that organic, free-range, and smaller farmers have no association with industrial activities. First of all, farming is not a hobby—small farmers are workers like you and me, who have material needs, dreams, and plans. Second, they purchase machinery, fuel, feed, antibiotics, cages, and other farming supplies produced on an industrial scale. Third, even small-scale, free-range, and organic farmers acquire animals from intensive breeding farms and large hatcheries. The field of genetic selection and production of pure strains of birds, for example, is highly technological and dominated by large multinational companies. In addition, the legislation does not require that organic-raised animals are *born* within organic standards, but rather that the animals are raised as organic from a certain age.

One can even buy birds online if aiming to keep some backyard laying hens or start a small broiler chicken or egg farm. For instance, on the US website www.strombergschickens.com, it is possible to buy birds of various types and ages and even fertilised eggs. The sentences 'Shipping is included with all prices,' 'The minimum order is 15 chickens,' and 'You can mix and match chicks,' demonstrate the commodification of animals.

We must understand that the use of animals is immoral and that the production and consumption of animal products are impacting the planet to the point of no return.

Catching, crating, transport, and slaughter

Whether the broiler chickens and laying hens are conventionally raised, free-range, or organic, they must be transported to the abattoir to be slaughtered. Only subsistence farms kill the birds on site. Typically, the birds are caught, crated, and loaded onto trucks. The animals can be caught mechanically by a harvesting machine or, most commonly, by hand (Mönch et al., 2020). Catching and crating are remarkably stressful and can result in painful injuries, from bruising to bone fractures and even death.

Many scientific papers report the prevalence of haemorrhages, bruises, fractures, and cranial trauma during pre-slaughter procedures (Knierim & Gocke, 2003; Nijdam et al., 2005; Saraiva et al., 2020; Mönch et al., 2020). Although upright catching is recommended for welfare reasons, workers typically grab the chickens by one or both legs and carry up to several birds upside down per hand. Some chickens are inadvertently caught by the neck (Mönch et al., 2020; Saraiva et al., 2020).

The chickens may be handed from person to person several times before loading into the transporting lorries. The birds get agitated in the catchers' hands, typically flapping their wings, kicking, and wriggling. Their fragile skeleton, already impaired by the sedentarism, overlaying, and fast growth rates, may not resist the impacts (De Lima et al., 2019; Saraiva et al., 2020). During quick and careless unloading upon arrival at the abattoir, the birds can also have their wings and combs torn when removed from the crates (HSUS, undated).

Mechanical catchers, such as the Apollo Universal (GTC Agricultural, the Netherlands) and the Chicken Cat Harvester (JTT Conveying A/S, Denmark), can collect 8,000–12,000 chickens per hour. These machines can capture the birds from the ground in an upright position, with no direct contact with operators, which can, in theory, be less stressful. Nonetheless, Nijdam et al. (2005) did not find differences in injury incidence and corticosterone plasma levels when comparing manual and mechanical loading methods. The prevalence of wing hematomas was 6.7–8.4%.

Knierim and Gocke (2003) reported a lower incidence of injuries in automatic loading, but no difference in the number of dead birds on arrival at the abattoir. Wing fractures, for instance, were found in 0.77% of broilers after manual loading versus 0.66% after mechanical loading.

Mönch et al. (2020) found a higher incidence of wing hematomas after mechanical loading versus manual loading, with respectively 7.19% of birds with at least one hematoma on the wing versus 1.49%. The number of dead broilers on arrival was also greater in mechanical (0.16%) compared to manual loading (0.06%). The authors observed that more staff during manual loading and shorter loading duration in manual catching resulted in a lower prevalence of wing hematomas. The occurrence of injuries was not affected by loading duration in mechanical loading.

Different comparison studies show contradictory results since several factors other than the loading method are relevant—e.g., individual attitudes of workers and number of personnel. Many workers simply show no compassion for animals. Rough handling and deliberate brutality add to the chickens' suffering. Regardless of the level of automation involved in pre-slaughter operations, the human factor *is always present* as machines do not operate alone. Insufficient training and pressure to meet deadlines and production targets can increase physical and psychological damage to the birds (De Lima et al., 2019; Mönch et al., 2020).

Several cases of animal abuse in chicken farms and abattoirs, including during pre-slaughter operations, can be found on Animal Equality's and Kinder World's websites (www.animalequality.org and www.kinderworld.org, respectively). Poultry farm workers catching and stacking birds into crates with violence, hitting their heads and limbs against the cage opening, kicking them, or throwing them are not unusual whatsoever. Google it. In the EU, an estimated 20 million chickens die every year before reaching the slaughterhouse (CIWF, 2020). In Brazil, an estimated 40% of the mortality related to the transport of birds occurs during catching and loading (Nunes, 2018).

Catching and transport are traumatic events for poultry, with studies reporting the elevated corticosterone (a stress hormone) levels associated with these operations (Kannan & Mench, 1996). As birds have difficulty thermoregulating their bodies, thermal discomfort (both cold and hot) is typical during transport. Heat stress is an important contributor to high corticosterone levels and mortality during the transport of chickens (Saraiva et al., 2020). Other stressors include noise, vibration, motion, luminosity extremes, overcrowding, as well as unfamiliarity with workers and other birds (HSUS, undated).

Mixing batches from different farms into the same truck is relatively common in many countries, such as Spain (Averós et al., 2020). Birds are submitted to fasting before transport and slaughter to avoid contamination of the feathers with faecal and gastric matter containing deteriorative and pathogenic microorganisms. Thereby, they can endure long journeys feeling hunger, thirst, pain, fear, and thermal discomfort.

During transport, the number of birds per crate, the water and feed withdrawal duration, and the distance and duration of the journey severely impact animal welfare. In a study with 64 mixed-sex broiler batches from 64 Portuguese farms, Saraiva et al. (2020) reported 0.02%–1.89% of birds dead on arrival per batch. The birds had been deprived of water for up to 17 hours. A third of the batches (32.81%) had been subjected to fasting for more than 12 hours. Transport duration and distance ranged from 22 to 184 minutes, and from 15 to 196 kilometres. Each crate contained 8–15 birds. The occurrence of bruise and death on arrival increased with journey distance, water withdrawal duration, and number of broilers per crate. Prevalence of bruises did not increase with transport duration, suggesting that bruises were associated mainly with catching, crating, and loading. Nearly 3% of carcasses were condemned due to dehydration, which occurred in 22 out of 64 batches.

Averós et al. (2020) evaluated the effect of transport conditions on the prevalence of death on arrival and carcass rejections of broiler chickens in Spain. The sample comprised 10,198,663 broilers from 2,284 flocks transported from 217 different farms to a single slaughterhouse. Longer journeys resulted in a greater carcass rejection. The results were worsened when the trucks were loaded with broilers from different farms. A higher risk of death on arrival occurred in extreme external temperatures during summer and winter. Death on arrival and carcass rejection had a mean prevalence of 0.26% and 0.77% of the flocks, respectively.

Upon arrival at the slaughterhouse, the crates must be unloaded into a safe area where the birds can be released. The time lapse between the arrival of the loaded trucks and the slaughter itself is called lairage. Lairage times must not exceed one or two hours to avoid further stressing the chickens, which impacts negatively on meat quality. However, lairage times can, in practice, range anywhere between zero to 18 hours, depending on the plant logistics and processing rate (Rodrigues et al., 2017).

Before being killed, chickens are hung by their feet on metal shackles attached to conveyor lines, which carry them to the stunning area. Being upside-down hung by the feet is obviously painful, or at least uncomfortable, and the chickens typically flap their wings frantically in an attempt to return to the upright position. As stunning efficiency can be impaired when the birds are agitated, a time gap between shackling and stunning is recommended (Bedanova et al., 2007). Nevertheless, studies outline that shackling increases the concentration of blood corticosterone in direct proportion to the shackling duration (Bedanova et al., 2007).

Conveyor lines take the chickens to an electrified water bath, into which they are dipped and electrocuted. Voltage and frequency requirements vary geographically. In EU member states, for instance, broilers are electrocuted using high voltage–low frequency (50–150 V, 50–350 Hz) alternating current. In the US, where welfare requirements are milder, low voltage–high frequency (12–38 V, ≥400 Hz) combinations are used, either employing direct or alternating current on water bath systems (Bourassa et al., 2017). Such conditions are used because they have a lesser impact on the physical integrity of the carcasses—again, the concern with profit margins prevails.

The purpose of the electrocution of live chickens is to induce irreversible loss of consciousness and sensibility before they are killed (Hindle et al., 2010). This is what the animal industry calls "humane slaughtering." Conveniently, unconscious animals facilitate subsequent operations, increasing productivity.

Scientists show that birds stunned with low voltage (as used in the US) can recover consciousness before having their necks cut (Hindle et al., 2010; Bourassa et al., 2017; NCC, 2013). Bourassa and co-workers reported that 50% of broilers stunned with 25 V for 60 seconds regained consciousness. The percentage increased to 75% for 15 V (Bourassa et al., 2017). When stunned by high voltage, as done in the EU, broilers usually die of cardiac arrest (Bourassa et al., 2017; NCC, 2013).

If the birds are not well-positioned in the shackles, are shorter than the average, or simply raise their heads above the water level, they will be fully awake having their heads chopped off or their throats cut, depending on the killing method (CWIF, 2020). Note that even when stunning is successful, the birds were still fully conscious whilst shackled upside down and electrocuted in a water bath. Very humane indeed.

Worth mentioning, the use of stunning is forbidden in kosher food but is sometimes allowed in halal food, provided the animals are still alive when they have their throats cut (Wright, 2016).

After stunning, the chickens can be killed manually with a cut in the throat or by machines that cut their heads off. A series of equipment to kill, defeather, eviscerate, and dismember poultry can be found at www.poultryprocessingequipment.com. If slaughter is done manually, the birds may not die from the cut itself, but rather from subsequent exsanguination. When stunning and throat slitting are not appropriately done, the bird may regain consciousness while bleeding to death. However, slitting efficiency is commonly hampered by high processing line speeds.

In the US, a typical poultry plant processes 140 birds per minute (Matthews & Pinkerton, 2020). The leading occupational cause of finger amputation in the country is poultry processing (Nevin et al., 2017). In Brazilian poultry processing plants, tendonitis rates are seven times higher compared to all other industries (Carne e Osso, 2011). Any human error or occupational injury along the processing line will directly affect the welfare of the animals being stunned and killed.

Following exsanguination, chickens are dipped into a scald tank for several minutes in preparation for the mechanical plucking of feathers. Now imagine the consequences of ineffective stunning or slitting. This was the case with the 81 chickens that were boiled *alive* in a halal slaughterhouse in 2016 due to an alleged equipment malfunction and an argument between workers (Wright, 2016; Webster, 2016).

One Stop Halal slaughterhouse in Eye, England, supplies major supermarket chains in Britain, including Tesco, Morrisons, Asda, and Sainsbury's. In the Eye plant, part of the birds is stunned in an electrically charged water bath, have their throats cut by the workers, and are finally placed in boiling water to facilitate feather removal. Another part is killed without stunning to be sold to the Islamic public as halal meat. However, a fault in the water bath forced the workers to slit the throats of up to 100,000 fully conscious chickens every day, even those not catered for Muslim consumers. To make things worse, on the 6th of July 2016, an argument between two workers on the quality of the cuts resulted in dozens of live birds following to the scalding stage. According to the company, the birds agonised inside the boiling water bath for around two minutes before dying (Wright, 2016).

After admitting having caused unnecessary suffering to the chickens, One Stop Halal was ordered to pay GBP 14,000 in fines and legal costs. A modest amount for the company's owner, Ranjit Singh Boparan, a British businessman with an estimated personal fortune of GBP 544 million. Boparan is also the founder and owner of 2 Sisters Food Group (remember the repurposed chicken meat mentioned earlier?) and many restaurants and farms involved with beef, poultry, and fish (Shetty, 2019). Animal exploitation is literally worth millions.

One Stop Halal declared the employees involved in the incident were dismissed from their jobs, while substantial investments were made in training and control systems since July of that year. Although the slaughterhouse stated the chickens were boiled alive in a *single* event, post-mortem tests by meat hygiene inspectors of the UK Food Standards Agency suggested otherwise. The inspection indicated that a further nine incidents happened over the next three months, with live chickens being plunged into scalding water. Morrisons and Sainsbury's pledged not to have sold chickens that were involved in the incidents (Wright, 2016).

I must add that nutrition is not limited to the mathematical intake of nutrients. Everything we ingest, physically and metaphysically, becomes part of our bodies. We integrate suffering into our very cells when we eat animal products.

Spent hens can suffer even further than broilers as they are lighter, more active, and get more agitated in the face of external disturbances. Therefore, they have much higher chances of missing the stunning water bath and, hence, receive painful shocks instead of being effectively stunned. Compared to broilers, spent hens also have weaker bones due to the high rates of osteoporosis (Webster, 2007).

Moreover, egg companies commonly sell their hens to a poultry processing company after their laying performance declines. Not surprisingly, egg companies do not take responsibility for the welfare of hens after they leave the farm. Take the word of poultry scientist Bruce Webster, from the University of Georgia in the US: '[...] egg companies typically do not see themselves as being responsible for a hen's welfare once ownership of the bird is transferred, and what happens to spent hens after that tends to fall under the radar' (Webster, 2007).

Alternative stunning methods exist, such as controlled atmosphere stunning (CAS) and low atmospheric pressure (LAP) stunning (Webster, 2007; Vizzier-Thaxton et al., 2010; NCC, 2013). CAS induces insensibility by exposing poultry to an inert gas or mixture of inert gases, such as nitrogen, argon, and carbon dioxide. The chickens lose consciousness due to a lack of oxygen. However, birds show aversion to these gases, typically gasping for air, flapping their wings, and convulsing (NCC, 2013). LAP also induces unconsciousness by hypoxia, but discomfort is reduced by the absence of toxic gases and the gradual adjustment to the atmosphere (Vizzier-Thaxton et al., 2010). Even though CAS and LAP are used in some processing plants, electrical stunning remains the classical method.

Production targets disguised as animal welfare concerns

Livestock farms, live transport companies, and animal processing plants are businesses, not charities. In the same way that you and I work to pay our bills, people behind animal agriculture are trying to earn a living. We must understand that the purpose of the animal industry is to *derive money from animals*.

Whenever capital is injected into a business, a return on the investment is expected. If the investment cannot be compensated, the company must charge more for their services or products. When measures to increase animal welfare are implemented in hatcheries, farms, and abattoirs, these businesses may have to offset costs with an increase in their products' final price. However, consumers look for the best quality for the lowest price. Being charged more for the same food is seen negatively by most shoppers.

Consumers of animal products do not always realise that improvements in animal welfare entail costs to the industry. For example, providing more space for birds in an indoor shed means fewer animals can be housed in the same area and, hence, a lower egg or meat yield is obtained. Increasing the team of catchers in a poultry farm aiming at gentler catching means higher costs with hiring and paying personnel. Thereby, it is not in the industry's best interest to invest in animal welfare—*unless they can profit from this.* 'Free-range,' 'pasture-free,' 'cage-free,' 'humanely raised,' and other labels can help add value to their products and thus charge more for them.

In the *worst-case scenario* for the animal industry, farms are mandated by legislation to increase animal welfare to meet minimal criteria in response to social pressure. The movements for cage-free hens and stall-free gestating sows are clear examples. In the *best-case scenario,* higher animal welfare results in improved productivity and product quality—and consequently, greater profit margins. Moreover, businesses can strengthen their brand by advertising their higher welfare standards compared to competitors or with their own brand over time. This is a perfect combination: higher revenue *and* improved brand reputation.

In the poultry industry, for instance, many performance targets are disguised as animal welfare concerns. Let us see some of them:

1. 'A time-lapse between shackling and stunning of chickens must be ensured to improve welfare levels.' Translation: Shackled birds flap their wings and wriggle nervously. Calmer birds are easier to manipulate, facilitating processing (Savenije et al., 2000).
2. 'Catching, loading, and transport conditions of broilers must meet minimal welfare requirements.' Translation: Stress and injury trigger metabolic reactions that impair poultry meat quality. Elevated corticosterone levels lead to glycogen depletion in the muscle, resulting in DFD or PSE meat (Lesiów & Kijowski, 2003). Besides, injuries may give rise to bruises, haemorrhages, and early death, which result in carcass downgrade or condemnation.
3. 'Broilers and layers are fed high-quality feed.' Translation: A balanced diet is crucial to ensure the flock's health and immunity and, therefore, the desired output. High-quality feed also ensures a proper FCR, while the nutrient balance affects product quality (e.g., fat content in meat and yolk colour and cholesterol levels in eggs).
4. 'Cage enrichment allows hens to engage in more of their natural behaviours, increasing their welfare.' Translation: Hens suffer horribly when confined in cages, which affects their laying performance. However, the negative effect of intense confinement is offset by the financial advantage associated with reduced housing area and workforce requirements. Were this offset nonexistent, hens would not be confined in tiny cages to begin with. In

addition, the existence of perches and other enrichments in cages reduces the hens' stress levels to a certain extent, which means better egg-laying performance in response to little investments.

5. 'Higher-welfare indoor systems ensure a better quality of life for broilers.' Translation: Poultry houses that offer more space per bird and environment enrichment (e.g., perches and straw bales) reduce stress and frustration levels in broilers. This translates into faster weight gain, lower injurious pecking and mortality rates, and a lower prevalence of condemned carcasses.
6. 'Measures to control ammonia concentrations in poultry houses increase animal welfare levels.' Translation: Ammonia concentrations from 11 ppm impact the animals' wellbeing, and consequently their *performance*. Some underlying reasons are the decreased intestinal surface area and the reduced resistance to oxidative stress, leading to less efficient nutrient breakdown and compromised immunity. Moreover, continuous exposure to high levels of atmospheric ammonia alters the expression level of genes related to lipid metabolism, reducing meat quality (e.g., increased fat content) and palatability (Yi et al., 2016).

Do you see? Investments in animal welfare are sometimes merely good business. I risk sounding like a broken record, but this message is essential: if animal welfare were a *legitimate* concern, no one would be using animals in the first place—neither industry nor consumers.

Poultry meat, eggs, and deadly plagues: speciesism, coronavirus, racism, and social inequality

We now know that slaughterhouses and meat processing plants are significant coronavirus clusters. But the mainstream media is not giving the topic the importance it deserves. Animal agriculture and governments are making a considerable effort to hide the relationship between meat and COVID-19 the same way it fails to inform consumers of the truth surrounding animal products and health. This is so evident that *meat* factories associated with coronavirus outbreaks are strategically referred to as *food* factories in the news, never mentioning the type of product being processed.

Take the case of O'Brien Fine Foods, a meat factory in Ireland that I cited in Chapter 2 as an example of coronavirus outbreaks in meat plants. Even though O'Brien is the primary supplier of meat products to the Irish market, supplying big supermarket chains such as Tesco, Lidl, and Dunnes Stores, well-known newspapers such as the *Irish Mirror* omitted which foods in particular the factory produces (Pattison, 2020a). The plant had had 86 workers tested positive for COVID-19 in May 2020. The word 'meat' was not mentioned once. The second piece of news by the same reporter in August 2020 about other cases in the cited meat factory "forgot" to mention the word 'meat' once again (Pattison, 2020b).

Remember 2 Sisters, the poultry company that was reported to repack and relabel chicken meat returned by a client supermarket? Yes, the same 2 Sisters that is owned by the multimillionaire businessman who is also the owner of the halal slaughterhouse that boiled chickens alive in 2016. Well, a 2 Sisters plant located in Llangefni, Wales reported over 200 cases of coronavirus among its workers as of June 2020. Llangefni is the smallest of the 12 2 Sisters plants across the UK and employs 500 people (*BBC News*, 2020).

In August 2020, a Scottish 2 Sisters plant in Perth and Kinross was temporarily closed due to coronavirus infections. It employs approximately 1,000 people (Taylor, 2020a). In parallel, a coronavirus infection cluster was reported among residents of Perth and Kinross (Perrett, 2020). Another 2 Sister plant in Coupar Angus, Scotland, which employs 1,200 staff, was closed for 2 weeks in August 2020 after 201 cases of COVID-19 were confirmed (Dickie, 2020). Many cases were reported in other 2 Sisters factories across the UK.

Disregarding that meat production and consumption are environmentally unsustainable, detrimental to human health, and are closely associated with COVID-19 outbreaks, the Scottish government awarded 2 Sisters Coupar Angus with GBP 1 million to improve their poultry processing facilities. The factory was the primary recipient of the GBP 5 million government aid that aims to strengthen the country's food supply chain and create new jobs (Taylor, 2020b). Let me tell you something top-secret: investments in horticulture also generate employment and strengthen the economy.

In any case, COVID-19 clusters in poultry processing plants are absolutely not limited to 2 Sisters or any other specific animal products company. Cases are being reported all over the globe in slaughterhouses and meat processing plants. Please note that meat refers to animal flesh in general, and therefore should not be confused with beef. Some KFC stores were even forced to close in Australia in 2020 due to a chicken meat shortage associated with a coronavirus outbreak in a Melbourne processing plant (Starkey, 2020).

We have already seen that the work conditions in meat factories are predisposing factors for bacterial and viral infections. Low pay and high rates of expatriates among the workers (many of them with irregular documentation and poor communication skills in the local language) aggravate the situation because these employees cannot afford to lose their jobs. CDC revealed that 87% of reported COVID-19 cases in US meat processing plants in April and May 2020 involved racial or ethnic minorities (CDC, 2020b).

In 2020, there were approximately 525,000 slaughterhouse employees in the US, spread over 3,500 factories (CDC, 2020b). The CDC study calculated a prevalence of 16,233 cases of COVID-19 and 86 related deaths among red meat and poultry processing facility workers *in April and May 2020 alone*. The actual numbers are likely higher, as only 28 of the 50 US states contributed data. Among the 9,919 infected workers who reported race or ethnicity, 56% were Hispanic, 19% non-Hispanic Black, 13% non-Hispanic white, and 12% Asian (CDC, 2020b).

According to a less conservative assessment carried out by the Food and Environment Reporting Network (FERN), 58,913 meatpacking workers were infected with COVID-19 in the US as of the 5th of August 2020, 297 of which died (Douglas, 2020).

Nearly 65% of the 91,278 farm and food processing workers who contracted the virus were meatpacking staff. Out of the 465 staff members who died, around 64% worked with meat processing (Douglas, 2020). Figure 35 illustrates the COVID-19 fatal cases among the US agri-food industry workers from the onset of the pandemic until early August 2021.

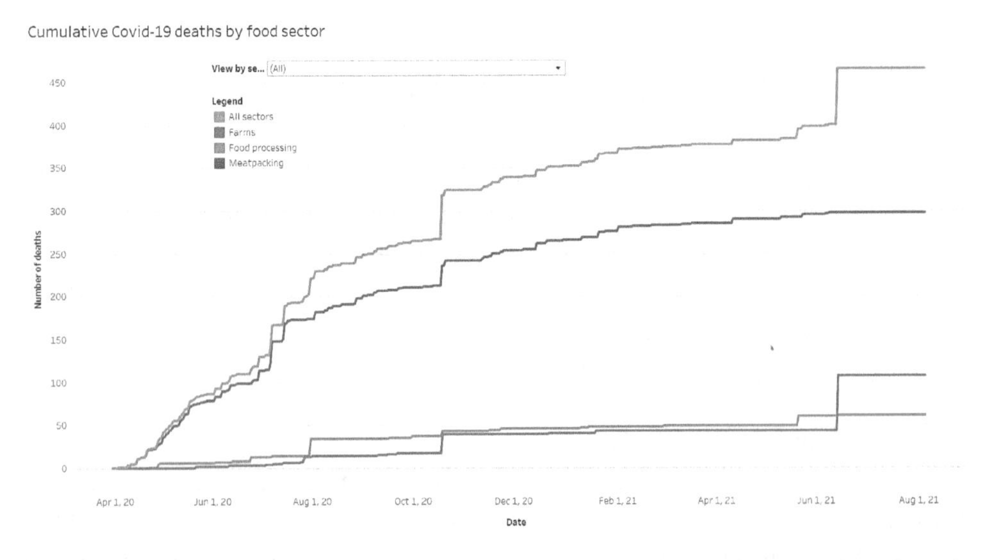

Figure 35. Total number of COVID-19 infections among US agri-food industry workers throughout the pandemic. Source: Douglas (2020).

In an interview to *Aljazeera*, pastor Reverend Willie Justis commented on the racial inequality in the poultry industry. Justis has lived his whole life in rural Virginia among poultry farmers and factory workers. He says: 'The poultry industry is built on a racist system. [...] It's very oppressive when you are extorting the labour of Brown and Black people for billion-dollar profits but you tell workers they can't speak to the media to tell them you are afraid to go back to work because COVID-19 is real. That's plantational: We can't say it because "massa" might get mad' (Aljazeera, 2020).

While 70% of processing line workers at US meat plants are people of colour, management teams are mostly white. White people represent 58% of the management employees at JBS and 73% at Tyson Foods, for example (*Aljazeera*, 2020). Not only do most meat processing plants worldwide continue to operate despite the worrying prevalence of COVID-19 cases, but also little, if anything, has been done to improve the labour conditions in this industry. On the contrary, the Trump administration urged the poultry industry to speed up processing from 140 birds per minute to 175 birds to offset production declines due to the pandemic (Matthews & Pinkerton, 2020). It is difficult to predict who is harmed the most: workers exposed to further pressure and occupational hazards, animals suffering (further) due to rough handling, faulty stunning and careless slaughtering, or the average consumer paying to become ill.

In the US, injury rates of slaughterhouse staff are higher than those of construction workers and coal miners (Bureau of Labour Statistics, 2018). In Iran, the prevalence of respiratory symptoms such as coughing, wheezing, and breathlessness is 3–4 times higher in poultry plant workers compared to office staff (Kasaeinasab et al., 2017). In Brazil, slaughterhouses rank first in occupational accidents in the agri-food sector. An inspection conducted by state agencies in 2015 at a chicken slaughterhouse in Rolândia, Paraná, confirms yet another aspect of the grim reality of animal agriculture. Interviews with 400 of the 4,000 employees revealed that 52.9% had to take medication or apply plasters to withstand pain during work in the previous 12 months (GGN, 2015).

The European animal exploitation industry likes to brag about their higher animal welfare and environmental standards compared to other major producers, such as the US, Brazil, and China. Nonetheless, the social issues surrounding animal food production are virtually the same everywhere: low remuneration, occupational diseases, work-related injury, stress, and discrimination.

Poultry workers in EU member states are commonly expatriates from Eastern European countries who are employed under short-term contracts, share accommodation with other migrants, and do not speak the local language well. In the Netherlands, 80% of meat factory workers are from Central or Eastern Europe. Most of them are subcontracted or employed on daily hires, which means they do not have a guarantee they will have a job tomorrow. Out of fear of losing their subsistence, many do not report COVID-19 symptoms (Marshall & Unger, 2020).

Again, we must understand that whenever we buy a product, we are supporting *the entire industry* and the underlying values behind it. Consuming animal products means not only paying for animal abuse, environmental depletion, and health hazards, but also voting for degrading work conditions and unequal rights. Each meal is a vote—*what is yours*?

Besides the direct health risks facing meat processing workers and the chances of coronavirus spread among their families and local communities, there is the probability that food items carry the pathogen throughout the food chain. Spokespeople of the food industry, national food authorities, and health agencies have declared that it is very *unlikely* that coronavirus can be transmitted through food. But note that *unlikely* does not mean *impossible*, especially considering that

SARS-CoV-2 (the virus that causes COVID-19) was an unknown pathogen until some years ago. In a guidance document on coronavirus and food safety dated April 2020, FAO & WHO (2020) stated:

- 'It is highly unlikely that people can contract COVID-19 from food or food packaging. COVID-19 is a respiratory illness, and the primary transmission route is through person-to-person contact and through direct contact with respiratory droplets generated when an infected person coughs or sneezes.'
- 'There is no evidence to date of viruses that cause respiratory illnesses being transmitted via food or food packaging. Coronaviruses cannot multiply in food; they need an animal or human host to multiply.'
- 'Recent research evaluated the survival of the COVID-19 virus on different surfaces and reported that the virus can remain viable for up to 72 hours on plastic and stainless steel, up to four hours on copper, and up to 24 hours on cardboard [(Van Doremalen et al., 2020)]. This research was conducted under laboratory conditions (controlled relative humidity and temperature) and should be interpreted with caution in the real-life environment.'
- 'Although COVID-19 genetic material (RNA) has been isolated from stool samples of infected patients [Ong et al., 2020], there are no reports or any evidence of faecal-oral transmission.'

Despite these guidelines, published in April 2020, meat imports were suspended over COVID-19 concerns. In June 2020, China banned imports of poultry from a US Tyson Foods plant in Arkansas following an outbreak in the factory. Products that had already arrived in Hong Kong were seized (Togoh, 2020). In August, China and the Philippines halted imports of chicken meat from Brazil after chicken wings from a major producer were tested positive for coronavirus (O'Kane, 2020; Cordero, 2020). The virus was also found on the packaging of frozen shrimp imported from Ecuador and sold in China (O'Kane, 2020).

We must consider we are dealing with a novel coronavirus with high pathogenicity and spread rate. Scientists all over the world are endeavouring to unveil how exactly SARS-CoV-2 is transmitted, how it infects live organisms, whether there is a seasonality, if patient's age, diet, and lifestyle have an effect, and what can be done to reduce infection and mortality rates. Since the beginning of the pandemic, research teams have been working tirelessly on vaccine development. Health and food authorities have been informing the public about the scientific findings on all issues surrounding COVID-19, as well as implementing lockdown and social distancing measures. However, the guidelines are continually changing following the latest research developments. In other words, *we are learning as we go along*. Rather than dismissing risks under blanket statements such as 'probabilities are low,' 'it is very unlikely,' or 'there is no evidence of,' all possibilities must be carefully considered for the global population's safety.

The reported number of COVID-19 cases are increasing across the world, despite the viral spread preventive measures implemented globally. In many countries, the gradual release of lockdown has been followed by an increase in infection and mortality numbers. Nations where the virus had been apparently eradicated are facing new outbreaks. New Zealand, Vietnam, and parts of China had experienced months with no reported cases in 2020 until new focuses emerged (Fisher et al., 2020). It is evident, therefore, that the transmission routes of SARS-CoV-2 have not been fully elucidated.

Day- to week-long persistence of MERS-CoV, SARS-CoV-1, and SARS-CoV-2 has been demonstrated by different studies on a wide variety of surfaces (e.g., plastic, cardboard, and paper) under different environmental conditions (Van Doremalen et al., 2020; Chin et al., 2020; Van Doremalen et al., 2013). In 2020, a study by Fisher et al. (2020) revealed that coronavirus remains active on the surface of frozen pork, chicken, and salmon for 21 days. The researchers inoculated slices of fresh meat and fish acquired from supermarkets in Singapore with SARS-CoV-2. The samples were stored at three temperatures, 4°C (standard domestic refrigeration), -20°C, and -80°C (industrial freezing), and monitored at 1, 2, 5, 7, 14, and 21 days after inoculation. Virus load and infectivity, two crucial factors for the onset of outbreaks, remained constant over the study period under all experimental temperatures. The research indicates that the virus can survive during transport and storage, even under harsh conditions (up to -80°C). The authors argue their findings can explain the re-emergence of COVID-19 in some countries.

Research demonstrates that it is unwise to rule out COVID-19 transmission through contact with food packaging and food surfaces, because these can act as vehicles of viable coronavirus in the same way as non-food surfaces (Van Doremalen et al., 2020; Chin et al., 2020; Oakenfull & Wilson, 2020; Fisher et al., 2020; Van Doremalen et al., 2013). Health agencies have been advising we constantly wash our hands and disinfect surfaces to avoid passing the virus from contaminated surfaces to our eyes, nose, or mouth. Why would food surfaces be an exception?

For all the above reasons, a comprehensive risk assessment by the UK Food Standards Agency has stated that transmission of COVID-19 through food and food-contact surfaces cannot be ruled out (Oakenfull & Wilson, 2020). But what about the transmission of coronavirus by *consumption* of contaminated food?

Oral ingestion has not yet been confirmed as a significant infection route for COVID-19. Nevertheless, researchers have discussed the possibility of faecal–oral transmission and foodborne transmission of SARS-CoV-2 (Aboubakr et al., 2020). But before we discuss their findings, let us cover some basic concepts for clarity purposes.

Foodborne pathogens are mostly pathogenic bacteria, viruses, and parasites that replicate in the gastrointestinal system of human and nonhuman animals and are excreted via faeces. If food or water contaminated with faecal matter are consumed, the virus, bacteria, or parasite can infect the host (the consumer) and give rise to foodborne diseases. This is a form of *cross*-contamination, as opposed to *direct* contamination when the pathogen is directly ingested via meat, eggs, and dairy products from infected animals. We have seen many examples of direct contamination throughout this book—e.g., *Clostridium perfringens* in poultry meat and *Salmonella* spp. in eggs, to name a few.

To the average reader, it might seem unlikely that food can be contaminated with faeces, but this is, in fact, a major route for foodborne infectious diseases. Let me give two classic examples. Scenario 1: sick animals defecate, their manure contaminates soil or underground water, and that contaminated water is used to irrigate crops or, alternatively, food is grown on contaminated soil. Scenario 2: sick people defecate, then a nonexistent or inadequate sewage treatment does not remove or inactivate the pathogen and, finally, people use contaminated water to wash their hands and utensils or prepare food.

The most common foodborne viruses are human Noroviruses (HuNoVs), hepatitis A virus, and hepatitis E virus. The first two are primarily transmitted through foods that are consumed raw, such as shellfish, vegetables, and fruit. Hepatitis E virus is associated with the consumption of pork products (Petrović & D'Agostino, 2016). Likewise, SARS-CoV-2 can be excreted in faeces, since it has been isolated from sewage in several cities across the globe (Aboubakr et al., 2020). Peccia et al. (2020), for instance, isolated SARS-CoV-2 RNA in sewage sludge in a US metropolitan area during the 2020 spring COVID-19 outbreak. All samples contained SARS-CoV-2 RNA in concentrations *highly correlated* with the coronavirus epidemiological curve.

Moreover, Greening and Cannon (2016) demonstrated that SARS-CoV-1 and MERS-CoV, other viruses from the coronavirus family, can be foodborne. Long before the emergence of the SARS-CoV-2 strain, Duizer and Koopmans (2008) had already discussed the possibility of foodborne transmission of SARS-CoV based on the virus's properties. The authors also cite several examples of viruses considered *unlikely* to be foodborne until cases of Avian Influenza and Nipah viruses attributed to food intake were reported. Therefore, it is reasonable to hypothesise that SARS-CoV-2 can be transmitted through food-related routes, whether by the handling of food by infected people, contact between food and infected surfaces, or the consumption of animal hosts.

Aboubakr et al. (2021) compared the characteristics of SARS-CoV-2 with the distinctive characteristics of enteric foodborne viruses, such as HuNoV and hepatitis A virus. SARS-CoV-2 was found to share four main traits that demonstrate their ability to transmit through food: (1) the virus is enteric and replicates in the gastrointestinal system; (2) its infectivity remains stable for long periods under a wide range of environmental conditions; (3) it is highly contagious due to the low infective dose; and (4) it tolerates the hostile conditions of the human gastrointestinal system.

Further investigations are needed to understand if the percentage of food-related COVID-19 cases are significant compared to person-to-person transmission and inhalation of airborne virus. Nevertheless, the cited studies clearly demonstrate that dismissing the role of foodborne transmission is unscientific.

The main route of SARS-CoV-2 transfer to food and food packaging recognised so far seems to be through cross-contamination from infected people—that is, food handlers. Well, what type of food factories are hotspots for COVID-19 outbreaks? Slaughterhouses and meat processing facilities. Therefore, by consuming meat and derivatives, we are increasing the demand for products that are spreading the virus throughout the food chain and the global community.

It is important to emphasise that even if the coronavirus pandemic had never emerged, animal products would still pose serious health risks, both microbiological and nutritional.

All public health agencies, food companies, and food professionals agree that cooking and handling of food products must be done correctly to avoid foodborne diseases. And two of the first things we learn in Food Engineering undergrad are: (1) the importance of avoiding cross-contamination between animal foods and plant-based foods, but also between cooked and raw food; and (2) animal products must be cooked until a certain temperature is reached in the core (typically 70°C). These food safety measures are not limited to industrial settings and catering services, but instead should be used in households too.

The high microbiological risks related to dairy, meat, eggs, fish, and shellfish explain many of the consolidated guidelines on domestic food preparation:

- Store red meat, poultry, and seafood products at low temperatures;
- Do not use the same utensils (e.g., knives, forks, and chopping boards) to cut fresh produce and meat products;
- Disinfect surfaces that were contaminated with animal products or their leaks (e.g., blood and water from chicken and red meat cuts) before preparing other foods;
- Do not consume raw meat and eggs or unpasteurised milk and dairy products;
- Store fresh meats separately in the fridge to minimise contamination risks;
- Cook red meat and poultry thoroughly to make sure the inside is not raw.

Is it sensible to gamble with our health just because we like how certain products taste?

INVESTIGATIONS

INVESTIGATION 1 – EGG-LAYING HENS IN INDIA

This 2019 investigation was carried out by Animal Equality in egg farms in the Indian states of Maharashtra, Gujarat, Andhra Pradesh, and Telangana. Figures 36a and 36b show pens filled with metal cages piled across three floors, each cage housing four to eight hens. In India, 80% of eggs are laid by caged hens, even though battery cages were banned nationwide in 2013 on the grounds of animal cruelty, with all cages to be phased out by 2017 (Animal Equality, 2020b; Madaan, 2017).

Figures 36a and 36b. Battery-caged hens in egg farms in India.

The layers suffer from sore, cracked, and deformed feet due to the tight space they must share with others and the configuration and material of the cages (Figures 37a and 37b). Their feet and claws are not anatomically adapted to step on metal wires all day long for months to years. Trampling, fighting, and overgrown claws are common in these cages, leading to further injuries.

Figures 37a and 37b. Typical living conditions of battery-caged hens (example from an egg farm in India).

The stress owing to captivity, overcrowding, and egg overproduction weaken their immune system and bones. High levels of ammonia inside the pens and the constant contact with excrements produce raw burns and blisters all over their bodies, in addition to respiratory problems. As they engage in abnormal behaviours such as feather pecking and toe pecking due to anxiety and stress, they leave their skin exposed and vulnerable to infections (Figures 38a to 38c).

Figures 38a to 38c. Health and welfare problems associated with battery cages (example from an egg farm in India).

The birds can go blind for different reasons: ammonia levels that damage the cornea, physical trauma to the eye, respiratory infections that affect the eyes, and Marek's disease. Nonetheless, they are often left untreated as long as they keep laying eggs (Figure 39).

Figure 39. Blind hen in an egg facility in India.

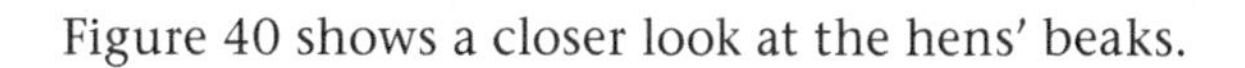

Figure 40 shows a closer look at the hens' beaks.

Figure 40. Beak-trimmed laying hens in an egg facility in India.

In the battery-cage system, the eggs roll to the front of the cage to be collected periodically. Faeces and urine fall underneath the stacked cages. Animals housed in the bottom layers are more affected by the urea and ammonia from the excrements (Figures 41a and 41b).

Figures 41a and 41b. Egg collection and manure accumulation in an egg farm in India.

The outstanding amounts of excrement and feathers produced in the pens pose a severe environmental issue. The investigators reported that the pile of dirt was only disposed of once every few weeks (Animal Equality, 2020b).

Animal Equality's team also witnessed sick hens left to die without proper veterinary care (Figures 42a to 42c).

Figures 42a to 42c. Negligence towards sick hens in an egg facility in India.

Hens were photographed trying to escape from their cages, another source of injuries and emotional distress (Figure 43). Like all living beings, they want to be free.

Figure 43. Hens trying to escape in an egg facility in India.

Please leave animals and their products off your plate. Let us create a world of peace.

INVESTIGATION 2 – BROILER CHICKENS IN THE UK

This investigation was conducted in different farms operated by Moy Park, one of the major producers of chicken meat in the UK. The farms, located in the East Midlands, supply big supermarket chains, such as Tesco and Ocado, as well as McDonald's (Animal Equality, 2020c). All of them are certified by Red Tractor, the UK scheme supposed to ensure high animal welfare standards in animal farms. As I have been showing throughout this book, animal welfare and animal exploitation are mutually exclusive concepts. Yet, people insist on denying that using animals as ingredient sources in the current times is immoral. But let us see some more evidence of that.

Figures 44a and 44b show tens of thousands of birds in a closed pen with no outdoor access or natural light. Many of them have deformed legs due to unnatural growth rates. Most of the chicks developed raw skin burns on their feet and body from the ammonia-soaked floor.

Figures 44a and 44b. Broiler chickens on Ladywath farm in the UK.

Unable to stand up or walk, some struggle to feed themselves. Many die of dehydration or starvation. The physical and emotional stress of these animals is worsened by witnessing dead birds in the pens. To be highlighted is the sanitary issue of mixing dead and live chickens in the same space (Figure 45).

Figure 45. Dead chicken on Ladywath farm in the UK.

On another farm operated by Moy Park, most chickens also suffer from skin abrasions and lesions. Their joints and hearts cannot support their artificially enlarged bodies, causing discomfort, pain, and even death. The dead ones are disposed of like garbage (Figures 46a to 46c).

Figures 46a to 46c. Injuries and mortality among broiler chickens on Mount farm in the UK.

Pictures taken on a third farm operated by the same group reinforce that agony and pain is the norm in broiler chicken farms. Figures 47a to 47c show birds suffering from deformed legs, blindness, and irritated, raw skin in an overcrowded shed. Please keep in mind that these are Red Tractor-accredited farms.

Figures 47a to 47c. Broiler chickens on Saltbox farm in the UK.

In Figures 48a and 48b, live broilers share space with dead birds of varying ages.

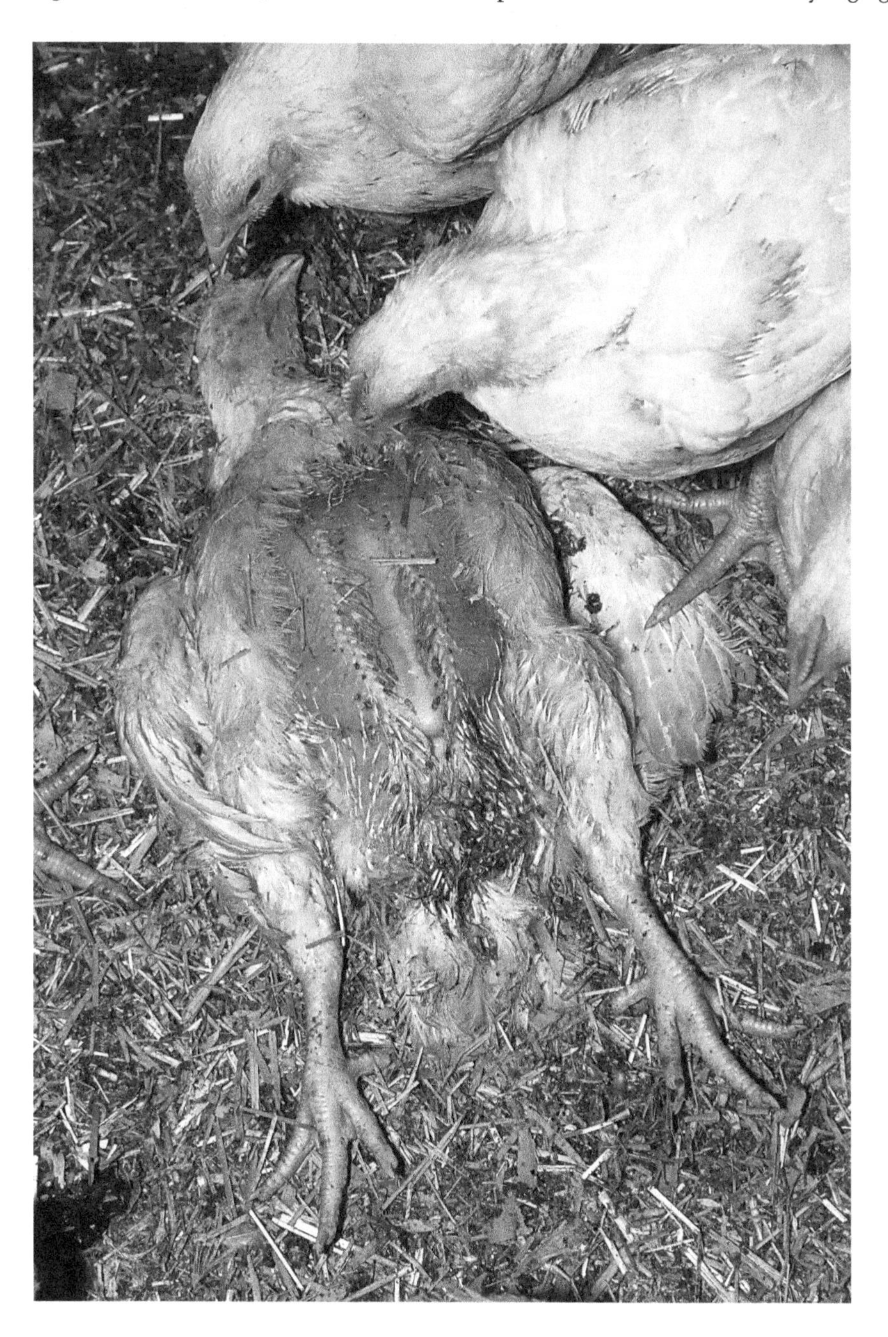

Figures 48a and 48b. Broiler chickens on Saltbox farm in the UK.

In these farms, the undercover investigators reported that hundreds of chickens were killed or simply left to die on-site when deemed too small or weak. A worker from one of Moy Park's farms said:

> I can look at a day-old chick and say that's going to make 1.8 kilos at 32 days or it's not. If it isn't, there's no point feeding it. It's cheaper to kill it and get rid of it. Because at the end of the day it's about making money. If I'm going to grow a bird for 30 days, feed it, keep it warm, let it drink the water and everything else, and then it not get processed at the factory because it's not big enough, it's just rejected, then I've just spent a pound feeding it. Well the more of those you can get out, the better the profitability of the farm will be; the better the profitability of the company will be. (Animal Equality, 2020c)

On many occasions, the birds were killed with the workers' bare hands—their necks were cut against sharp edges of nearby buckets or feeders. Animal Equality's team also witnessed workers tossing chickens from a certain height onto the floor, thereby hurting and even killing some of them (Animal Equality, 2020c). When confronted with the photo and video shoots, a Moy Park spokesperson declared that 'the overall flocks are displaying natural behaviours and appear in good health in most of the footage' (Animal Equality, 2020c). That is their level of regard towards the animals from which they make a living.

Investigation 3 – Egg-laying hens in Germany

This investigation was conducted in two *organic* farms in Northern Germany in early 2015 (Animal Equality, 2020d). Both farms held up to 30,000 egg-laying hens at the time. One of the farms supplies eggs to Deutsche Frühstücksei GmbH, the country's largest egg producer and one of Europe's leading egg producers. Deutsche Frühstücksei GmbH supplies all of Germany's leading retail chains and supermarkets.

Figures 49a to 49g show hens living in poor hygiene conditions in overcrowded, dark barns. The birds shared the barn with several dead hens.

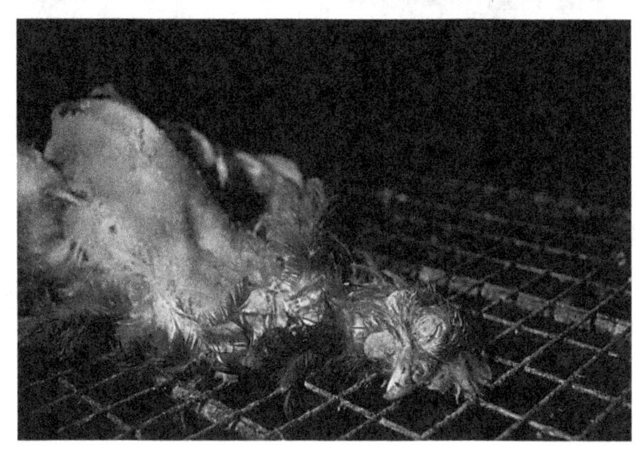

Figures 49a to 49g. Living conditions of laying hens on organic egg farms in Germany.

Figures 50a and 50b show a hen that died a dreadful death, possibly from egg-bounding or egg yolk peritonitis, without veterinary assistance. These horrific pictures show abundant blood in the vent area and protrusions on the abdomen. We can also see a broken egg inside her oviduct and another bloody spot on one of the abdominal protrusions. This may indicate that the eggshell punctured the hen's body. Not being able to lay an egg is extremely painful to birds, and egg-bound hens can die in a few days if untreated.

Figures 50a and 50b. Dead hen with severe injuries on an organic egg farm in Germany.

Most hens had featherless areas due to stress-induced feather pecking (Figures 51a to 51e). The raw, sometimes injured skin offers perfect conditions for inflammations and microbial infections.

Figures 51a to 51e. Poor feathering in hens on organic egg farms in Germany.

Bear in mind that the two farms investigated supply *organic* eggs. Worldwide, 200,000 to 300,000 hens in a shed still count as free-range, cage-free, or organic production. In the wild, however, chickens typically form groups of *12* animals. Figures 52a to 52c show that the birds are indeed cage-free, but is this enough?

Figures 52a to 52c. Hens living amongst dirt and faeces on organic egg farms in Germany.

We feel deeply offended about prejudice against women, yet we enslave non-human females for taste gratification.

Motherhood is sacred in all species. Cows do not produce milk because they are cows; they produce milk because they are *mothers*. Hens do not lay eggs because they are hens, but because of their reproductive cycle.

Do we have the right to subjugate females over an *ingredient*? Is it ethical to dominate females for *any* reason? In the end, it seems justice is only valid when it benefits us. Think about this.

CHAPTER 4

FISH

Earth's oceans could be completely devoid of fish by 2048—that is what a *Science* paper predicted in 2006 (Worm et al., 2006). Data from 64 large marine ecosystems and 48 protected ocean areas were analysed in 12 coastal regions from the US to Europe over long periods. The researchers attributed fish depletion to overfishing, pollution, habitat loss, and climate change. I am not a fan of the word "overfishing" because it gives the idea that *industrial-scale* fishing is the problem, rather than fishing itself. However, I will use this word throughout the chapter to emphasise the colossal rate at which we are destroying aquatic life—we are not just fishing, we are *over*fishing.

Important to highlight is that life on Earth is dependent on marine health. The ocean is greatly responsible for regulating the global temperature, other than producing 50–80% of the planet's oxygen, primarily by algae and cyanobacteria. *Prochlorococcus*, a genus of marine cyanobacteria measuring 50–100 μm, produces 5–20% of the atmosphere's oxygen. Although *Prochlorococcus* is the most abundant photosynthetic organism known to date, the genus was isolated for the first time less than two decades ago (Pennisi, 2017).

We know very little about life on this planet. At the same time, we know *enough* to cure many of the problems afflicting humanity. By insisting on animal use, despite the obvious harm it is causing on multiple levels, we are missing valuable opportunities.

In November 2020, around 50,000 fishes escaped a salmon farm in Tasmania, Australia, called Huon Aquaculture, after an electrical fire melted part of their pen (Rachwani, 2020a). For those who are not familiar with the term 'farm' used for fish, aquaculture consists of raising aquatic animals (fish, molluscs, crustaceans, and even frogs) in captivity—for instance, in net tanks submerged in rivers and seas, in artificial ponds or in tanks inside sheds.

While you might have smiled at the thought of thousands of fishes gaining freedom—I surely have—what you might have overlooked is the environmental impact of this incident. The loss represented merely 1% of Huon Aquaculture's fish stocks, and the company was quick to claim that there is limited evidence of ecosystem alteration due to farmed fish escapes (Rachwani, 2020a). Untrue. Numerous papers report that fish farm escapees impact wild populations in three primary ways: long-term genetic alteration, competition for food and habitat, and the proliferation of sea lice, all leading to demographic decline (Larsen & Vormedal, 2021; Bradbury et al., 2020; Glover et al., 2020; Weitzman et al., 2019; McGinnity et al., 1997).

Fish can escape from farms for various reasons, including net biting, storms, vandalism, predators, outdated or damaged pens, and human error. Fish escapes have been reported in virtually all regions where they are farmed. In many studies, researchers have identified escaped salmon in rivers and seas hundreds of kilometres away from fish farms (Jensen et al., 2013; Morris et al., 2008; Hansen & Jacobsen, 2003; Hansen et al., 1993).

Farmed fish are traced back to their farms of origin by DNA-based methods (Glover et al., 2020). Genetic interactions through breeding between wild and escaped Atlantic salmon have been repeatedly confirmed in North America and Europe (Wringe et al., 2018; Sylvester et al., 2018, Glover et al., 2017; Bourret et al., 2011; Crozier, 1993).

The dispersion of farmed fish over rivers and oceans is driving significant genetic alteration. For example, Norwegian farmed genomes were detected in wild salmon in Norway, Ireland, England, and Scotland rivers. Therefore, many fish sold as wild contain farmed DNA (Kurlansky, 2020). In Norway, the farmed fish population is *800 times* greater than the wild fish population (Kurlansky, 2020).

A group of researchers published a risk assessment about Norwegian farmed salmon's introgression in native populations (Glover et al., 2020). Ten out of 13 aquaculture zones exhibited a moderate to high risk of introgression of salmon escapees, which is particularly important since salmon travel long distances (Glover et al., 2020). Mitigation initiatives, such as the OURO program, are dedicated to removing fish escapees from rivers. The program, financed by the aquaculture industry, was working on 60 of the country's 448 rivers as of 2020 (Glover et al., 2020). Other mitigation strategies consist of reducing fish escapes through husbandry practices and technologies that hinder the fishes' ability to survive and reach rivers. One of the methods is the use of genetic selection to reduce early maturation in domesticated strains. A third strategy is farming sterile fish (Glover et al., 2020). Reducing, let alone extinguishing, fish farming was not contemplated in the study. The research was funded by the Norwegian Ministry of Trade, Industry and Fisheries.

One week after the Tasmanian fish escape episode mentioned before, another 130,000 salmon swam to freedom from the Huon fishery—this time due to a tear in a net. In response to conservationists' fears of environmental damage, a company spokesperson claimed the fishes would hardly survive given their small size (Rachwani, 2020b). Once again, an unscientific statement. Outraged by the company's claims, co-chair of the Tasmanian Alliance for Marine Protection Peter George told *The Guardian Australia* that the issue is being overlooked by regulatory agencies and politicians (Rachwani, 2020b).

Despite the prolific scientific evidence of environmental damage attributed to fish farming, new fish farms continue to open worldwide. Tassal, a 28-pen Tasmanian salmon farm, gained federal approval to function near a world heritage area in Okehampton Bay, back in 2017. Activists had their legal action against Tassal rejected by the court on the grounds of employment generation (Australian Associated Press, 2018). Tasmanian Atlantic salmon is Australia's largest seafood sector, with 58,000 tonnes of wild and farmed salmon produced in 2018–2019 (FRDC, 2020).

Tasmania's fishing industry is dominated by three major fisheries: Tassal, Huon, and Petuna (Morton, 2018). In 2018, the aquaculture industry lost hundreds of thousands of fish in the Macquarie Harbour due to illness, low dissolved oxygen levels, or both. Infections by pilchard orthomyxovirus (POMV), which are innocuous to humans but have no cure in salmon, are growing in number. The deficient water oxygenation is ascribed to the sector's rapid expansion in the area—an eight-fold increase over a decade (Morton, 2018). Hypoxia in Tasmanian waters due to anthropogenic stressors is putting many species at risk, including the endangered Maugean skate (*Zearaja maugeana*) (Morash et al., 2020).

Huon Aquaculture's chief executive, Frances Bender, accused rival Tassal of exacerbating the environmental damage in the Macquarie Harbour by filling pens with salmon of different ages. She claims this practice increased the POMV spread rate across Tasmanian fisheries, which sit in proximity to each other (Morton, 2018). Apparently, environmental preservation is only a priority to the fish industry when their business is financially affected.

A third of the Macquarie Harbour, which occupies 315 km^2, sits in the Tasmanian world heritage area. In 2005, the harbour's fish production was 2,000 tonnes. Ten years later, state and federal governments approved the expansion of aquaculture to 29,500 tonnes of fish per year (Morton, 2018). The result? In 2016, at a production of approximately 16,000 tonnes, evident signs of environmental collapse started to show, including a large dead zone. The Environment Protection Authority was repeatedly forced to revise the biomass limit until reaching 12,000 tonnes (Morton, 2018). We are gambling with ecosystems.

Bacteria that survive in low-oxygen conditions have multiplied to such an extent that, in February 2018, the Macquarie Harbour's West Coast beach was closed to swimmers due to high bacterial levels. The West Coast Council declared the water was unsafe for human contact due to gastrointestinal illness risk (Morton, 2018; McBey, 2018).

Intensive fish farming results in increased nutrient and heavy metal concentrations in surrounding waters, and generates anthropogenic organic carbon (OC) deposition (Sweetman et al., 2014; Schenone et al., 2011; Norði et al., 2011). Carbon and nitrogen losses in fish farms reach 67–84% and 68–81%, respectively (Norði et al., 2011). A study conducted in a trout farm in the Faroe Islands, a North Atlantic archipelago located between Scotland and Iceland, demonstrated that the farm's OC and nitrogen footprints covered an area ten times the farm's size (Norði et al., 2011). In addition, organic sedimentation beneath fish farms can be up to 27 times higher than in other sites (Norði et al., 2011).

OC and excessive nutrient input into water bodies pose major threats to the ecosystem balance, including benthic organic loading (Sweetman et al., 2014). The benthic zone comprises the sediment surface of oceans, lakes, and streams.

This zone is inhabited by benthos, which include bacteria, fungi, amphipods, crustaceans, polychaetes, and other large invertebrates. As many of these organisms have adapted to live on the ocean or lake floor, they cannot resist the lower pressures of the upper parts of the water column. While benthos feed on decaying matter sedimented on the ocean or lake bottom, they are prey for other creatures, such as sharks. Each species has a vital role in the aquatic ecosystem, both due to their participation in the food web and other ecological functions they play.

The oversupply of nutrients, primarily phosphorus and nitrogen, in a water body promotes excessive algae growth (i.e., eutrophication). The so-called algal blooms reduce light penetration, killing plants that use sunlight for photosynthesis. When the algae ultimately die, bacterial degradation may result in oxygen depletion, causing fishes to die. What is more, the anaerobic digestion of dead algae releases GHGs, such as CO_2 and methane. Besides, algae can outcompete other organisms and promote the occurrence of low-diversity algal communities. Algal blooms can also release toxins, odours, and taste substances into the water, killing aquatic organisms and reducing water quality for human use.

But please do not get the wrong idea, algae are not villains. Algae use sunlight and carbon dioxide to produce organic compounds, releasing oxygen back into the atmosphere. Algae also provide food for the aquatic fauna and remove pollutants from the water (Krienitz, 2009). Disruption of this delicate aquatic equilibrium can affect the food web, the CO_2 sequestration potential, the oxygen cycle, and the symbiotic relationships between algae and other organisms, such as corals and sea sponges (Frankowiak et al., 2016). For example, microscopic algae supply organic matter to corals, which in turn serve as a host to the algae. The corals also emit ammonium as waste, which is consumed by algae. Coral reefs are built ten times faster when symbiosis occurs between corals and algae (Princeton University, 2016).

As you can see, fish escapes and waste from fish farms unchain massive ecosystem disruption. According to a recent study on fish escapes in Norway, roughly 300 escape incidents involving Atlantic salmon (*Salmo salar*) or rainbow trout (*Oncorhynchus mykiss*) were reported to the Norwegian Directorate of Fisheries between 2010 and 2018. The estimated number of escapees nears two million fishes. Ninety-two per cent of the escapees derived from sea-based fish farms versus 7% inland fish farms. The remaining 1% of the fishes escaped during transportation between sites (Føre & Thorvaldsen, 2021).

Sea-based farms are obviously not the only source of fish. Fishes and other aquatic creatures can be either farmed or wild-captured, both from marine environments and freshwaters. As we have only one chapter dedicated to aquatic animals, I will focus on fishes.

Let us now look at the different types of fishing and the associated animal welfare and environmental issues.

Wild-caught versus farmed fish

Simply put, wild-caught fishes are captured from their natural habitat (sea, river, or lake), while farmed fishes are raised either in sea pens, artificial ponds, or inland tanks. Notably, the term 'fishing' includes aquatic animals other than fishes, such as crustaceans (e.g., shrimp and lobster), molluscs (e.g., scallops and oysters), cephalopods (e.g., squid), and echinoderms (e.g., sea cucumbers). The term 'fishing' usually does not apply to farmed fish and aquatic mammals, such as whales.

Fishes are consumed in such large numbers that fish captures are traditionally measured by the tonne. Table 3 puts numbers to the violence inflicted on animals used as food in the world. While around 298 million cattle, 1.3 billion pigs, and 57 billion birds are slaughtered every year, up to 110 billion farmed fishes and 2.7 *trillion* wild fish are killed. But, unlike land animals, slaughter and welfare regulations for fish are vague.

Table 3. Number of animals killed for food globally every year, by species.

Species	Number of animals (millions)
Farmed animals	
Chickens	52,887
Ducks	2,556
Turkeys	669
Total birds	56,769
Pigs	1,313
Sheep	527
Goats	398
Cattle	298
Total mammals	2,572
Fish (estimated)	6,400–110,000 (31,927,813 tonnes)
Wild fish (estimated)	
Peruvian anchovy	300,000–870,000
Atlantic herring	3,600–22,000
Bombay duck or bummalo	2,600–38,000
European pilchard	8,300–15,000
Atlantic mackerel	1,400–1,700
Total wild fish	970,000–2,700,000 (77,388,322 tonnes)

Source: Mood (2010).

You may be wondering, but where are all those animals? Well, factory farms confine a vast number of animals, often indoors, which can go unnoticed by most people. This is especially true for fish farms—we may spot some metal poles or maybe the top rim of pens poking through the seawater and never realise this is a fish farm. Did you know fish pens can go down 50 metres deep?

Fishes have been the last class of animals used as food to be studied in terms of sentience. Nonetheless, fishes have been scientifically proven to feel pain, fear, and distress based on multiple behavioural studies and assessments of sensory systems, brain structure, and functionality (Yue et al., 2008; Volkoff & Peter, 2006; Kihslinger & Nevitt, 2006; Dunlop & Laming, 2005; Gilmour et al., 2005; Huntingford & Adams, 2005; Chandroo et al., 2004; Yue et al., 2004; Acerete et al., 2004; Arends et al., 1999; Chervova, 1997; Bradford, 1995; Denzer & Laudien, 1987).

Legislative acknowledgement that animals are sentient beings and, as such, deserve ethical treatment is recent. The European Convention for the Protection of Animals kept for Farming Purposes dates back some decades only (1976). The Convention's rules reflect the principles of the Five Freedoms, a guideline based on a 1965 British parliamentary investigation into the physical and emotional needs of animals. The Five Freedoms have been used extensively by industry, governments, and certification schemes. The Five Freedoms correspond to animals:

1. Free from hunger and thirst;
2. Free from discomfort, who have access to an appropriate environment, including shelter and a comfortable resting area;
3. Free from pain, injury, or illness;
4. Free to express their natural behaviour;
5. Free from fear and distress.

Paradoxically, the Five Freedoms are the antithesis of what animal agriculture represents. Animals raised for food are simply not free. Freedom from hunger, thirst, discomfort, pain, injury, fear, distress, and disease, as well as the freedom to express normal behaviours, are violated on a daily basis in animal farming, capture fishing, transport, and slaughter.

In 2006, the European Convention set the Recommendation Concerning Farmed Fish, which specifies husbandry methods, housing standards, and other provisions (Council of Europe, 2006). And only in 2009, the Convention was amended by the Lisbon Treaty to finally recognise fish sentience. I am focusing on the European regulations for practical purposes, but the conclusions apply elsewhere.

While these are critical advances in how we perceive animals, there is a long distance between *acknowledging* that animals are not things and *stopping to treat them as such*. Our poor understanding of what it means to respect our fellow sentient beings is strongly reflected in scientific papers on animal welfare. For example, 'Fish are an important biological *resource for humanity* as food through fisheries and aquaculture that allow for sustainable production and are *useful study models* for scientific research' (Toni et al., 2019). Our deeming animals as 'resources for humanity' and 'useful study models' says more about us than about animals.

The main difficulty in establishing specific welfare standards for fishing and aquaculture is that too many species are exploited. Husbandry methods, feed, housing, equipment, and slaughtering techniques must be tailored to each species according to their biological and behavioural characteristics. If violations against the welfare of terrestrial animals, who have been extensively investigated for decades, are still *the rule* rather than the exception, let alone against fishes.

Due to the behavioural and biological similarities with other animals and humans, the use of teleost fish has increased significantly among the scientific community (Toni et al., 2019). Zebrafish, for example, have been employed in neuroscience and behavioural and pharmaceutical research because over 71% of human genes have one or more zebrafish orthologues (Howe et al., 2013). Orthologue genes are genes (not necessarily identical) found in different species that originated from a common ancestor. Many psychiatric drugs have been tested in zebrafish before approval to human trials (Giacomini et al., 2016). Animal experimentation is one more reason why eating plant-based is not enough if we deem other animals worthy of moral consideration.

Capture fishing

Global wild catches vary between 86 million and 93 million tonnes per year. Approximately half of the world's capture is represented by 7 countries: China (15%), Indonesia (7%), Peru (7%), India (6%), Russia (5%), the US (5%), and Vietnam (3%), in this order. Wild marine species represent 87.4% of the global catch, versus 12.6% of freshwater animals. FAO reports marine catches for more than 1,700 species, with finfish representing 85% of them (FAO, 2020).

The fishing method depends on several factors, including fishery production, catch size, target species, marine versus freshwater capture, and tradition. Bycatch rates vary substantially among different fishing methods, but at least 10% of the global catch is wasted (FAO, 2020). Some studies estimate global marine bycatch rates at 40% or over (Davis et al., 2009).

Trawling

Trawling is a fishing method in which cone-shaped nets are dragged through the water by one or more fishing vessels called trawlers (Figure 53). Trawl nets can be pulled in midwater at a certain depth or along the ocean floor, sweeping everything they meet—fish, shellfish, mammals, turtles, corals, algae, and so on.

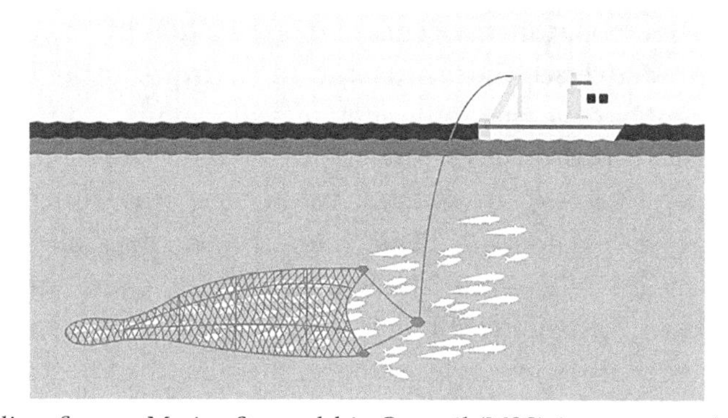

Figure 53. Trawling. Source: Marine Stewardship Council (MSC) (www.msc.org).

Each trawler can operate one or more trawl nets simultaneously. The catch is maintained under chilling conditions until unloading to slow down deterioration. Part of the fish processing is done on board the trawlers, including sorting, washing, gutting, and filleting. Trawlers vary in architecture, power, and size, with factory trawlers measuring up to 80 metres long. Figure 54 gives a glimpse of the dimension of commercial trawl nets.

Figure 54. A factory-sized trawl net. Source: Quora (www.quora.com/What-is-the-difference-between-gill-netters-and-trawlers).

In this fishing method, fishes are chased to exhaustion by the trawl nets. Once overrun by the trawler vessel, the fishes reach the narrower, closed end of the net, where they commonly have their gills compressed, making it difficult to breathe. The net is usually hauled on deck after hours of tow. Throughout the process, the animals experience fear, stress, exhaustion, and may suffer from injuries, circulatory failure, and suffocation. The fishes can also die from decompression, especially those caught at depths of 20 metres or over. The pressure may cause the abdomen to prolapse, forcing the gut out of the anus and mouth and the eyes from the orbits. Many of those who survive catching and landing will die of suffocation *before* slaughter.

As fish stocks decline due to overfishing, trawlers resort to deeper and deeper waters, damaging ecosystems and menacing species. The smaller the mesh size, the greater the bycatch rates. Bycatch in shrimp trawling is particularly substantial due to the small mesh size required.

In a study conducted in the Arabian Gulf's Saudi waters, bycatches from shrimp trawler fisheries included 104 fish and shellfish species (Abdulqader et al., 2020). Data was collected from 37 large trawlers during 2 consecutive fishing seasons (2013–2015). Each 500–755 tonnes of shrimp were accompanied by 281–563 tonnes of bycatch and 114–339 tonnes of discard. A total of 1,492–2,018 turtles were accidentally captured.

Bycatch rates can be reduced by tailoring the mesh size to the target species and using various devices—e.g., turtle excluder devices and sea lion excluder devices. However, marine animals can still be injured and even die when escaping from trawl nets (Mood, 2010).

Other impacts of trawler fishing, beyond habitat destruction and bycatch, are water pollution, global warming, eutrophication, and acoustic pollution (Vázquez-Rowe et al., 2010). The vessels pollute the waters with fuel, fishing gear, and other debris, in addition to stressing marine life with noise levels above the damage thresholds. Cetaceans (e.g., whales and dolphins) are significantly affected, with declined populations and reduced biological diversity resulting from noise exposure. For these reasons, scientists recognise that trawler fishing is the *primary* anthropogenic threat to marine life (Daly & White, 2021).

Purse seining

In purse seining, a school of fishes is surrounded by a long wall of netting called a seine net. The net is suspended vertically in the water, floating attached to buoys. To catch the fishes, the net is pulled tight to form a bag and hauled on board the vessel (Figure 55). Unlike trawlers, purse seining ships travel at a low speed to avoid alarming the fishes away from the net. Purse seines can be as long as 2,000 metres and reach deeper than 200 metres. Fish schools are located by observing movement underwater, fast-moving dolphins, or birds flying near the water surface (Mood, 2010).

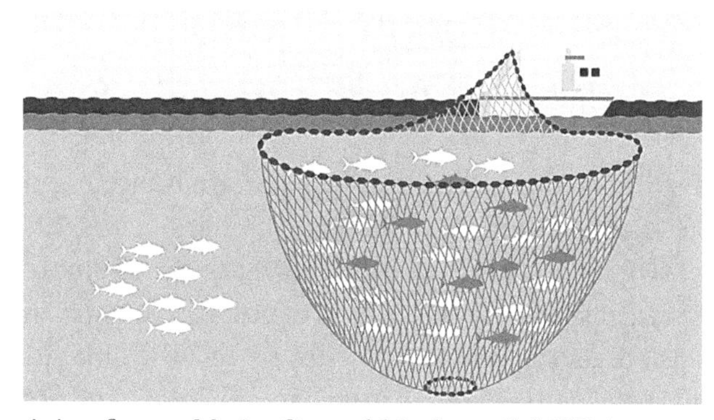

Figure 55. Purse seining. Source: Marine Stewardship Council (MSC) (www.msc.org).

Commonly, turtles, seabirds, and mammals are trapped in the nets along with the target species and can sustain serious injuries. Smaller animals (e.g., turtles, otters, and birds) can even be crushed to death under the tow's weight. Seine nets are typically associated with the incidental capture of dolphins and whales because these animals chase fish. In fact, dolphin pods have been long used as a cue for the proximity of fish schools.

The total fishing duration in purse seining is generally shorter respective to trawling, with significantly lesser discard and bycatch rates. Nonetheless, the numbers are high in both fishing methods. A 2019 FAO report estimated the annual bycatch and discards by marine fisheries worldwide between 2010 and 2014 (Pérez Roda et al., 2019). Around 10% of the yearly catches were discarded, corresponding to 9.1 million tonnes. Approximately 46% of total annual discards were from bottom trawls (4.2 million tonnes), 11% from purse seines (1.0 million tonnes), nearly 10% from midwater trawls, and 9% from gillnet fisheries (0.8 million tonnes) (Pérez Roda et al., 2019).

At least 20 million animals are unintentionally captured and discarded every year by the fish industry, including 10 million sharks, 8.5 million sea turtles, 1 million seabirds, 650,000 marine mammals, and 225,000 sea snakes (Gray & Kennelly, 2018). Purse seine fisheries commonly release the catch with undesirable size or species. Still, the mortality rates are substantial, increasing with exposure time and tow weight (Pérez Roda et al., 2019). The actual figures are probably much higher since these estimates did not include smaller-scale fisheries and did not cover every single region around the globe.

Although there are regulations intended to reduce the destruction of marine life, especially regarding endangered, threatened, and protected species, those are often inconsistently enforced (Pérez Roda et al., 2019). Modifications to fishing gear and fishing practices, fishing area restrictions, discard bans, bycatch limits, and economic incentives put some halt to bycatch and discards by fisheries. Nevertheless, incidental captures are not at all the only threat to marine health.

Besides, governments and fisheries are interested in ocean preservation for the industry's survival and productivity. Overexploited oceans are simply not profitable.

Gillnetting

Gillnetting is a fishing method that employs a curtain of netting that hangs in the water from a line attached to spaced floaters. The net has weights on the bottom to keep it vertical (Figure 56). Mesh size and floater-to-weight ratio can be manipulated to catch the target species at the desired depth. As the net is invisible to fishes, they swim into it and get trapped by the gills, body, teeth, maxillaries, and other protrusions when they try to reverse. As the name suggests, the fishes are generally gilled, which is very painful.

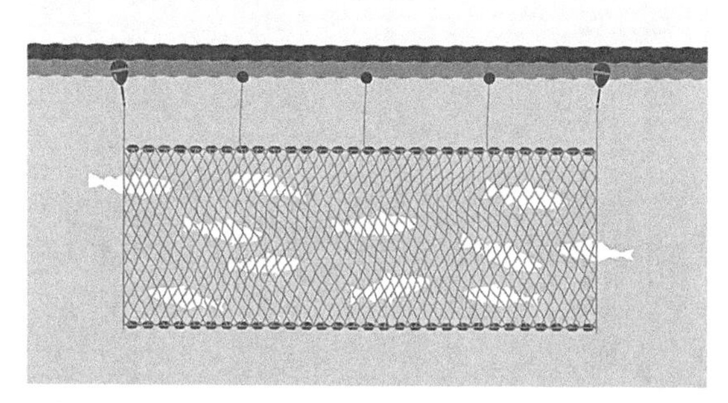

Figure 56. Gillnetting. Source: Marine Stewardship Council (MSC) (www.msc.org).

Gillnets, the most used fishing gear in the world, can range from 90 metres to 11 kilometres in length. The fishing vessel is called a gillnetter. Although gillnetting can have a lesser impact on the seafloor than other fishing methods, it is traditionally associated with bycatch and ghost fishing. Ghost fishing occurs when the net is discarded or lost in the water body, entrapping animals who can then sustain painful injuries, suffocate, or starve to death. The nets can also reach the ocean bottom and harm the benthic environment. Additionally, net fragments can be swallowed by the fauna, causing injuries and mortality.

Replacing nylon gillnets with biodegradable ones has been widely proposed as a mitigation strategy against ghost fishing and ocean pollution. However, large fisheries and small-scale fishermen alike are often resistant to change due to the lower catch efficiency and higher cost of biodegradable nets (Standal et al., 2020). Also, biodegradable nets do not degrade magically after being abandoned or lost in the ocean. They decompose gradually, with the degradation rate depending on the net composition, current strength, water temperature, frequency of storms, and microbial action.

In a study with biodegradable nets made of 82% polybutylene succinate (PBS) and 18% polybutylene adipate-co-terephthalate (PBAT), the catch rates for several species were only slightly lower compared to nylon gillnetting (Kim et al., 2016). That seems excellent, right? However, the nets *began* to lose their ghost fishing capacity after *two years* in seawater.

And before anyone says that the problem is simply one of formulation optimisation, note that a compromise between durability and biodegradability must be reached. A gillnet that dissolves promptly in saltwater is obviously useless to the fishing industry.

Trolling

In trolling, lines bearing lures or live baited hooks are towed through the water by a low-speed vessel. Trolling lures are designed to look and behave like prey fish (Figure 57). The type and size of the hook and bait are primarily determined by the target species. Caught fishes are landed quickly, sometimes with the help of a club or hook. Trolling is common in the commercial fishing of tuna, salmon, mackerel, and other pelagic fishes (i.e., those that swim nearer the surface).

Figure 57. Trolling. Source: US National Oceanic and Atmospheric Administration (NOAA).

As the capture duration is short compared to other fishing methods, trolling can be less harmful to fishes. Despite the *shorter* suffering, the fishes sustain serious injury nonetheless, as they can be hooked by the mouth, gills, eyes, or any other part of the body during fishing and landing (Mood, 2010).

The suffering of live fishes used as bait is obviously not to be ignored. Besides, bycatch levels are lower in this method, but cannot be neglected.

Pole and line fishing

In pole and line fishing, fishes are attracted to the water surface by live bait scattered at the vessel's side. This process, known as chumming, creates the illusion of an active school of prey. As the target fishes start to bite everything they see, the fishers throw poles with lines and barbless hooks into the water (Figure 58). The hooked fishes are then brought on board one at a time and disengaged from the hook.

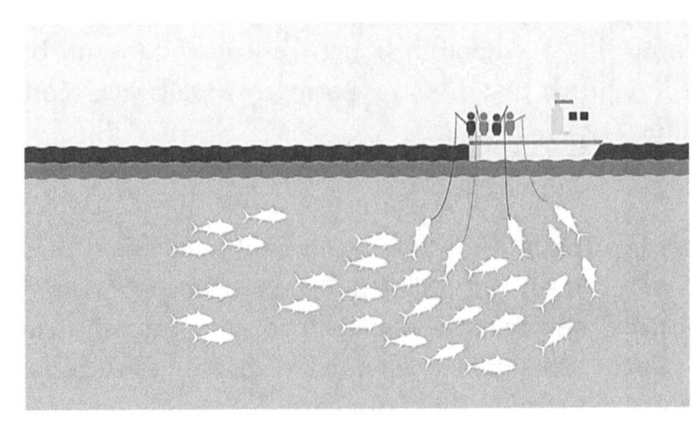

Figure 58. Pole and line fishing. Source: Marine Stewardship Council (MSC) (www.msc.org).

Pole and line fishing is commonly used to catch tuna and other pelagic species. Suffering levels can be lower compared to other fishing methods due to the shorter time the fishes remain on the hook. There is also little bycatch. Nonetheless, fishes can be impaled with gaff hooks to bring them aboard. And of course, the smaller fishes used as live bait suffer during fishing and over the confinement period too.

Longlining

In longline fishing, a long line carrying hundreds or even thousands of baited hooks is trailed behind a vessel (Figure 59). Live fishes are commonly used as bait. The line can measure anywhere between 50 metres and many kilometres. Pelagic (surface) and demersal (bottom) fishes are both common targets in longlining. The capture duration varies from hours to days, with the caught fishes suffering from fear, pain, and hunger the entire time. Some are attacked and even killed by parasitic crustaceans or predators.

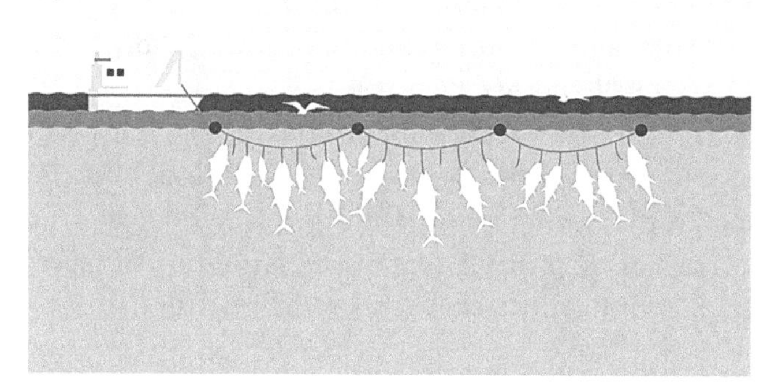

Figure 59. Longlining. Source: Marine Stewardship Council (MSC) (www.msc.org).

Birds, turtles, rays, and sharks are common bycatch in longlining. Seabirds (e.g., albatrosses and gulls) are common victims because they are attracted by the bait, get entangled by the lines, and end up drowning. Almost half (47%) of all seabird species is threatened, largely due to fisheries (Croxall et al., 2012; Anderson et al., 2011).

An estimated 160,000–320,000 birds are killed every year in longline fisheries worldwide. Due to unobserved and unreported bird bycatch cases, the actual levels can be up to 50% higher (Anderson et al., 2011). Exclusion devices can reduce seabird bycatch by 89%; however, implementation costs are still a drawback to fisheries (Avery et al., 2017).

Bycatch sharks and rays may survive capture depending on the gear type and capture duration (Oliver et al., 2015). However, when finned and thrown back into the sea (which is commonplace), sharks will die an agonising and slow death. According to the Shark Research Institute, 100 *million* sharks are slaughtered every year, especially for their fins, used as a delicacy food. Ironically, fins have no nutritional value and contain high mercury levels (Shark Research Institute, 2021).

As shark catches are largely unmonitored, the figures mentioned are likely underestimations. Although shark finning is prohibited in many parts of the world (e.g., EU, US, Brazil, South Africa, and Costa Rica), the global fin demand moves a multibillion-dollar industry, with profits only behind those of the illegal drug trade (Shark Research Institute, 2021).

Other methods

Fishes, crustaceans, and other aquatic creatures can be caught by various other techniques, such as harpooning, trapping, and dredging. Like all other fishing methods, they involve luring and killing sentient beings.

Harpooning is commonly used to catch large species, such as swordfish. Basically, a barbed spear is fired at the fish, who is hoisted aboard after exhaustion. This method results in low bycatch rates but high levels of suffering and stress.

Trapping involves using cages of varying sizes, shapes, and materials (e.g., wood, wire netting, or plastic) to catch marine life. Lobsters, crabs, and fishes are typical targets in this method. Traps can be laid individually or, more commonly, in large numbers pending from a long rope (Figure 60).

Although the time between setting and retrieving the traps is generally short, entrapped fishes can sometimes wait for many hours and even over a day until being hauled aboard the fishing vessel. During this period, they may suffer from hunger, fear, and distress and may be attacked by predators that manage to enter or destroy the trap. Additionally, fishes can sustain painful injuries while trying to escape the trap.

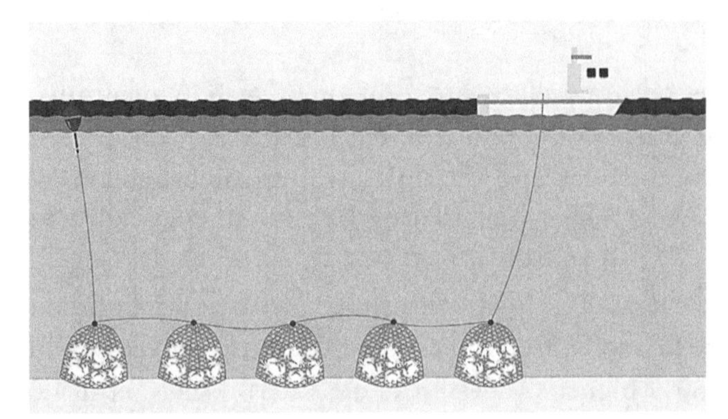

Figure 60. Trapping. Source: Marine Stewardship Council (MSC) (www.msc.org).

The traps may be dragged by tides along the seabed, damaging habitats. Like gillnets, traps that get lost in the water can continue ghost fishing for long periods. Bycatch rates are usually lower in trapping respective to other fishing methods. However, marine life can still become entangled in the lines connecting the traps to the buoys. Mesh size and exclusion devices can prevent marine animals from becoming entangled while foraging for food.

Bivalves like scallops, clams, and oysters are commonly caught with dredges: rigid structures, typically triangular-shaped, that are dragged along the seafloor (Figure 61).

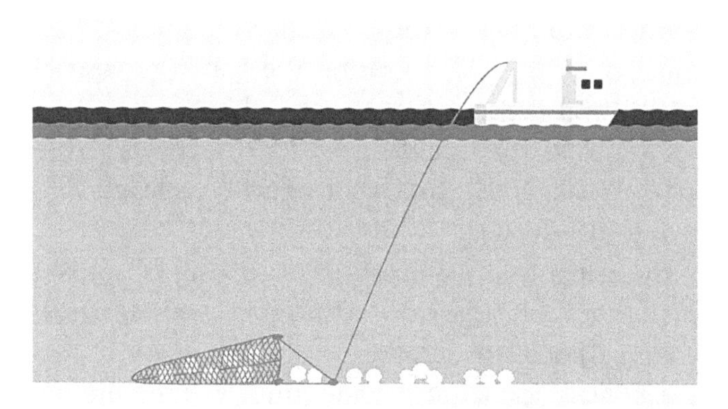

Figure 61. Dredging. Source: Marine Stewardship Council (MSC) (www.msc.org).

Dredge types and fishing areas are sometimes regulated to minimise environmental damage. For example, the Marine Stewardship Council (MSC) is a non-profit organisation operating in Europe, North America, and Australia that issues certifications to fisheries that comply with their sustainability requirements. The MSC Fisheries Standards bases on three pillars—namely, sustainable fish stocks, minimised environmental impact, and effective fisheries management (MSC, 2021). Nonetheless, animal welfare and ethics are not part of MSC's scope.

Dredges are also associated with bycatch and ghost fishing. Besides, some animals collide with the ship or escape from the gear and thus are not brought on board. These so-called "pre-catch losses" can result in high mortality rates due to the injuries sustained (Pérez Roda et al., 2019).

As we have seen in the last few pages, capture fishing is linked to multiple animal welfare issues, which affect the target fishes, bycatch species, and baits. Not to be dismissed is the suffering inflicted on fishes *after* they are on board the vessel. They can be clubbed several times before finally dying, be simply left to suffocate in air, be eviscerated or mutilated while conscious, and be thrown in iceboxes while still alive. When put onto ice, fishes undergo muscle paralysis and cannot show signs of pain and distress. But the reality is that they suffer a slow death due to asphyxiation and thermal shock (Mood, 2010).

The slaughter method depends on the fishing company and species, with little or no national or state regulation to be observed. Animal welfare academics propose some methods for a more "humane" slaughter—for example, percussive stunning and spiking (Mood, 2010; Boyland & Brooke, 2017). Percussive stunning is defined as a blow to the head, using a club, so the fish bleeds to death before regaining consciousness. Spiking consists of inserting a spike in the fish's brain so that the animal instantly becomes unconscious. Both concussion and spiking may potentially kill the fish directly. Spiking combined with food-grade sedatives, as permitted in Chile, Australia, and New Zealand, is another potentially "humane" method (Fishcount, 2019). Other recommended techniques are free-bullet slaughter for tuna and electrical stunning for carp, eel, and salmonids (OIE, 2019).

These methods' efficiency in preventing the fish from reawakening during exsanguination largely relies on the fisherman's skill. Although there are automatic stunning devices available, they are designed for specific species (e.g., pneumatic club for salmon) (Poli et al., 2005). To the best of my knowledge, automated spiking systems are still to be developed.

In any case, the slaughter by percussion or spiking is not exactly practical, especially for small fishes and large catches. Imagine clubbing or spiking hundreds of thousands of animals one after another...

A combination of electrical stunning and nitrogen exposure under conditions specific to each species is also recommended by researchers (Poli et al., 2005). Sadly, the interest in "humane" slaughter for fishes is frequently driven by flesh quality issues, since stress prior to and during killing affects the muscle tissue just like in terrestrial animals (Poli et al., 2005).

Notably, all the "humane" methods described are the exception in commercial fisheries and subsistence fishing alike (Mood, 2010). In most cases, the fishes are simply tossed, clubbed, pierced, placed on ice, and opened alive as if they were objects.

To be noted is that approximately half of the fishes and shellfishes that are eaten nowadays is *farmed.* Aquaculture is one of the fastest-growing "food" production sectors in the world (FAO, 2021). While many defend that aquaculture is the fishing industry's future, the modality poses serious environmental problems, in addition to the ethical issue of eating animals.

Farmed fish

Aquaculture comprises sea pens (i.e., marine aquaculture), coastal ponds and gated lagoons adjacent to the sea (i.e., coastal aquaculture), and inland fish farms, which include tanks, natural ponds, and excavated ponds. Inland fish farms can use freshwater, brackish water, or saline water, but most commonly the former. Nearly 600 aquatic species are currently farmed worldwide (FAO, 2021).

Asia is the major continent in terms of aquaculture production, representing almost 90% of the world's farmed fish in 2018. China alone accounted for nearly 60% of the 2018 global farmed fish production (FAO, 2020).

Dozens of countries produce more aquatic animals from farming than fishing. For instance, in China, India, and Bangladesh, aquaculture provided 77%, 57%, and 56% of the fish consumed countrywide in 2018, respectively (FAO, 2020).

In inland fish farms, hundreds and even thousands of fishes are packed together in tanks or ponds, sometimes saturated with excrements, availing of little space to swim and minimum opportunity to express their natural behaviours. Colossal amounts of water are used to fill the pens, process the dead fishes, and clean the facilities. The large tracts of land and water required in aquaculture reduce the agricultural yield from the surrounding areas. Small and subsistence farmers are exceptionally affected. Despite common belief, fish farms do put pressure on wild marine life because farmed fishes are commonly fed with wild-caught, smaller species.

The same pattern is observed in marine aquaculture, where sea pens are filled with unnatural amounts of fishes. The fishes have insufficient space to swim, which is especially disturbing for such migratory species as salmon. Feed and drug residues from the pens pollute the marine environment, posing a threat to the ecological balance and the coastal communities.

A Scottish study showed that lipids from the diet fed to salmon in fish farms were present in high concentrations in the sediment surface layer. The lipids included triacylglycerols, free fatty acids, sterols, polar lipids, and a combined fraction of hydrocarbons, wax esters, and cholesterol esters (Henderson et al., 1997). In addition, fish manure deposits on the seafloor beneath the pens, leading to coastal eutrophication, water quality decrease, and disease spread (Weitzman et al., 2019).

Welfare and stress indicators encompass different behavioural and physiological parameters that can be observed or measured. These indicators include eye and skin colour alterations, morphological changes (e.g., bitten fins and weight loss), presence of injuries, signs of disease (e.g., parasites, delayed growth, and reduced appetite), mucus production, opercular beat frequency, aggressive behaviour, and changes in swimming patterns (Toni et al., 2019). Multiple factors affect the welfare of farmed fishes, including feed, water physicochemical parameters, water volume, stock densities, tank design, handling, and others.

Like pigs, cows, and chickens, fishes suffer from boredom, stress, and depression when confined. Researchers suggest using environmental enrichment in tanks and ponds to simulate the natural habitat and reduce stress levels (Pounder et al., 2016; Giacomini et al., 2016). Nevertheless, there are way too many stressors in captivity other than the environmental barrenness.

Fishes are social creatures who establish positive relationships with other individuals and within groups, as well as hierarchy and territoriality interactions. When packed in unnatural numbers in ponds and tanks, aggressive behaviour can occur between the fishes, leading to stress, injury, and even death.

Disease incidence in aquaculture is so high that fish farmers spend large sums on prevention and control. We have seen the example of sea lice, which costs farmers an estimated EUR 305 million per year (Erkinharju et al., 2020). Sea lice were not a significant problem before fish farms existed. These tiny crustaceans would roam the ocean looking for salmon, a small fraction of the fish population. The salmon would carry a few parasites until they entered a river, when the lice would die and fall off as they do not tolerate freshwater. However, salmon farms are the perfect spot for sea lice—tonnes of fishes enclosed in a pen become vulnerable hosts.

As salmon lice eat the host fish's skin, they aim at their head and neck, which are scale-free. Sea lice can skin salmon to death, resulting in the death of up to 25% of the fishes, which are found dead at the bottom of the pen. The environmental problem is even more severe: lice spread quickly and attack wild salmon. The issue is more evident in the Atlantic than in the Pacific, where the number of wild salmon still exceeds that of farmed fishes (Kurlansky, 2020).

But there are way more parasitic and bacterial diseases of concern in the fish industry: for instance, internal parasitosis (e.g., *Philometra fasciati, Huffmanela ossicula,* and *Cymothoa exigua*) and varied infections (e.g., *Vibrium* spp., *Aeromonas* spp., *Mycobacterium* spp., *Streptococcos* spp., *Edwardsiella* spp., *Renibacterium salmoninarum, Pseudomonas anguilliseptica, Pseudomonas fluorescens, Clostridium botulinum, Flavobacterium* spp., *Piscirickettsia salmonis, Hepatobacter penaei, Francisella noatunensis,* and *Chlamydia* spp.) (Haenen, 2017).

Some of these pathogens pass from fishes to humans by direct contact, infecting farmers, technicians, and processors (Haenen, 2017). The parasites feed on the gills, organs, and blood of fishes, leading from discomfort to extreme pain and even death. *Cymothoa exigua,* for example, enters fishes through the gills and causes their tongue to atrophy by severing the area's blood vessels. To contain the spread of parasites and bacterial diseases in inland tanks and sea pens, larger and larger amounts of antiparasitics and antibiotics are employed, driving antibiotic resistance in fishes, land animals, and humans as well (Santos & Ramos, 2018).

High levels of ammonia, nitrates, and suspended solids, as well as inadequate water temperature, reduce animal welfare and productivity. Nonetheless, investigations around the world repeatedly reveal the unsanitary conditions and animal suffering that exist on fish farms. Fishes swimming in filthy water and sick animals with massive tumours, bacterial infections, open wounds, or covered with parasites are commonly reported.

In fact, salmon farmers are familiar with "loser fishes" or "dropouts," smaller-sized fishes exhibiting apathetic behaviour, floating near the surface as if already dead. Recent research showed that the depression state results from elevated brain serotonergic activation combined with increased cortisol production (similarly to what happens to humans) as a response to prolonged stressors. The problem affects nearly *a quarter* of the fishes in salmon aquaculture (Vindas et al., 2016).

In 2018, finfish was the predominant farmed aquatic animal, amounting to 54.3 million tonnes, valued at USD 139.7 billion. Of the total farmed finfish, 47 million tonnes were raised inland, while 7.3 million tonnes were raised in marine and coastal farms. Molluscs were the second most farmed aquatic creatures, with a production of 17.7 million tonnes, valued at USD 34.6 billion. The list continues with crustaceans (9.4 million tonnes, USD 69.3 billion), marine invertebrates (435,400 tonnes, USD 2 billion), aquatic turtles (370,000 tonnes, USD 3.5 billion), and frogs (131,300 tonnes, USD 997 million) (FAO, 2020).

These figures might have you wondering what would happen to livelihoods and the global economy should we switch to a plant-based diet. Well, the production of grains, legumes, seeds, vegetables, nuts, leaves, mushrooms, and fruit is also profitable. Notably, the direct profits would be accompanied by numerous financial and intangible benefits: improved physical health, increased longevity, lower rates of substance abuse and mental disorders, decreased pollution, reduced risks of new pandemics, lesser environmental impacts at the local, regional, and global scale, mitigation of hunger and food insecurity, reduced social conflicts surrounding water and land access, and ultimately a new paradigm based on harmony and peace. We will see more details about the relationship between animal exploitation, drug use, and mental health in the next sections.

I am not saying the switch could be done overnight. Still, the change is unavoidable for all the reasons outlined in this book. While industry and policymakers advance at small steps, I kindly invite you to encourage and support this change with a growing demand for ethical food—before it is too late.

Farmed fishes are usually slaughtered within months to two years. Notably, neither farmed nor wild-caught fishes are covered by the US Humane Slaughter Act. The same applies to Brazil and elsewhere, where the state and local regulations surrounding fish slaughter are vague or nonexistent (Ferreira et al., 2018). Although the European legislation includes fishes in the general principle of avoiding unnecessary suffering to farmed animals, it recognises that more specific rules are required to protect them (Boyland & Brooke, 2017).

The World Organisation for Animal Health (OIE) provides standards for the treatment of aquatic animals through all stages of production until slaughter. The recommendation is that the stunning and killings methods should consider species-specific information, where available (OIE, 2019). However, not only "humane" slaughtering techniques have been deemed little attention from the scientific community, but also the existent recommendations are greatly ignored by fisheries worldwide. In addition, the OIE guidelines do not have the force of law.

Most studies on fish slaughtering methods focus on meat quality, with the enhancement of welfare levels serving economic purposes. Severe pre-slaughter and slaughter stress deplete the muscle energy stocks, accelerating lactic acid production, thereby reducing the muscle pH and speeding up the *rigor mortis* process. This will reduce shelf-life and hinder texture (Poli et al., 2005). As stated by animal welfare specialists Jeff Lines and Jade Spence, 'Even now, with a broad scientific consensus that fish are sentient and have the capacity to suffer, there is little evidence of improvement in slaughter methods on the majority of fish farms around the world' (Lines & Spence, 2014).

In fact, the vast majority of farmed fishes worldwide are killed with little or zero consideration for their welfare, including "inhumane" killing methods, excessive fasting periods, and stressful transport (Lines & Spence, 2014).

You might have noticed I use the words 'humane' and 'inhumane' in quotes whenever they are followed by 'slaughtering' or 'killing.' Well, killing can only be considered humane, or merciful, when the purpose is to alleviate someone's suffering—for instance, euthanising someone who is in extreme pain and cannot be helped by current medicine.

Killing sentient beings for taste gratification is evidently not humane, regardless of the method employed.

Similar to land animals, many fishes are fasted before slaughter to facilitate handling, transport, and processing, as well as enhance meat quality and shelf life by minimising microbial contamination. The fasting period varies from one to five days, depending on the species and the water temperature, which affects the fish metabolism and hence the time required to empty their gut (Lines & Spence, 2014). However, exceedingly long periods are widely applied for the sake of tradition and convenience.

Fishes and shellfishes can be transported over variable distances based on the location of the slaughterhouse. As the animals are gathered for loading, they may sustain injuries, over-inflation of the swim bladder, and oxygen shortage. Gathering or crowding is achieved by lifting part of the net, moving grids, or using a seine net to catch the fishes, depending on the farm type. The fishes are then transferred by brailing (i.e., using a large net) or pumping (i.e., using centrifugal or vacuum pumps that suck both water and fishes). Pain and stress can result from bruising, puncturing, abrasion, crushing, collisions between fishes, swimming against the flow, and swimming in unexpected turbulent waters. Road transport of fishes out of water is regular, with the animals suffering from anoxia, physical shock, thermal discomfort, and noise exposure (Lines & Spence, 2014).

At the abattoir, the fishes can be left to recover in a lairage for up to days, with no food, to reduce exhaustion and maintain the flesh quality. Excessive handling adds to all these stressors.

The overwhelming majority of both farmed and wild-caught fishes are simply left to die by suffocation in air or ice, with no prior stunning. Killing fishes this way serves several purposes: it is effortless, cheap, and preserves the flesh for longer when ice is used. Nevertheless, a body of research shows that fishes can remain conscious for up to 90 minutes in ice slurry (Rahmanifarah et al., 2011; Roth et al., 2009).

Some species do not require exsanguination before processing, but most do to avoid residual blood in the tissue. Fishes are generally bled without stunning. Some are beheaded, mutilated, and eviscerated while fully conscious. Percussive stunning (which we have seen as potentially more "humane") is widely used in salmon aquaculture but hardly employed for smaller species for practical reasons.

Other violent and common slaughter methods applied to both farmed and wild fishes are asphyxiation in the air followed by percussive stun (e.g., carp), chilling with ice in holding water, CO_2 in holding water (e.g., salmon and trout), gill cutting without prior stunning (e.g., turbot), and immersion in salt or ammonia baths followed by gutting of conscious animals (e.g., eels) (Fishcount, 2019; OIE, 2019).

In February 2021, Animal Equality released an undercover footage shot at a slaughterhouse operated by the Scottish Salmon Company, a major UK and international fish supplier (Animal Equality, 2021a). The investigation was published by *The Times*. The investigators recorded many salmon being killed or having their gills torn out (sometimes with the worker's fingers) while fully conscious. Many animals were seen flapping, wriggling, and gasping for air after inadequate stunning. Some fishes who had their gills cut after stunning regained consciousness and were re-stunned with a club. The footage also documented fishes falling or being thrown to the floor and left to suffocate.

In the UK alone, more than 2 fishes are killed per *second* (around 77 million individuals per year). The Scottish Salmon Company supplies more than 20 countries, from North America to the Far East, and major supermarket chains (e.g., Waitrose and Co-op), retailers, hotels, and restaurants around the UK. The company spreads across Scotland in 12 freshwater sites (where the eggs are spawned and the fishes grow to the smolt stage), 41 marine sites (where they are raised to slaughter age), 3 processing plants, and 1 head office. From spawning through alevins to parrs, the fishes live 10–14 months in freshwater tanks. The animals are then transferred to sea pens, where they spend a further 16–20 months, from the smolt stage to adulthood.

On their website, the company talks about their passion for people, salmon, and the planet. Animal treatment, environmental issues (e.g., salmon are fed on wild-caught species), and health problems associated with fish consumption are not addressed; however, there is a section dedicated to the benefits of omega-3.

> Our marine farming site locations are steered by Environmental Impact Assessments (EIAs) led by our Environmental Department. We work collaboratively with the Scottish Environmental Protection Agency (SEPA), Marine Scotland, Scottish National Heritage (SNH) and other governmental and non-governmental organisations to make sure all environmental factors are considered. Our goal is to grow the optimum number of salmon while ensuring that our farming activities do not have an adverse long-term effect on the ecology of our sea lochs. (The Scottish Salmon Company, 2021)

Most average readers who visit their website might have the same impression conveyed by other animal agriculture sectors: animals raised for food live a happy life and the activity is in perfect harmony with the environment.

A recent investigation carried out by Animal Equality in the Indian fish industry exposed the animal abuse, childhood labour, and unsustainability behind their activities (Animal Equality, 2021b). Shrimp farms, hatcheries, and fish markets in West Bengal, Andhra Pradesh, Tamil Nadu, and Telangana were inspected from February 2019 to May 2020. Overcrowded ponds were home to thousands of fishes, stimulating the spread of diseases. Fishes were simply pulled from the water, thrown on ice, and left to suffocate to death. Many animals were killed by the weight of the others during catching and relocation. Workers were reported to club conscious catfishes several times before finally dying. Other fishes had their gills cut without stunning, resulting in a slow and painful death. All this is very similar to what was documented in the Scottish fish company.

Additionally, the staff included children, who were seen slaughtering fish. Finally, the investigators recorded a brutal procedure called 'fish milking,' in which female fishes have their eggs extracted by hand. All practices outlined conflict with India's Prevention of Cruelty to Animals Act, which aims to prevent unnecessary pain and suffering on animals (Animal Equality, 2021b).

Aquatic creatures suffer tremendously in fish markets too. Carps, the most numerous farmed finfish species, are traditionally sold alive in many parts of the globe, including Asia and Europe. After the stress of transport and excessive handling, the animals will be displayed in unacceptable conditions and killed with methods that do not conform to the OIE's fish welfare guidelines.

Fishes do not fall from trees

Many pescatarians eat fish but not red meat due to animal welfare concerns. Although I sincerely appreciate the pescatarians' effort to avoid eating land animals, I kindly encourage them to include aquatic creatures in their circle of compassion.

Other people not concerned with animal use often replace red meat with fish or poultry flesh due to health and environmental reasons. I am sorry to disappoint them, but fish consumption is neither healthy nor sustainable. Seafood production is not only linked to ocean pollution, antibiotic resistance, and aquatic biodiversity loss, but is also associated with *terrestrial* resources.

As the aquaculture industry progressively resorts to crop-based feed ingredients to replace fish meal, a greater portion of arable land worldwide is destined to the animal exploitation industry, instead of being used to feed people directly with plant-sourced food. While feed-to-flesh conversion ratios are generally higher in fishes than in mammals, a colossal amount of resources is still lost by using animals as nutrient "filters." In addition, in using more crops to feed fishes, aquaculture is increasingly associated with pesticide and fertiliser runoff from crop production (Fry et al., 2016).

In a third category, there is the well-meaning consumer of local products. 'Eating local' has been widely advocated by many to evade the environmental damage caused by animal agriculture. To begin with, let us agree that eating *local animals* does not help animals —they are enslaved and killed as much as any other farmed animal.

Although the 'eating local' pledge does not solve animal exploitation, it can surely help reduce the distance our food travelled (the so-called 'food miles') and support local farmers. Nevertheless, food transport's contribution to the GHG emissions associated with food is generally *minor*, especially for animal-sourced foods. What is more, as I have been highlighting several times throughout this book, carbon emissions are far from being the only environmental emergency the planet is currently facing. Beyond the problem posed by GHG emissions, we must also worry about resource use (e.g., land, energy, and freshwater), eutrophication, and biodiversity loss. Therefore, what you eat is *far more* critical than the distance your food travelled.

Scientist Dr. Hannah Ritchie broke down the carbon emissions of different foods throughout the supply chain (Richie, 2020) based on data from Dr. Poore's and Dr. Nemecek's study across 38,700 commercial farms in 119 countries (Poore & Nemecek, 2018). The calculations are published on *Our World in Data*'s website (ourworldindata.org), along with a detailed discussion on the environmental impacts of food production. Dr. Ritchie is Senior Researcher and Head of Research at *Our World in Data*, a project dedicated to presenting evidence on multiple global development issues. *Our World in Data* was founded by Dr. Max Roser, the Programme Director of the Oxford Martin Programme on Global Development at the University of Oxford.

Figure 62 shows the negligible fraction of the carbon emissions attributable to the transport stage in the animal foods supply chain. The chart also illustrates the immense difference in carbon footprint between animal and plant-based products. Transport typically accounts for 0.4% of beef's GHG emissions, 0.9% of pig meat's, 1.1% of farmed fish's, and 2.3% of poultry meat's (Ritchie, 2020).

The transport share is far more significant for plant-based products, ranging from less than 1% for dark chocolate and coffee to as high as 22.9% for cane sugar and 27.3% for bananas (Ritchie, 2020). Bottom line: if you are concerned about the climate crisis, give preference to local food but, *more importantly*, avoid animal products—even if animal suffering is not an issue to you.

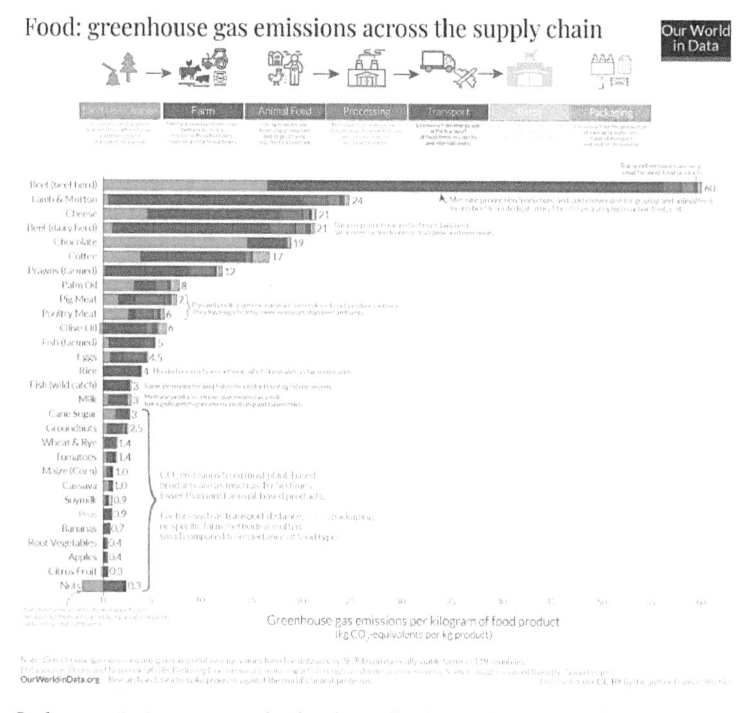

Figure 62. Carbon emissions across the food supply chain. Source: Ritchie (2020).

In another study, Sandström et al. (2018) calculated the GHG emissions from the average *diet* (instead of that of individual products) across different EU countries (Figure 63). The emissions were segmented into food production, land-use change, and transport or trade for each food group. The calculations corroborate the estimations of Ritchie (2020), showing that *what* people eat is way more important than *where* their food comes from in terms of carbon footprint. From 49% to 64% of EU dietary emissions were attributed to meat and eggs. Dairy products contributed to a further 16–36% (Sandström et al., 2018). On average, only 6% of the GHG emissions of the average European diet were related to international transport. The research did not include fish and seafood due to insufficient country-level data on the origin and type of fish consumed and trade flows between countries.

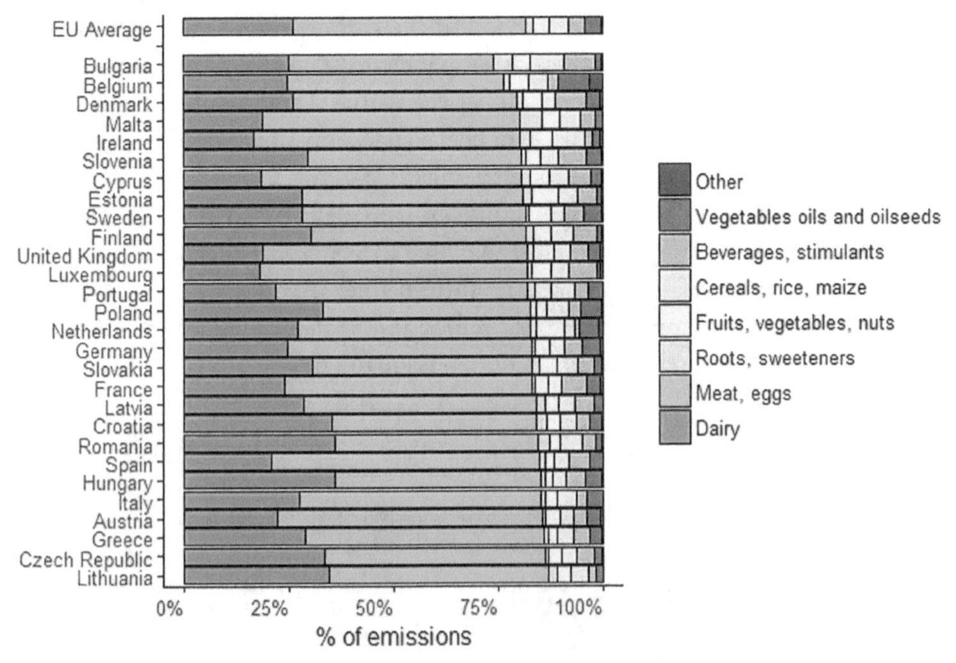

Figure 63. Percentual dietary emissions of different food groups across EU countries. Source: Sandström et al. (2018).

Science has consistently shown that animal products have in general a much larger environmental impact than plant-based foods. Typically, red meat production emits more GHGs than fish and poultry production, primarily due to land-use changes ascribed to deforestation. However, depending on the type of animal flesh under consideration, fish and beef can have comparable ecological footprints, especially if we include other equally important environmental metrics (e.g., resource use) (Guzmán-Luna et al., 2021).

Guzmán-Luna et al. (2021) compared the resource use of aquaculture to that of livestock operations, using extensive, semi-intensive, and intensive tilapia (*Oreochromis* sp.) farming in Mexico as a case study. The research included the water, energy, and land footprints of all stages from broodfish to processing and transport. Worth mentioning, Mexico ranks ninth in the global list of major tilapia producers, with a 2017 production of 163,293 tonnes. Besides, aquaculture accounts for 91% of the country's tilapia production.

The *energy* and *land* footprints for producing a gram of protein were comparable among tilapia, poultry, and pork raised in intensive systems. Beef was the most energy- and land-intensive protein (Table 4). However, the *water* footprint was significantly larger for tilapia, with 126 litres of water per gram of fish protein produced, versus 14 L/g for poultry, 33 L/g for pork, and 51 L/g for beef. Water is considered the most critical resource due to its fundamental role in maintaining life on the planet and its complex management (Lele et al., 2013). It seems the statement 'fish is more sustainable than other animal meats' does not hold true—at least not for tilapia.

Table 4. Energy (EF), land (LF), and water (WF) footprints of the industrial production of animal protein.

Animal product	EF (MJ/g protein)	LF (m^2/g protein)	WF (L/g protein)
Beef	0.15–0.55	0.14–0.25	51
Poultry	0.06–0.43	0.04–0.05	14
Pork	0.10–0.54	0.06–0.08	33
Tilapia	0.12	0.05	126

Source: Guzmán-Luna et al. (2021); Mekonnen & Hoekstra (2010); Williams et al. (2006); Kramer & Moll (1995); Gerbens-Leenes et al. (2002); Gerbens-Leenes et al. (2013); De Vries & De Boer (2010); Ibidhi et al. (2017).

You might be wondering if the method under which tilapia is produced has any influence on the environmental impact. After all, some people argue that intensifying production (which is achieved mainly by confining animals in a smaller space) reduces the resource requirements. Well, Guzmán-Luna's team (2021) not only made this comparison, but also segmented the water footprint (WF) of tilapia production into green, blue, and grey WF (Table 5).

Simply put, green water represents rainwater; blue water is the surface and ground freshwater used for irrigation of feed crops, refilling or refreshments of fish pens, transportation of fingerlings (juvenile fish), and other purposes; and greywater is wastewater from domestic activities, agriculture, and other industrial uses. Greywater can be treated and used to irrigate crops, depending on the type of contamination.

Absolutely contrary to what factory fish farming defenders advocate, *extensive* systems had the smallest blue WF, green WF, and grey WF (Table 5). The blue WF, for instance, was 14 and 4.5 times larger in intensive farms compared to the extensive and semi-intensive counterparts, respectively. The large blue WF is owed primarily to the daily requirements of pond water refreshment in intensive farms (Guzmán-Luna et al., 2021).

The extensive system also required significantly less energy, especially electricity. Finally, the land footprint (which includes direct land use for reservoirs and ponds, and indirect land use related to feed) was nearly ten times lower in extensive farms (Guzmán-Luna et al., 2021).

Table 5. Water, energy, and land footprint of tilapia filet produced under different systems.

Metric	Extensive	Semi-intensive	Intensive
Blue water footprint (m^3/ton)	927	2,909	13,027
Green water footprint (m^3/ton)	5	7,827	7,831
Grey water footprint (m^3/ton)	398	1,873	1,873
Energy footprint (MJ/ton)	4,241	14,451	22,200
Land footprint (m^2/ton)	1,193	10,723	10,711

Source: Guzmán-Luna et al. (2021).

The results by Guzmán-Luna et al. (2021) show that tilapia is not environmentally friendly. The production requires considerably more water and has similar land and energy requirements to other types of animal flesh, regardless of the production system. Citing the authors: 'Tilapia fillet not only requires more freshwater than beef, pork and poultry, but also pollutes larger amounts of water than terrestrial animals due to constant effluent loads coming from the ponds. From a freshwater perspective, it is more sustainable and efficient to obtain animal protein from terrestrial animal sources' (Guzmán-Luna et al., 2021).

At this point, you might be asking yourself if fish is at least healthier than other types of animal flesh. Spoiler alert: it is not. But let us see why.

Omega-3: the fish industry's darling

The animal industry has consistently used the "nutrient-rich" tactic to sell its products. Even illiterate people seem to know the associations between calcium and milk, protein and iron and meat, lean protein and eggs, and omega-3 and fish. Curiously, an insignificant fraction of the world's population is aware of the relationships between animal-sourced foods and cholesterol, cancer-promoting substances, antibiotic resistance, allergens, hormones, foodborne diseases, and zoonotic-origin pandemics. Fishes and other aquatic animals, in particular, contain cholesterol, saturated fat, and high concentrations of heavy metals and dioxins. Our system strongly relies on ignorance, manipulated information, and biased research to keep people consuming violence foods that harm everyone, not just animals.

Furthermore, fish and other animal products have been genetically modified too. Curiously, people are terrified of GMOs in plant foods but have no concerns about genetically engineered *animals*. In 2015, the US FDA approved the national sale of genetically modified salmon produced by AquaBounty Technologies. The federal regulator does not mandate any special labelling, meaning consumers cannot identify the product. The FDA claimed the salmon was approved for sale based on robust scientific evidence. The salmon was modified with genes from an eel-like ocean pout to grow twice as fast as wild salmon and consume 25% less feed to reach market size (Milman, 2015).

AquaBounty's engineered salmon is raised in land-based farms in the US and Canada. The company's chief executive, Ron Stotish, claimed their salmon is "responsibly raised" because it does not harm ocean habitats and uses fewer resources to generate the same amount of protein (Milman, 2015). The company had waited three decades to launch the product in the US.

Well, there is another, much more efficient way to increase sustainability in the agri-food industry: replacing animal protein with plant protein. No need for expensive technology and decades of research. If we are serious about our survival on the planet (and ethics), we must go vegan—this is a no-brainer.

Further information about AquaBounty's salmon production is available on their website. Let us have a look at some of the company's claims:

1. 'Always Free of Antibiotics and Other Contaminants—The waters our Atlantic salmon swim in never come in contact with harmful diseases and toxins that can be a concern with traditional sea-cage farms and net pens. When we say our Atlantic salmon is a safe choice, we mean it.' It amazes me how the animal exploitation industry shoots itself in the foot. In this statement, they imply that the consumption of fish from sea-based farms poses health problems due to the drugs used to keep the fishes free of diseases.

2. 'Fresh Salmon is Closer than You Think—Eating fresh, local salmon isn't just for coastal living anymore. We're bringing Atlantic salmon closer to seafoodies throughout the U.S. and Canada, without the high carbon footprint.' The company forgot to mention that the transport stage represents a tiny fraction of the carbon footprint of animal-sourced foods, as we have seen previously in this chapter.

3. 'AquaBounty can help reduce pressure on wild salmon stocks.' And what about the other wild fishes used to feed salmon? The company's claim disregarded that salmon are carnivores, and that fish meal and fish oil required in salmon farming are obtained from wild-captured small fishes, such as anchovies and sardines. On average, 1.7 kg of wild-caught fishes translates into 1 kg of Atlantic salmon (RSPCA Australia, 2019).

4. 'The contained, land-based systems used by AquaBounty are endorsed by most seafood certifying organisations and environmental groups as a more

> environmentally friendly and responsible alternative to traditional sea-cage farming of salmon.' The fact that inland salmon farms can cause *less* environmental damage than sea-cage salmon farms does not mean it is 'environmentally friendly' whatsoever.

But let us go back to the "nutrient-rich" argument. The fish industry relies on two main nutritional marketing strategies to promote its products: protein and omega-3 fatty acids. Omega-3 (n-3) fats are polyunsaturated fatty acids (PUFAs) necessary for the human body's proper functioning. Omega-3s are known to exert protective cardiovascular effects and improve brain function and mental health (Lange, 2020; Valenzuela et al., 2014; Poudyal et al., 2013; DeFilippis & Sperling, 2006).

Fishes and other marine animals are rich in two primary omega-3 fatty acids, namely eicosapentaenoic acid (EPA) and docosahexaenoic acid (DHA). However, fishes are far from being the only source of omega-3 fats. First, because these PUFAs come originally from algae, the primary producers. Second, because nuts, seeds, vegetable oils, and other plant-based foods are rich in alpha-linolenic acid (ALA), an *essential* omega-3 fatty acid. In other words, ALA *must be obtained from our diet* because it cannot be synthesised by the human body (Valenzuela et al., 2014).

Such omega-3s as EPA, DPA, and DHA can be produced in our organism from precursor ALA, although the conversion rate depends on several factors. The research on the protective health effects of individual PUFAs and the conversion rates from precursor ALA to other omega-3s is still in its infancy. For instance, some studies indicate that the bioconversion from ALA to EPA and DHA in mammals is limited, while others demonstrate the conversion rates are significant once enough ALA is ingested (Lane et al., 2021; Valenzuela et al., 2014; Poudyal et al., 2013). Due to the limited scientific data, the omega-3 daily recommended dose varies broadly across countries and health agencies. On the other hand, tolerable upper intake levels are consolidated because excessive consumption of n-3 PUFAs can impair the immune system, hinder the lipid and glucose metabolism, increase lipid peroxidation, and cause bleeding (EFSA, 2012).

Dietary recommendations from regulatory agencies worldwide range from 115 mg to 610 mg/day for EPA and DHA, and 0.8–2.0 g/day for ALA for adults. The recommended intake for children drops to 40–200 mg/day for EPA and DHA, and 0.5–1.6 g/day for ALA, depending largely on age (EFSA, 2012).

Let us compare the omega-3 content of different food items using the data published by physicians DeFilippis & Sperling (2006) in the *American Heart Journal* (Table 6).

You may find it difficult to compare food items on a 100-gram basis. So let us understand how much omega-3 fatty acids we find in common servings of different products and how we can meet the dietary recommendations (Table 7). Keep in mind that the daily intake for adults should be at least 115 mg/day for EPA and DHA, and at least 800 mg/day for ALA (EFSA, 2012).

Table 6. Omega-3 fatty acid content of different food items, in grams per 100 grams of food.

Food item	EPA	DHA	ALA
Catfish	Trace	0.2	0.1
Cod	Trace	0.1	Trace
Mackerel	0.9	1.4	0.2
Farmed salmon	0.6	1.3	Trace
Wild salmon	0.3	1.1	1.3
Canned salmon	0.9	0.8	Trace
Swordfish	0.1	0.5	0.2
Bluefin tuna	0.3	0.9	–
Canned white tuna in oil	Trace	0.2	0.2
Canned white tuna in water	0.2	0.6	Trace
Mussels	0.2	0.3	Trace
Shrimp	0.3	0.2	Trace
Butternut	–	–	8.7
Walnuts	–	–	9.1
Canola	–	–	9.3
Flaxseed	–	–	53.3

Adapted from: DeFilippis & Sperling (2006).

Table 7. Omega-3 fatty acid content associated with selected food items, in typical portions.

Food item	n-3 in a typical portion	Portion to meet minimum daily requirement (adults)
Walnuts	2,542 mg ALA in 28 g	Five halves
Chia seeds	4,915 mg ALA in 28 g	Half tablespoon
Flaxseed	6,388 mg ALA in 28 g	One teaspoon
Hemp seeds	6,000 mg ALA in 28 g	One teaspoon
Brussel sprouts, cooked	270 mg ALA/cup	Three cups
Algal oil supplements	200–600 mg EPA + DHA in 1-2 soft gels	One soft gel
Salmon	1,400–1,900 mg of EPA + DHA in 100 g	6–8 g
Bluefin tuna	1,200 mg of EPA + DHA in 100 g	10 g
Cod	100 mg EPA + DHA in 100 g	115 g
Tuna, canned in water	800 mg EPA + DHA in 100 g	14 g
Mussels	500 mg EPA + DHA in 100 g	23 g
Shrimp	500 mg EPA + DHA in 100 g	23 g

Sources: Carey (2018); Link (2017); DeFilippis & Sperling (2006).

When we stick to fish and seafood as dietary sources of n-3 PUFAs, we also ingest toxic metals, antibiotic residues, microplastic, and cholesterol. Yes, cholesterol! But what about fish being a healthier alternative to red meat, as we have all heard? Well, although some fish do have substantially lower cholesterol levels than beef and pork, this by no means translates into 'fish is low in cholesterol.' Like all animal products, fish and seafood are packed with cholesterol, with levels varying vastly across species.

Most studies that find benefits in adding fish to our diet make the comparison between the standard meat- and dairy-rich Western diet and one that replaces some of these animal products with fish. Not surprisingly, any replacements of saturated fat–packed products with "leaner" foods will yield benefits. In contrast, research has consistently shown that balanced vegan diets outperform others (e.g., Mediterranean, pescatarian, vegetarian, and omnivorous) by reducing the incidence of cardiovascular diseases, hypertension, chronic kidney disease, inflammatory bowel disease, total cancer, obesity, diabetes, and all-cause mortality (Sakkas et al., 2020).

Claiming that eating fish is healthy by comparing fish–rich diets to red meat–rich diets is as logical as defending that everyone should smoke nicotine-free cigarettes because they are less harmful than regular cigarettes.

While a 150 g portion of cooked cod contains approximately 150 mg of n-3 fatty acids (predominantly DHA), it is also packed with 177 mg of *cholesterol* (Carey, 2018; DeFilippis & Sperling, 2006). Bear in mind that the maximum cholesterol intake recommendation is 200–300 mg/day for adults (Bellows & Moore, 2012).

Another example: 150 g of cooked salmon contains 2,100–2,850 mg of n-3 fatty acids (EPA and DHA) and 92 mg of cholesterol. And the worst option: 150 g of shrimp contains 750 mg of EPH and DHA combined and 284 mg of cholesterol (Carey, 2018; DeFilippis & Sperling, 2006).

You know something is very wrong about a product when food and health agencies advise pregnant women and children against its consumption. For example, the US FDA and the Environmental Protection Agency (EPA) released official advice on eating fish due to the risk of mercury intoxication (FDA, 2019). Many state and local agencies in the country have issued similar guidance for fish consumers, for instance, the Office of Environmental Health Hazard Assessment of California (OEHHA, undated).

Basically, children younger than six, pregnant or nursing women, and women expecting to get pregnant within a year should avoid bigeye tuna, swordfish, seabass, yellowtail, mackerel, marlin, shark, orange roughy, striped bass, pikeminnow, white sturgeon, and tilefish altogether (FDA, 2019; OEHHA, undated). Pregnant and nursing women should also limit consumption of yellowfin tuna, albacore tuna, white tuna, carp, bluefish, and other fish to one serving a week (FDA, 2019; OEHHA, undated). The FDA recommends that other fish and crustaceans with lower mercury levels should be limited to 2–3 servings a week, such as anchovy, catfish, butterfish, pollock, sardine, trout, tilapia, scallop, shrimp, oyster, and lobster (FDA, 2019). Please note that the same species can pose different mercury intoxication risks—for example, bigeye tuna (high risk) versus yellowfin tuna (medium) versus skipjack tuna (low), according to the agency (FDA, 2019). EFSA gives similar precautionary advice, asking European consumers to select fish with lower mercury levels and, hence, avoid large predatory fishes (EFSA, 2004).

Predatory fishes (e.g., salmon, swordfish, and tuna) have typically higher mercury levels because the contamination builds up in their flesh over their lifetime as they eat other fishes. Therefore, small fishes such as sardines, anchovies, and scallops often contain lower levels of toxic metals, dioxins, and polychlorinated biphenyls (PCBs).

Toxic metals occur naturally on the planet's surface. They can be released into the atmosphere (in the elemental form) through volcanic activity and volatilisation from the ocean. However, most metal exposure is attributed to industrial emissions, residential heating systems, coal-fired power stations, and mining. The metals make their way into rivers, lakes, and the ocean through rain, wastewater, sewage discharges, and runoff. Direct, occupational exposure (e.g., miners) to the inorganic form can also occur.

Fishes literally filter the water through breathing. All substances dissolved in the water accumulate in their tissues over time, especially in the gills. When we eat fish and seafood, we are therefore ingesting pollution. Mercury, for instance, builds up in fish and shellfish as methylmercury, *the most toxic form*. Our gastrointestinal system absorbs nearly 95% of the methylmercury ingested (Rice et al., 2014). Cooking temperatures do not degrade toxic metals; they instead can increase the metal concentration in the food through moisture loss (Chiocchetti et al., 2017).

Other heavy metals that can accumulate in seafood include arsenic, lead, and cadmium. Even small amounts of heavy metals can induce serious health problems in adults, children, and foetuses, ranging from malignant tissue growth to brain damage (Carver & Gallicchio, 2017). To be noted is that the absorption of toxic substances by fishes has been increasing with global warming as a result of higher water temperatures, since absorption is proportional to temperature (Manciocco et al., 2014).

The leading cause of methylmercury intoxication in the world is fish and shellfish consumption (EFSA, 2004). This substance is included in the WHO's list of ten major public concern chemicals and is listed as a Group 2B carcinogen (WHO, 2010; WHO & IARC, 2020). Instead of advising people against the consumption of marine animals altogether, FDA, EFSA, WHO, and other advisory agencies encourage consumers to eat fish due to the omega-3's benefits, provided they consult local advisories on the species they should or should not eat.

Now every time you order a dish containing aquatic animals, you just need to grab your fact sheet, consult the "okay" species and the safe number of weekly servings, ask the waiter what type of flesh precisely your food contains, and make some calculations. Very practical!

Studies have associated high mercury plasma levels with neural and cognitive problems in children, weakened immune system, cardiomyopathy, increased lipid peroxidation, ulcers and other digestive issues, high blood pressure, endocrine system impairment, renal cancer, liver cancer, gastric cancer, and gallbladder cancer (Carver & Gallicchio, 2017; Rice et al., 2014; WHO, 2010). Foetuses exposed to methylmercury through fish and shellfish consumption by the mother may have their memory, attention, language, cognitive thinking, and fine motor and visual-spatial skills affected after birth (WHO, 2016a).

The US Committee on the Toxicological Effects of Methylmercury of the National Academy of Sciences and the US EPA recommend keeping the blood mercury levels below 5.0 µg/L and the hair levels below 1.0 µg/g. EFSA proposes a tolerable weekly intake (TWI) for methylmercury of 1.3 µg/kg body weight, while the WHO suggests a TWI of 1.6 µg/kg. An individual's mercury dietary exposure is largely dependent on their seafood consumption pattern (species and the amount consumed) and the fish origin (e.g., proximity to mining, agricultural and industrial activities, and wastewater and sewage treatment). Therefore, mercury levels vary broadly among individuals and across studies. Nonetheless, higher mercury blood and hair levels are consistently associated with greater seafood consumption, regardless of the value range found (Crespo-Lopez et al., 2021).

Riverine Brazilian Amazon populations were recently shown to exhibit two to six times the international reference mercury levels (Crespo-Lopez et al., 2021). Mercury exposure in the region has increased steadily in the few last years due to mining intensification, dam construction, and deforestation (Crespo-Lopez et al., 2021). The United Nations Environment Programme (UNEP) estimates that nearly 38% of the global mercury emissions are attributable to artisanal and small-scale gold mining (UNEP, 2019). Stationary combustion of coal (~21%), cement production (~11%), and non-ferrous metal production (~10%) are also significant emitters.

Interestingly, biomass burning (including deforestation) accounts for 2% of global mercury emissions (UNEP, 2019). Mercury vapour from mining and other activities are deposited on trees and leaves and later return to the atmosphere upon deforestation and burning of the vegetation. In the Amazon, biomass burning constitutes the *second leading cause* of mercury emissions, only after mining (Crespo-Lopez et al., 2021).

Both legal and illegal deforestation in the Amazon is driven primarily by mining and animal agriculture activities. A 2020 paper published in *Science* magazine has shown that soy and beef exports to the EU alone account for 62% of the Brazilian Amazon's illegal deforestation (Rajão et al., 2020). And, of course, fewer trees mean lower mercury sequestration from the atmosphere, enabling the metal to reach the food chain (Crespo-Lopez et al., 2021).

Approximately 80% of the protein intake by Amazon riverine people is sourced from fish, mainly predatory species (Crespo-Lopez et al., 2021). The metal exposure risk is further increased by the presence of dams near these communities. Hydropower dams intensify the accumulation and biotransformation of mercury by favouring conditions for converting inorganic mercury into methylmercury (Crespo-Lopez et al., 2021). Extremely high mercury exposure has been reported in populations near Amazonian dams, with a study in the Tucuruí dam revealing that 30% of the participants exhibited hair levels over 20 μg/g (Arrifano et al., 2018).

In fact, tropical riverine populations (especially in the Amazon) have been identified as one of the four populations of concern regarding mercury intoxication, along with Arctic populations who consume fish and sea mammals, coastal communities that depend on seafood, and people who work or live nearby mining sites (Basu et al., 2018). Nonetheless, people are exposed to mercury *everywhere in the world*, with differences in the *level* of exposure across and within countries (Basu et al., 2018). Populations that consume large amounts of seafood (e.g., Spain, Brazil, and Faroe Islands) typically exhibit blood mercury levels above 10 μg/L (Basu et al., 2018).

For example, a study with 89 patients conducted in San Francisco, California, revealed that 89% of the subjects exceeded the mercury maximum limit. Blood levels ranged from 2.0 to 89.5 μg/L. Most patients followed diets rich in fish (Hightower & Moore, 2003). Another research carried out in Sweden showed that half of the 143 participants exhibited mercury hair levels higher than 1.0 μg/g. The study was conducted in angling communities. One-third of them consumed freshwater fish at least once a week. The highest mercury level, 18.5 μg/g, was found in an individual who consumed pike and perch several times per week (Johnsson et al., 2004).

Fish and shellfish are also carriers of cadmium, arsenic, and dioxins, which are included in the WHO's top ten chemicals of concern (WHO, 2010). What is more, arsenic, cadmium, and dioxins are listed as Group 1 carcinogens (i.e., proven carcinogenic potential). By comparison, remember that methylmercury compounds are in the 2B Group (possibly carcinogenic to humans) (WHO & IARC, 2020).

Arsenic intake is associated with skin lesions, diabetes, gastrointestinal symptoms, peripheral neuropathy, renal dysfunction, cardiovascular diseases, and many types of cancer (e.g. lung, bladder, skin, colon, gastric, prostate, kidney, and nasopharyngeal) (WHO, 2010; Carver & Gallicchio, 2017; Zimta et al., 2019). The primary dietary source of arsenic is aquatic foods. Arsenic concentrations in these products vary broadly, with reported values from ≤ 1 mg/kg of fresh weight in canned tuna to 26.2 mg/kg in shellfish and as high as ≤ 100 mg/kg in cod (Chiocchetti et al., 2017).

Multiple studies have shown that exposure to inorganic arsenic from fish consumption poses a considerable carcinogenic risk (Jia et al., 2018; Hull et al., 2021). In Norwegian blue mussels, for example, the percentage of inorganic arsenic comprised up to 42% of the total arsenic (i.e., 5.8 mg/kg of 13.8 mg/kg) (Sloth & Julshamn, 2008).

As most fish and shellfish are high in the *organic* form (mainly arsenobetaine), which is less toxic, the risk of arsenic exposure through fish and shellfish consumption has been neglected. Nonetheless, scientific evidence points out that dietary arsenic (even inorganic) can be metabolised to more toxic forms in the human body (Watson, 2015).

Seaweed is also an important dietary source of arsenic (Chiocchetti et al., 2017). Interestingly, the consumption of antioxidant-rich foods is recommended to lower the carcinogenic effects of arsenic ingestion (Carver & Gallicchio, 2017). Veggies and fruits are the most abundant sources of antioxidants.

Cadmium is a carcinogenic metal that affects the renal and skeletal systems and is associated with the onset of leukaemia and breast, prostate, teste, oesophagus, stomach, intestines, and lung cancer (WHO, 2010; Zimta et al., 2019). Cadmium accumulates in seaweed and aquatic animals, especially molluscs and crustaceans, but also in plants. Other non-dietary exposure routes are direct inhalation in the metal industry and active or passive smoking (WHO, 2010).

Varying cadmium levels are found in fish and shellfish, ranging from 0.1 mg per kg of fresh weight in fishes to up to 62 mg per kg in crabs (Chiocchetti et al., 2017). Crab consumption poses exceptionally high health risks when the liver, hepatopancreas, and gonads are included (Chiocchetti et al., 2017; Wiech et al., 2020).

Dioxins and PCBs are persistent organic pollutants (POPs) that travel long distances from the source of emission, contaminating the soil and water bodies. Coal combustion, steel smelting, chlorine bleaching of paper pulp, and incineration of waste from electrical equipment are typical dioxin and POPs emitters (WHO, 2010). These substances bioaccumulate in animal tissues and secretions and, to a lesser extent, in plants. The toxic effects of dioxins and PCBs include immunotoxicity, physical developmental and neurodevelopmental problems, diabetes, changes in thyroid and steroid hormones, and reproductive dysfunction (WHO, 2010).

Dioxins are some of the most toxic chemicals in existence. As PCBs act like dioxins, they are included in the term 'dioxin-like substances.' The most potent dioxin—2,3,7,8-Tetrachlorodibenzo-para-dioxin (or 2,3,7,8-TCDD)—and PCBs are considered Group 1 carcinogens (WHO & IARC, 2020). Tolerable intakes for 2,3,7,8-TCDD set by health agencies worldwide vary from 2 pg TEQ per kg of body weight a *week* to 1–4 pg TEQ per kg a *day* (Fürst, 2010; EFSA, 2018). The TEQ (toxic equivalent) of different dioxins is calculated in terms of their percentual toxicity in relation to 2,3,7,8-TCDD. For example, a dioxin exhibiting 50% of the 2,3,7,8-TCDD's toxicity will count as 0.5 TEQ.

A survey among the US population revealed that meat contributes the largest share of dioxin intake in the *adult* diet, followed by dairy, and then vegetables. Among individuals aged 1–19, the largest share is from dairy foods, followed by meat, and lastly vegetables (Schecter et al., 2001). In fact, over 90% of human exposure to dioxins and PCBs is through food consumption, with around 80% of the dietary intake attributed to animal-sourced foods (Fürst, 2010).

The consumption of aquatic animals poses health risks directly associated with the pollution type and level in their environment. A Polish study indicated unsafe dioxin and PCB concentrations in freshwater fish, according to the EU Commission

Regulation 1259/2011/EU (Mikolajczyk et al., 2020). Fishes from fisheries located in urban areas were especially affected. Regular consumption of some species exceeded the tolerable weekly intake set by EFSA.

Note that dioxins are so toxic that the maximum intake levels are given in picograms—one picogram equals one trillionth (or 10^{-12}) of a gram. The average intake of dioxins in developed countries (1.2–3.0 pg TEQ/kg/*day*) is 4 to 10 times higher than the EFSA's safe intake (2 pg TEQ/kg/*week*) (EFSA, 2018). No wonder why cancer rates are growing rampant worldwide. In contrast, a US study of dietary dioxin exposure indicated that vegans had the lowest dioxin blood levels compared to the general population (Schecter & Päpke, 1998).

The fact that dioxins are lipophilic explains the rapid accumulation in animal tissues, especially fatty ones. As dioxins are highly stable, their half-life in the human body is 7–11 years (half-life being the time required for a substance to reduce to half of its initial value) (WHO, 2016b).

But there is more to worry about concerning fishing and fish consumption. As plastic pollution disperses in the aquatic environment, microplastic enters the food chain in larger and larger numbers. A study published in the *Marine Pollution Bulletin* in 2020 revealed that fishing gear comprises 41–94% of marine litter weight in the Wetlands Reserve in Al Wusta Governorate, Oman (Van Hoytema et al., 2020). Trash was collected along 100-metre intersects of 7 beaches. After fishing-related items, the predominant litter consisted of water bottles and caps and food packaging. Plastic was the dominant material, accounting for 71–99% of the litter weight.

A 2015 research published in *Science* estimated that 4.8–12.7 million tonnes of plastic enter the ocean every year (Jambeck et al., 2015). Another paper featured in *Science* in 2020 estimated that 710 million tonnes of plastic waste will accumulate in the environment between 2016 and 2040, even in the face of coordinated global action to reduce plastic consumption and increase reuse and recycling rates (Lau et al., 2020). Marine and river pollution is so pervasive that over 50 freshwater species and almost 700 marine species were reported to have ingested or become entangled in debris, especially plastic debris (>80%) (Gall & Thompson, 2015; Rochman et al., 2016).

Another 2020 research evaluated the presence of microplastics in the digestive tracts of commercial fish in China (Wu et al., 2020). The sampling included 125 fishes from 13 sites belonging to 24 different species and 3 distinct feeding habits. Microplastic was found in nearly 40% of the fishes, predominantly cellophane and cellulose. Another paper reported the presence of microplastics in the organs of 60% of the fishes analysed. The sample comprised 198 species captured in 24 countries (Sequeira et al., 2020).

Due to bioaccumulation and biomagnification, carnivore fishes are more contaminated than omnivore fishes (Sequeira et al., 2020). More research is needed to estimate the microplastic intake by consumers of fish and shellfish. But as an example, Van Cauwenberghe and Janssen (2014) found the presence of microplastics (<1 mm) in commercially grown blue mussels (*Mytilus edulis*) and oysters (*Crassostrea gigas*) in Sweden. The researchers estimated that European shellfish consumers could ingest up to 11,000 microplastics each year exclusively from this source.

Bioaccumulation and biomagnification are some of the main reasons why our insisting on eating high on the food chain is unclever. At the pyramid base are the producers, plants, and phytoplankton that derive energy directly from the sunlight and inorganic nutrients through photosynthesis. Primary consumers eat the producers, secondary consumers feed on the primary, and so forth, magnifying the concentration of toxic substances. As they grow and eat, these substances accumulate in their tissues. Therefore, heavy metals, pesticides, antibiotics, and other drugs tend to be consumed in small amounts by primary consumers and in large amounts by the following.

Again, another important reason for going vegan is food efficiency: resources are lost between each trophic level. Up to 90% of the energy in a food material is lost as we move to the next level in the food pyramid. That is one of the primary rationales for the planetary shift away from animal products. We cannot feed 7.8 billion people eating at the top level, but we can provide everyone with a nutritionally complete plant-based diet. The more people consuming meat, dairy, fish, and eggs, the more we fuel hunger and food insecurity.

COVID-19 and seafood processing

Just like meat processing facilities, seafood processing plants have been facing numerous coronavirus outbreaks. In June 2020, American Seafoods declared that three-quarters (92 people) of its crew on board the American Dynasty ship tested positive for COVID-19. The boat fishes for pollock and cod off Alaska and for hake off Washington and Oregon. A few days later, 25 crew members on 2 other American Seafoods ships tested positive for the virus (Herz, 2020). In July, a 285 foot–long American Seafoods' trawler travelling from Unalaska to Seward (both in Alaska) was found to have 85 COVID-19 positive workers on board. Several outbreaks took place on the same trawler, the American Triumph, around that time. The trawler had been fishing from Washington to Alaska and had docked in several ports along the journey to offload the catch (Hagenbuch, 2020a).

Alaska is home to dozens of large-scale fisheries and fish processing plants, with pollock being one of the main targets. Alaska pollock (*Gadus chalcogrammus*), a member of the cod genus *Gadus*, inhabits a broad niche across the northern Pacific Ocean, with notable prevalence in the Bering Sea (between Russia and Alaska). Alaska pollock ranks second as the most important fish species in volume globally, with 3.1 million tonnes in 2018 (FAO, 2020). Approximately half are caught from the Bering Sea by US fisheries (NOAA, 2017). Alaska pollock is found in many products, from surimi to high-grade fillets to McDonald's sandwiches. All pollock is wild-caught, primarily with mid-water trawl vessels. Some vessels serve both as catchers and processors, while others deliver their catch to sea-processing ships or to inland seafood processing facilities.

Also in June 2020, Pacific Seafoods announced that 124 workers from 5 processing plants in Newport, Oregon, were infected with COVID-19. This had been the largest single outbreak in Oregon state since the start of the pandemic. Seafood processing factories like Pacific Foods hire hundreds of temporary workers during the high season. Such workers are hired from other states and even countries through the US's H2B visa program, which aims to meet labour needs (Smith, 2020).

In December 2020, 24 of the 25 people on board the United States Seafoods' Legacy trawler tested positive for COVID-19. By that time, the 132-foot-long trawler had been perch fishing off Unalaska (Hagenbuch, 2020c). In January 2021, a Trident Seafoods factory in Akutan, Alaska, reported that 135 of 307 tested workers were positive for the virus (Matthews, 2021). The factory was expected to double the approximately 700-employee team for the pollock A season, which runs from late January to late April. However, given these events, Trident Seafoods announced a three-week shutdown at the plant as of the 21st of January (White, 2021).

UniSea, one of the world's largest seafood processing facilities and located in Unalaska, had 54 employees test positive for COVID-19 in mid-January 2021. By the time I am writing this page, 475 people were tested out of the 700 processing workers. UniSea President Tom Enlow said: 'Not knowing how and where the virus is getting introduced will make everyone on edge until we've got everyone inoculated and have achieved herd immunity' (McKenney, 2021). This was the second outbreak in the same plant. In September 2020, 5 workers had tested positive for the virus, 4 of whom were part of a group of 100 processing employees recently hired to fill a gap in the market (McKenney, 2020). The factory had to shut down part of its operations for a few weeks (White, 2021).

In early December 2020, 9 of 28 workers on board a trawler owned by the O'Hara Corporation tested positive for coronavirus (Hagenbuch, 2020b). In the same month, six positive coronavirus cases were reported by the City of Unalaska. All the infected people had arrived by plane to work in the fish industry—five were employed by Alyeska Seafoods and the other worked at UniSea (Hagenbuch, 2020c).

By the 4th of December 2020, Alaska state had reported 2,214 coronavirus cases in Unalaska (Hagenbuch, 2020b). The town has approximately 4,500 residents and remains at a high coronavirus risk level. On the 5th of December, Alaska experienced its record daily rate, with 933 infections (Hagenbuch, 2020b).

According to US official statistics available at www.covid19.alaska.gov, Alaska had 75,654 *resident* coronavirus cases as of the 12th of August 2021 (over 10% of the state's population), resulting in 392 deaths. Among *non-residents*, there were 3,426 infections, 1,766 hospitalisations, and 8 deaths. While residents are being urged to respect movement restrictions and additional safety guidelines, the seafood industry is free to bring workers in.

And please bear in mind that a similar trend has been seen worldwide in animal-flesh processing plants. In Chile, multiple fish processing plants were forced to shut down operations in January 2021. In Quellán, for example, Salmones Austral, Cultivos Yadran, and Marine Farm temporarily closed their facilities that month (White, 2021). Quellán, in southern Chiloé Island, is a port city in the Los Lagos Region and a hub for aquaculture and fishery operations. As of the 2nd of February 2021, Los Lagos was the fourth region with the most coronavirus cases in the country (43,026 infections and 422 deaths).

According to a spokesperson of Salmones Austral, Milton Castaing, the company's processing plants follow strict safety protocols, and therefore the rising COVID-19 cases are attributed to the employees getting infected *outside* work, especially during the holiday period. Commenting on the government calls to temporarily close the salmon factories in the region, Castaing claimed this would be counterproductive as their workers would be at higher infection risk having more free time to roam around. He added:

> [...] all the preventive proactive control tests that we carry out on the plants and that have managed, until now, to identify possible positives early and thus prevent further infections are stopped. In this sense, salmon farming companies are a great ally of the Ministry of Health, generating thousands of Covid-19 tests in the communes where we operate. Each test of our worker means a tested home and a family, neighbours and community, in general, more protected. (Garcés, 2021)

How generous of them—they offer the disease and the testing at the same time.

Mowi, from Norway, is the world's leading seafood company and the primary producer of Atlantic salmon. The company employs 14,537 people in 25 countries. Coronavirus cases have been detected in several Mowi facilities worldwide (Riise, 2021; Njåstad & Riise, 2020).

Fish is not essential, and yet fish processing plants remain open despite the multiple coronavirus outbreaks worldwide. Please reflect if that is the ethics you want to support. We can shape a new future through our purchases.

Violence reverberates

The current worldview sustains that one species has moral prevalence over the others. This *belief system* opposes the widely recognised interconnectedness of all living species and its role in our survival. Therefore, a lack of compassion towards other sentient creatures is not just a moral issue; it is also profoundly unwise. Deciding to ignore other creatures' pain is choosing to get caught in our own web of selfishness and carelessness—to our own disadvantage.

For those who are not concerned about the ongoing sixth mass extinction, animal use, and human pandemics, maybe there is still a window of opportunity for a mindful life: *human suffering.*

Fishing is a demanding job, both physically and psychologically. Long shifts, extended voyages, variable working hours, low living standards on the vessel, strict social isolation, occupational hazards, weather adversities, and low pay contribute to dissatisfaction at work (Szymańska et al., 2006; Fort et al., 2016). Just like slaughterhouse workers, fishermen tend to injest alcohol and drugs to handle occupational stress, which in turn increases their vulnerability to HIV and other sexually transmitted diseases (STDs) (Kissling et al., 2005).

Numerous studies have indicated a staggering prevalence of substance abuse and HIV among fishermen (West et al., 2014; Kissling et al., 2005). The extended fishing network, comprising people working in fish processing and trading, is also at higher risk of HIV infection due to sexual connections with fishers (Kissling et al., 2005). Demographic variables (social status, educational level, income, access to health services, and mobility) alone do not fully explain the high rates (Fort et al., 2010).

In a survey among individuals aged 15 to 49 years in African, Asian, and Latin American countries, the HIV prevalence was 4–14 times higher in the fishing community relative to the national average. The estimate comprised the Democratic Republic of Congo, Kenya, Uganda, Indonesia, Myanmar, Malaysia, Cambodia, Thailand, and Honduras. In most countries, the HIV prevalence among fisherfolk nears or surpasses that of recognised high-risk groups, such as truck drivers and military personnel. For example, in Malaysia, HIV rates are ten times higher among fishermen than the national population average (Kissling et al., 2005).

In a paper on injected drug use among Malaysian fishermen, West et al. (2014) reported that social relationships and occupational culture on fishing boats induced substance use and increased HIV risks. Among the risk factors for HIV infection was syringe and needle sharing. Crewmembers were aware of drug use in the ship, which commonly took place in groups, with the captain often providing drugs to the crew to increase their performance. Out of the 406 fishermen included in the study, 37% had a history of injected drug consumption, which correlated positively with the HIV seroprevalence. Heroin was the most frequently used drug, although others were also consumed: ecstasy, buprenorphine, *pil kuda*, methamphetamine, methadone, benzodiazepines, cannabis, and even glue.

Chinnakali et al. (2016) assessed the alcohol consumption pattern of 304 fishermen aged 18 years or older in the coastal village of Puducherry, India. Eighty-two per cent of the sample had been working as fishermen for over a decade. A total of 79% of the respondents had consumed alcohol in the past year. Of those, a shocking 89% reported alcohol use *during* fishing. Another study conducted in the same village provided further details on drug abuse among fishermen (Manoj et al., 2018). A house-to-house survey revealed that harmful use of alcohol, understood as violent behaviour towards other people, was noted in 77% of the fishermen. Approximately 12% of the sample was characterised as having a 'probable alcohol dependence.'

In Malawi, HIV prevalence in fishing communities was estimated at 9.6% in 2017, considering people aged 15–49. MacPherson et al. (2020) evaluated the social factors driving HIV infection rates among young fishermen of two fishing communities in Lake Malawi. The authors depict the young fishermen's lifestyle and discuss the factors leading to risky behaviour, namely the power dynamics within the fishing industry, the high mobility of fishermen, gender norms, economic insecurity, and low educational levels, among others. Alcohol consumption among crew members was common and was associated with episodes of verbal and physical violence in bars and residential areas.

To be highlighted is the role of transactional sex in impoverished fishing communities. Female fish traders secure access to fish at lower prices or at no cost in exchange for sex with fishermen, increasing their vulnerability to HIV and other STDs. In interviews with sex workers conducted by the researchers, they stated that young fishermen rarely used protection. They added they were afraid of refusing to have unprotected sex as the men could become violent.

But before one claims that fishing-related drug abuse and HIV prevalence are limited to developing countries, let us highlight that fishermen from Indonesia, Bangladesh, China, India, and Vietnam account for more than two-thirds of the planet's 36 million fishermen (Kissling et al., 2005). Therefore, a seafood dish served in any part of the world is likely associated with the portrayed issues. Furthermore, substance abuse and unsafe sexual behaviour is observed *worldwide* among fishermen.

In the face of dire poverty and adverse climate conditions, fish is often the easiest nutrient source in some regions. However, this is not the reality in developed countries, where we have plentiful options. We do not eat animals because there is no other choice; we do it out of pure selfishness—taste preferences, habit, social status, you name it.

But let us go back to our analysis of substance abuse and STDs among fisherfolk.

Smolak (2004) analysed 44 peer-reviewed articles and the grey literature from 1992 to 2012 on the HIV-risk behaviour among fishermen around the globe. Nearly half of the fishermen did not use protection with female sex workers. The review showed that alcohol abuse, the fishing culture, and mobility played an essential role in HIV infection risks.

Fort et al. (2010) assessed the tobacco, drug, and alcohol use among fishermen and merchant seamen working in 19 French ports. Smoking, daily alcohol consumption, and nicotine dependence were more prevalent in fishermen, while the use of cannabis and other drugs were predominant in merchant seamen. Higher consumption of alcohol and tobacco among fishermen was attributed to the working conditions.

The same authors (Fort et al., 2016) evaluated the prevalence of cannabis and cocaine consumption among French fishermen. Out of the 1,000 respondents, 63% were current smokers and 51% smoked 11 to 19 cigarettes a day. One-third of the fishermen were at risk of excessive drinking according to the Alcohol Use Disorders

Identification Test (AUDIT-C), a questionnaire approved by the WHO. In addition, 58% had already experimented with cannabis, and 16% had already used cocaine. Urine tests found that 28% of fishermen were positive for cannabis and 4.5% for cocaine. The results of the Cannabis Abuse Screening Test (CAST), developed by the French Monitoring Centre for Drug and Drug Addiction, revealed a strong level of cannabis dependence in 43% of the fishermen.

In Poland, suicide rates among fishermen were shown to be significantly higher than the national average (Szymańska et al., 2006). Suicide rates among Polish fishermen and seamen between 1960 and 1999 ranged from 7 to 9%, versus 0 to 2% among the 20-year or older general male population. The authors ascribe the high suicide prevalence to the work conditions and lifestyle in the marine and fish industries, which include substance abuse to cope with stress and depression. Acute and chronic mental illnesses can also explain the high suicide rates. Deep-sea fishermen were the most prone to suicide, possibly due to loneliness and stress related to long voyages. Deaths attributed to suicide corresponded to 6 to 18% among deep-sea fishermen, 3 to 14% among merchant seamen, and 2.5% in boat fishermen.

In a paper published in the *Psychological Medicine* journal, Roberts et al. (2013) compared the suicide rates among different professions in the UK over a 30-year period. The study was divided into two phases: (1) 1979–1980 and 1982–1983; and (2) 2001–2005. During phase (1), veterinarians and merchant seafarers ranked first and second in suicide rates, with 77 and 76 suicides per 100,000 worker-years, respectively. Although the 30 top occupations in terms of suicide rates differed significantly among the two study periods, merchant seafarers remained in the second position from 2001 to 2005, with 68 suicides per 100,000 worker-years. High suicide rates were partly attributed to the easy occupational access to guns, pharmaceuticals, drowning, and other methods of committing suicide.

The animal exploitation industry heavily relies on instruments to trap, restrain, mutilate, stun, kill, and eviscerate animals. How can we evolve as a society while consistently recurring to violence?

While some (sadly) might claim that some forms of violence are "necessary," I kindly invite you to question this perverse paradigm. We must understand that violence is still violence, regardless of the victim involved. Violence against women, against children, against foreigners, against people of colour, against animals, and so forth are all rooted in one same thing: a lack of empathy.

Throughout this book, we have seen how deforestation, habitat destruction, biodiversity loss, the climate crisis, animal exploitation, sub-employment, human rights abuse, dietary-based illnesses, zoonotic diseases, and pandemics are all interrelated. All activities based on violence reverberate back to us.

In the fish industry context, for example, there are strong links between fishing activities, drug and gun trafficking, human smuggling, and human slavery (Stringer & Harré, 2019; Belhabib et al., 2020). These associations have prompted the emergence of the term 'fisheries crimes.' According to Stringer & Harré (2019),

'Fisheries crime is an emerging paradigm for conceptualising the range of illegal activities taking place at sea, including illegal fishing activities, document fraud, and human trafficking.' Although the term has no legal definition, it has been broadly used by scientists and policymakers.

Fisheries crimes

Firstly, I would like to address the most elemental fisheries crimes: illegal, unreported, and unregulated fishing (IUU). IUU occurs whenever fishing exceeds legal catch limits, include protected or endangered species, are done in protected areas, employs illegal gear, or misreports catch volumes. National and local measures have been implemented worldwide to deter and prevent IUU—for instance, coastal surveillance, regulations, and policies (May, 2017).

However, the first international agreement developed with this intent, the Agreement on Port State Measures to Prevent, Deter, and Eliminate Illegal, Unreported, and Unregulated Fishing (PSMA), entered into force only in 2016 (FAO, 2020). Deterrence of IUU by PSMA is mainly done by preventing ships involved with IUU fishing from using ports and unloading the catch, hence blocking the entrance of the fishery products into national and international markets (FAO, 2020).

Despite the PSMA's efforts, the yearly revenue from IUU fishing is valued at USD 15.5–35.6 billion, versus USD 109.2 billion for the global legal capture. The actual figures are likely to be even higher, as this estimate does not include unregulated fishing and IUU in inland fishing zones (May, 2017).

In addition, FAO recognises that the seafood sector is highly vulnerable to food fraud. A recent action orchestrated by the European Commission, INTERPOL, and Europol in 11 EU countries confiscated over 51 tonnes of tuna due to fraudulent practices, including selling tuna meant for canning as fresh tuna and substituting tuna with cheaper species (FAO, 2020).

A US nationwide survey by Oceana, a non-profit ocean conservation organisation, found that nearly half (43%) of all salmon sold in restaurants and grocery stores over the 2013–2014 winter were mislabelled. Sixty-nine per cent of the salmon labelled as "wild-caught" were proven by DNA tests to be farmed Atlantic salmon. Others were falsely labelled as higher-value species of salmon, such as Chinook (Oceana, 2015).

But beyond fish stocks depletion, IUU, and food fraud, there are other concerning issues surrounding the fishing sector. IUU captures are commonly associated with serious offences, such as the smuggling of migrant workers and the trafficking of people and illicit substances. The economic and societal impacts of IUU fishing are so prominent that IUU is listed in the Global Financial Integrity's report on transnational crime, along with such activities as drug trafficking, weapon trafficking, human trafficking, illegal organ trade, illegal wildlife trade, illegal logging, illegal mining, and crude oil theft (May, 2017).

Belhabib et al. (2020) analysed the link between drug trafficking and fisheries from worldwide data on vessel interdiction between 2010 and 2017. According to the authors, the transhipment of drugs involving fishing vessels is already responsible for 15% of the global retail value of illicit substances. In the face of the growing interdiction, fishing vessels carry ever-smaller drug loads, while the trafficking network expands. Drug trafficking is becoming more and more attractive to small-scale fishers in financial vulnerability due to overfished seas and marine conservation measures (Belhabib et al., 2020).

In a paper on fisheries crimes in New Zealand's waters, Stringer & Harré (2019) elaborate on the relationship between illegal fishing and human smuggling for forced labour. The study was based on a series of reports of human rights abuses and illegal fishing practices by Indonesian fishermen working on South Korean ships in New Zealand's marine territory. In fact, economically vulnerable people are commonly lured into jobs in the fisheries sector only to end up trapped in inhumane living conditions. Other times, they are literally trafficked for labour exploitation, as repeatedly reported in Southeast Asia (Kadfak & Linke, 2021; Wilhelm et al., 2020; May, 2017; Marschke & Vandergeest, 2016).

In a report on transnational crimes, May (2017) explains how smugglers in Thailand and Indonesia act as brokers, coordinating recruitment and transportation of migrants, especially from Myanmar. With vague or nonexistent employment terms, the fishermen are frequently forced to pay back the recruitment fees. Slave labour is tightly linked to IUU fishing and environmental degradation (Sparks et al., 2021; May, 2017).

Modern slavery is well recognised in the global fish supply chain. Tickler et al. (2018) discuss how gradually smaller catches due to overfishing are driving modern slavery as a way to reduce labour costs and offset diminishing returns. Reports of severe labour abuse among seafood workers have emerged in the last years from multiple countries. The abuses include human trafficking, debt bondage, forced confinement, physical abuse, and even murder. Typically, individuals from developing countries and war zones are recruited voluntarily or smuggled. Numerous allegations exist on human trafficking of African, Eastern European, and Asian crew among the US, New Zealand's, English, Scottish, and Irish fisheries (Murray, 2014; Mendoza & Mason, 2014; Stringer et al., 2016; Lawrence & McSweeney, 2017; MRCI, 2017; NCA, 2020).

In 2014, officers from the Scotland human trafficking squad freed dozens of labourers from several countries working under appalling conditions in fishing vessels (Murray, 2014). In 2017, nine men from Ghana, India, and Sri Lanka maintained on a slavery regimen were liberated by the British police from two national scallop trawlers (Lawrence & McSweeney, 2017). The UK's National Crime Agency regularly reports on cases of people smuggling in several sectors, including the fishing industry. In November 2020, a fishing boat was intercepted off East Anglia's coast with suspicion of immigration and labour legislation offences. Three men (from Ukraine and Latvia) were charged and remanded in custody for the alleged exploitation of 69 Albanian workers (NCA, 2020).

And the story is not much different in North America. While Hawaii is recognised by the US fish industry for its strict sustainability regulations, labour conditions remain an unsolved issue. In 2016, an investigation on Hawaiian fishing ships revealed that hundreds of underpaid, skilled foreign fishermen work with no documentation. Many are victims of verbal and physical abuse, forced labour, and debt bondage. Some of them reported having taken years to pay off alleged costs with airline tickets, recruitment fees, and others. Despite the high unemployment rates in Hawaii, few US nationals are willing to apply for the demanding and underpaid work as fishermen (Mendoza & Mason, 2016).

In 2017, the Migrant Rights Centre Ireland conducted an in-depth investigation among 30 fishermen from Egypt and the Philippines working on Irish boats. Although most of them (85.7%) were qualified and experienced fishermen, they worked more than a hundred hours a week for an average pay below EUR 3 an hour. Poor occupational safety, discrimination, verbal and physical abuse, and documentation issues were also reported. The workers claimed to be paid less than their Caucasian colleagues, given more difficult tasks, and getting explicitly racist insults (MRCI, 2017).

The Irish fishing industry received EUR 147.6 million from EU funds for the period 2014–2020 through the European Maritime and Fisheries Fund (EMFF). EMFF's purpose is to increase the industry's productivity and generate jobs. Further EUR 92 million were invested by the Irish government to strengthen the fishing sector. Still, none of the funding was used to address labour issues (MRCI, 2017). I wonder how bright the future would be if these investments were relocated to plant-based crops.

IUU fishery products entering the food chain are widespread and sizeable (Donlan et al., 2020; Ganapathiraju et al., 2019; Agnew et al., 2009). In Hawaii, for example, catches associated with the labour abuses previously mentioned are widely commercialised in fish markets, restaurants, hotels, and supermarkets across the state (Mendoza & Mason, 2016). In Japan, an estimated 24–36% of the 2015 wild seafood imports were linked to IUU fishing (Ganapathiraju et al., 2019). What is more, in a highly globalised seafood trade, large volumes of fish are imported from countries with lax labour regulations. Therefore, the fish consumed anywhere in the world has a great chance of being connected to human exploitation—beyond the inherent animal exploitation.

In fact, a recent paper published in *Nature Communications* using the Global Slavery Index (GSI) showed that 'all seafood exported by a given country, whether caught by domestic fleets or processed from imports, carried the same risk of potentially involving slavery' (Tickler et al., 2018). The study comprised figures from the Sea Around Us (a research initiative by the University of British Columbia) and the UN's COMTRADE databases.

Industry and governance frequently stress that the fishing industry's legitimacy is not to be overshadowed by criminal activities in the sector (International Organization for Migration, 2016). While this reasoning applies to illicit businesses operating under the cover of fishing vessels (e.g., drug and weapon trafficking on fake fishing vessels), it is imprudent to ignore the severe labour breaches among *legitimate* fishing vessels and widespread IUU catches entering the global market. The poor management of the supply chain erodes consumer confidence in the fish industry as a whole.

Animal exploitation industry: a dead end

As hard as it might be to hear, all occupations relying on finite resources are due to collapse when demand reaches unsustainable limits. The fact that fisheries are one of the world's oldest industries, and that fishing remounts to ancient times, does not justify our blind attachment to eating animals and deriving money from them. *Our history is not our future*; we must look ahead. Animal agriculture and capture fishing are unsustainable because the damages they cause are far greater than the benefits. Wild fishes cannot multiply at the same rate we are consuming them. The exploitation must stop one way or the other—either through lucidity and ethics or through dramatic socioeconomic and environmental crises that will leave no time for a plan B.

Fishers and fishing communities have been adapting to numerous constraints over history, owing to natural disasters, trade policies, legal catch limits, oil spills, ocean dead zones, overfished areas, climate change, and recently, the COVID-19 pandemic (Smith et al., 2020; Coulthard, 2012). In the US, for example, 70% of seafood is consumed in restaurants (Smith, 2020).

Smith et al. (2020) analysed the adaptation mechanisms adopted by fishers in the Northeastern US between March and June 2020. A total of 258 commercial fishers were surveyed. The alternatives to which the respondents turned for income included: engaging in extra or alternative activities in other sectors (e.g., landscaping, carpentry, and real estate), shifting economic responsibility to a family member, relying on unemployment assistance and social security payments, fishing different species, fishing the usual species with other gear types, and engaging in alternative marketing (home delivery and direct sale from the vessel).

As complex as finding the best resilience mechanism to socioeconomic, political, and environmental challenges can be, many of the fishers in this and other studies have complemented the fishing revenue with other activities or have abandoned the fishing industry altogether (Smith et al., 2020; Himes-Cornell & Hoelting, 2015; Marshall & Marshall, 2007).

As important as individual adaptability and resilience are, we are more than ever faced with a vital dilemma: insisting on an outdated industry or rising out of ignorance towards a brighter future. The wise choice is clear.

INVESTIGATION – TUNA FISHERY IN ITALY

This investigation was conducted in a tuna fishery in Carloforte, Sardegna, by Animal Equality's team (Animal Equality, 2012). Atlantic bluefin tuna is the largest tuna, weighing up to 680 kg and measuring approximately 180 to 300 cm. There are three species of bluefin tuna (Atlantic, Pacific, and Southern), of which Atlantic is the largest in number and also the most endangered. The largest tuna fisheries in the world are concentrated in the Mediterranean Sea (WWF, 2021).

The pictures below might look like scenes from hell but are just standard routine in fisheries worldwide. Figure 64 shows fishes desperately trying to escape the fishing net.

Figure 64. Tuna trying to escape a fishing net in Sardegna, Italy.

The fishes are hauled aboard the ship by hooks pierced through their mouth or face, causing severe pain (Figures 65a to 65e).

Figure 65a to 65e. Hooked fishes being brought on board fishing vessels in Sardegna, Italy.

In Figures 66a to 66d, note how the fishes bleed abundantly and gasp for air.

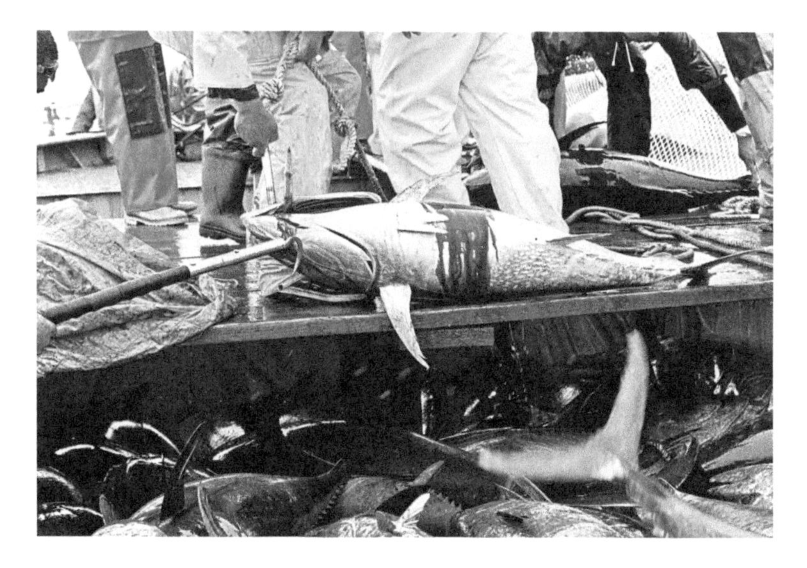

Figures 66a to 66d. Fishes bleeding during a fishing operation in Sardegna, Italy.

Once on board the fishing vessel, the fishes are either left to die by suffocation or are harpooned in one or more regions of their bodies (Figures 67a to 67c). While "humane" slaughter could be ensured by killing the animal with a concussion to the head, this is not what happens in real life, where the fishermen's priority are the production targets.

Figures 67a to 67c. Tuna slaughter in Sardegna, Italy.

The fishes may be sliced open and eviscerated while still conscious (Figures 68a and 68b).

Figures 68a and 68b. Evisceration of tuna in Sardegna, Italy.

Work in a fishery can be exhausting, unpleasant, and underpaid, and is associated with high rates of substance abuse (Stoll et al., 2020). Figures 69a to 69c show the fishermen's routine.

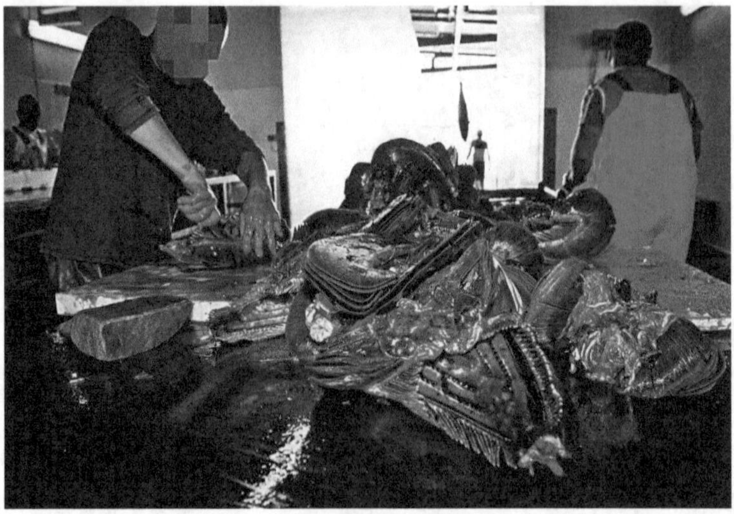

Figures 69a to 69c. Fishermen at work in Sardegna, Italy.

Fishing operations pollute the ocean with nonbiodegradable or long-term biodegradable materials, such as nets, harpoons, hooks, and others (Figure 70). Pollution with oil and missing parts from the boats is also typical. Sea animals can get entangled in the nets, or these can "sweep" the ocean's bottom, destroying delicate ecosystems.

Figure 70. Nets, lines, and other fishing gear in a fishery in Sardegna, Italy.

Please consider veganism.

CHAPTER 5

AN INVITATION

If you have reached this page, you probably wish to break free of all this horror and destruction. And this means you have considered veganism. Contrary to popular belief, living vegan is far from being restricting, complicated, or expensive. And most importantly, being vegan is not limited to what we eat because animals are exploited in many other ways: as sources of raw materials to produce clothes, shoes, cosmetics, vaccines, and medicines, in scientific research, entertainment, transport, and as cheap labour. Being vegan means, above all, awakening to the fact that no one needs to suffer for us to be happy. Quite the opposite, as we are all connected. Therefore, vegans are committed not to support animal exploitation of any kind. Commonly, we also make efforts to reduce our environmental impact and reject companies that follow unsustainable practices and abuse their employees. Being vegan is ultimately about being just and having empathy.

If you are not a vegan yet, you might be looking for guidance on how to make a transition, starting from your meals. Although there is no single way of doing it, you may find the following tips helpful.

Transitioning to a vegan diet

Let us be honest: the average Western diet is deeply reliant on processed foods—white bread, refined sugar, salami, ham, sausages, cheese, chicken nuggets, meat burgers, soft drinks, doughnuts, cookies, ice cream, and so forth. Therefore, many people have the erroneous idea that eating vegan equals replacing meat, eggs, fish, and dairy products with animal-free analogues on a heavily processed diet.

In terms of animal ethics, there is no harm in consuming plant-based and fungi-based alternatives that mimic animal products, but ideally these should not be our staples. Vegan processed food can be harmful to our health too, so we should enjoy Nature's abundance: fruits, leaves, tubers, shoots, flowers, legumes, cereals, seeds, mushrooms, truffles, and algae yield infinite nourishing and delicious combinations.

When we go vegan, we generally pay more attention to what we put into our bodies and naturally lean towards a more natural diet, at least with time. Furthermore, a whole-food, plant-based (WFPB) diet is not only substantially healthier, but also cheaper and more environmentally friendly than an "ultra-processed" vegan diet. If you can afford to eat organic, you will reap even more benefits. In any case, all kinds of vegan diets have a much lower carbon footprint than the traditional, animal-based way of eating, for all the reasons outlined in the previous chapters.

When we wake up to the ramifications of animal agriculture and fishing, many of us cannot contemplate having another meal involving violence foods. But others opt to change their diet gradually. Whatever your case is, I recommend a transition plan where you replace animal products with healthy, plant-based foods.

In this chapter, I suggest a 28-day transition plan. But why 28 days specifically? Four weeks is long enough for us to adapt to a new routine, but short enough for us not to give up midway. Although studies show that the time for habit formation varies largely among individuals (Lally et al., 2009), we are not moved exclusively by neural pathways created by the repetition of actions and thoughts (Volpp & Loewenstein, 2020). *We are not machines.* Our values are drivers of sustained behaviour change.

But please do not take the 28 days as a rule. These are just suggestions, so feel free to adapt the timeframe. You can go vegan overnight and implement the tips you find most helpful from day one, without any particular order. Alternatively, you can remove the animal products gradually over four weeks.

Week 1

Pick one of your three main daily meals and make it animal-free. You can either look up new recipes online or in books or simply adapt your old recipe with cruelty-free ingredients. For example, replace cow's milk with plant-based milk. There are plenty of options in the market. Alternatively, you can make your own plant-based milk at home with cereals, fruits, seeds, or nuts of your preference. It is super easy; it just requires water, the chosen base, a blender, and a clean cloth to filter the resulting mixture.

When baking cakes and dessert breads, replace eggs with mashed bananas or ground flax seeds soaked in water. When preparing dishes and snacks, replace dairy butter with vegan butter, hummus, aubergine caponata, pine nut pesto, sun-dried tomato pâté, or other plant-based spreads and dips. If you want to try something completely different, replace that old breakfast made of white bread, eggs, and sausages with avocado wholegrain toasts, chia pudding with berries and granola, or a bowl of oatmeal with fruit, pumpkin seeds, cinnamon, and plant-based milk.

It might also be useful to remember this simple advice: by adding more fruit and veggies to your meals, you have less room for processed food.

Week 2

Now follow to the second meal. Let us suppose you chose lunch or dinner. You can adapt your old recipes or try something brand new. For example, replace beef burritos with burritos made with your favourite beans, brown rice, guacamole, mixed peppers, sweetcorn, onions, lettuce, cilantro, and fresh lime juice. Mix and match your favourite veggies and a plant protein source (e.g., beans, tofu, or tempeh) and you will have several types of veggie burritos.

Alternatively, prepare a colourful plate of delicious and whole plant-based foods. Explore the flavour, texture, and colour of legumes (e.g., lentils, chickpeas, beans, etc.), which are great protein and fibre sources. You can soak them overnight so they release the phytic acid (which can cause flatulence and reduce nutrient absorption) and cook faster (rendering your meal even more sustainable). Delve into vegetables (e.g., potatoes, peppers, aubergine, carrots, squash, cabbage, kale, chicory, lettuce, etc.), cereals (e.g., rice, sorghum, corn, etc.), and seeds (e.g., pumpkin, sunflower, sesame, hemp, chia, flaxseed, etc.).

Try ingredients that are not so common in the omnivore diet—quinoa, amaranth, buckwheat, tofu, tempeh, seaweed, jackfruit meat, etc. Spice up your cooking with herbs, pepper, garlic, leek, and onions. Try new salad dressings using olive oil, mustard, balsamic vinegar, different seasonings, maple syrup, or any ingredients of your liking. The sky is the limit!

Week 3

Now that two-thirds of your main meals are vegan, discover new ingredients. Go food shopping with a curious mindset. Pay attention to that different-shaped vegetable or fruit you had never noticed before. Taro roots, pak choi greens, oca, romanesco, kohlrabi or turnip cabbage, sunchoke or Jerusalem artichoke, jicama or Mexican potato, yacon roots, physalis or golden berries, jackfruit, mulberries, dragon fruit, and so forth. Go to local or ethnic markets and buy pulses and grains you have not tried or never knew existed.

Start preparing varied salads, with more than the good old lettuce and tomatoes. Add texture and flavour to your salads by using olives, sundried tomatoes, fruit, shoots, nuts, and seeds. Experiment!

Try dark chocolate with a higher cocoa percentage, maybe with a hint of fruit or simply pure. With time you will realise your taste and smell get more refined. You can notice flavours you were not able to before.

And how about giving plant-based yoghurts a try? There are so many great quality options nowadays that we cannot complain. Yoghurt, custard, ice cream, condensed milk, chocolate dessert, spreadable cheese, cheese blocks, grated cheese, and so forth are all available in plant-based versions. Be open and find your favourite brand—or make them at home to keep them as natural as possible.

If you are in a mood for comfort foods, make pasta al pomodoro or delicious focaccia with different toppings. And how about pumpkin "ricotta" gnocchi? You should also explore the potential of cashew nuts as an excellent ingredient for preparing white creamy sauce and cheese. Nutritional yeast will give that extra cheesy flavour to your sauces and soups, besides being a great source of iron, magnesium, B_{12}, and protein. It also tastes great sprinkled over pasta and popcorn.

When you do not feel like cooking, dining out or ordering something 100% plant-based has never been easier. Almost every restaurant will have at least one vegan-friendly option or will tailor the dish for you—for example, by removing the cheese. Moreover, restaurants of every type (even fast-food chains) are increasingly offering plant-based options. You can download Happy Cow, a brilliant online app that lists sources of vegan, vegetarian, and gluten-free restaurants worldwide.

Week 4

Go to the farmers market more often or even visit local farms. You will be supporting local, smaller-scale farms and consuming fresher, tastier products. Explore the power of peanut butter, almond butter, pecan butter, nuts, dates, apricots, and other dried fruit in snacks or as dessert ingredients. Replace traditional custard and sweetened condensed milk with plant-based commercial alternatives or prepare your own at home—just search the Internet.

And if you are a cheese lover, how about taking a course in plant-based dairy products? Search in the region where you live or take an online course.

Eat more fresh fruit, drink more water! Learn how to prepare healthy pre- and post-workout snacks. Sugar-free granola with coconut yoghurt or a fruit smoothie are great before the gym. After your workout, you can have a pea protein shake or baked beans on a toast with roasted tomato and mushroom. Look it up; there are many ideas on the Internet.

If you feel you should reduce your alcohol intake, replace alcohol with non-alcoholic beverages, such as natural juices, teas, kombucha, and flavoured waters. Worth mentioning, not all alcoholic beverages are vegan because they can use animal-origin substrates in the manufacturing process, even if the product formulation is 100% plant-based. And that is because the liquid can be clarified or filtered through chitin, egg albumin, or isinglass, a material obtained from the swim bladder of fishes. Fortunately, these substrates are by no means necessary in the production process, as they can easily be replaced by bentonite clay, potato protein, and other materials of plant origin. Many beer and wine brands are already removing animal substrates from their clarification process. Still, it is worth checking before buying the beverage as ingredients not contained in the final product do not need to be labelled.

And perhaps start a veggie garden, why not? Even if you live in an apartment, you can plant herbs in small pots and leave them on a windowsill. Basil, oregano, cilantro, parsley, fennel, thyme, and chives are easy to grow and make all the difference to your meals. You can also grow peppermint, chamomile, and lemon verbena to make your own herbal tea.

The purpose of this section is not at all to be a cooking guide. But I wanted to show you how easy, diversified, and delicious a vegan diet is by giving you clear ideas of ingredients and dishes. You can also use the vegan food pyramid (Figure 71) to understand the various food groups and the number of servings you should consume daily. The pyramid does not need to be followed to the letter, but it can be helpful.

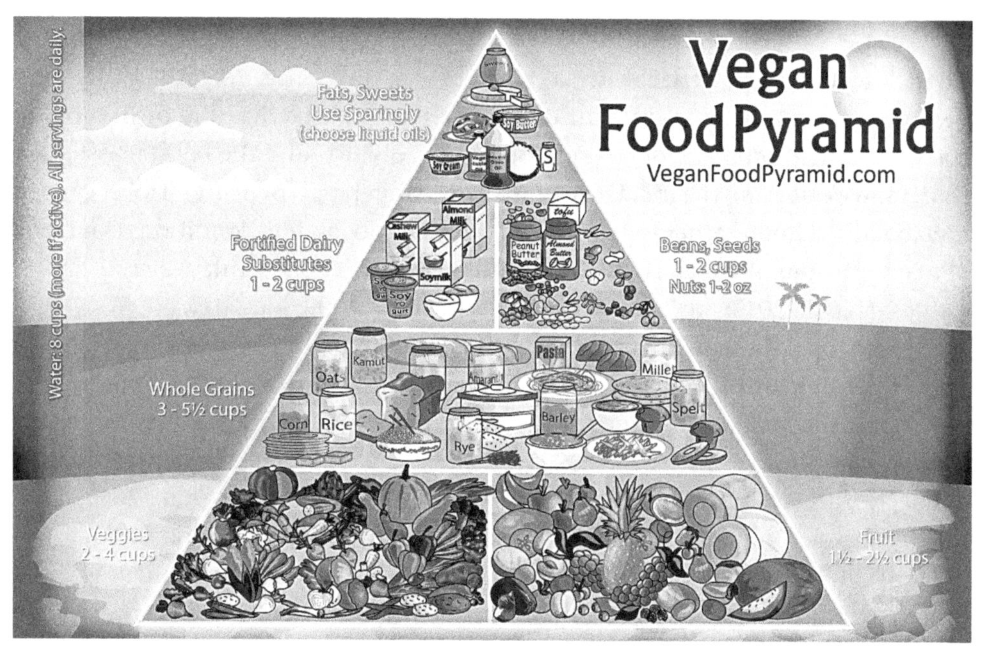

Figure 71. Vegan food pyramid. Source: www.veganfoodpyramid.com.

By following these simple recommendations, you will see how preparing food will become great fun.

Also, there are many vegan starter kits available on the Internet. You can order a free kit with everything from animal rights to nutrition, recipes, and tips on eating out from PETA's or the International Vegan Association's website. If you prefer something in your native language (other than English), virtually every country has a vegan or vegetarian society to provide you with similar information. You can also sign up for a vegan challenge online. There are plentiful options on the Internet, with different duration—seven days, two weeks, one month, etc. Some examples are www.veganchallengediet.com and www.onemonthveganchallenge.com.

You can also count on apps to support you in this transition. For example, the 21-Day Vegan Kickstart app developed by the Physician's Committee (PCRM) is free to download on your phone. This app was developed by medical doctors, nutritionists, and chefs and provides 21 days of nutritionally complete meal plans in 2 languages (English and Spanish). The PCRM's vegan kickstart app includes an interactive grocery list and many videos on health and nutrition by prominent health professionals, including leading physician Dr. Neal Barnard, President of PCRM. According to Dr. Barnard, followers will start to see improvements in their health within three weeks.

If you have any pre-existing health condition, look for a doctor or nutritionist specialising in plant-based eating to help you transition. If you have diabetes, for example, your doctor may need to adjust your medication as your insulin levels normalise with the new diet.

Health benefits of eating vegan

There are multiple benefits to a plant-based diet. The scientific literature demonstrates that balanced vegan diets are the most effective way of improving the plasma lipid profile, reducing excess weight, obesity, and the body mass index (BMI) and decreasing the incidence of diabetes, high blood pressure, stroke, cancer, metabolic syndrome, atherosclerosis, and other cardiovascular conditions (Marrone et al., 2021; Aljuraiban et al., 2020; Viguiliouk et al., 2019; Kahleova et al., 2018; Kahleova et al., 2017; McCarty, 2014; Barnard et al., 2009; McCarty, 1999).

Healthy plant-based diets can reduce the risk of hypertension by 75%, coronary heart disease by 40%, and cerebral vascular disease events by 29%. The risk of developing metabolic syndrome and type 2 diabetes is 47–78% lower among vegans (Kahleova et al., 2017; Le & Sabaté, 2014). Balanced vegan diets can also *reverse* type 2 diabetes, atherosclerosis, and coronary heart disease (Kahleova et al., 2018).

Diets excluding animal products are generally rich in fibre, mono and polyunsaturated fats, plant sterols, high-quality proteins, polyphenols, and other compounds that have anti-inflammatory and anti-cancer effects. They are also low in inflammatory and pro-cancer substances, such as animal protein, saturated fat, lactose, whey, choline, heavy metals, and hormones. This results in improved gut bacteria composition and lower inflammation of the gastrointestinal system, reducing the incidence of cancer, digestion problems (e.g., bloating and constipation), and irritable bowel syndrome (IBS) and eliminating lactose intolerance symptoms. WFPB diets also diminish the occurrence of rheumatoid arthritis, joint pain, cataract, and asthma (McCarty, 2021; Willett et al., 2020; Alwarith et al., 2020; Sakkas et al., 2020; Nath & Singh, 2017; Kahleova et al., 2017; MacDermott, 2007).

Notably, weight loss and improvements in lipid profiles when shifting to a plant-based diet are not exclusively related to calory input, but also to changes in the gut microbiome and the intake of specific nutrients triggering health benefits (Sakkas et al., 2020; David et al., 2014). Studies show that the gut microbiota in vegetarians and vegans is abundant in beneficial bacteria that provide anti-inflammatory properties, for instance, *Prevotella* spp., *Bifidobacterium* spp., and *Lactobacillus* spp. On the other hand, the gut microbiota of omnivores typically contains harmful microorganisms, such as *Bacteroides* spp., *Alistipes* spp., *Clostridium* spp., and *Bilophila* spp., which are involved in inflammation and infections and can provide antimicrobial resistance to a variety of drugs. Beneficial microorganisms are also reduced in the gut of omnivores (Sakkas et al., 2020). The good news is that the gut microbiome responds rapidly to diet changes (David et al., 2014).

For all the reasons outlined, vegan diets boost our immune system, which includes reducing the severity of COVID-19 symptoms (Kim et al., 2021). In fact, a recent study conducted with 2,884 frontline healthcare workers from 6 countries concluded that vegans had 73% lower odds of moderate-to-severe COVID-19 symptoms. Compared to vegans, people following keto diets had over three times more chances of moderate-to-severe COVID-19 (Kim et al., 2021).

Reduced stool transit and straining during defecation in vegetarians and vegans lessen bowel disorders and diverticular diseases (Nath & Singh, 2017). Colorectal cancer and pancreatic cancer have a lower incidence in people following a plant-based diet due to improved gut transit and low intake of pro-inflammation nutrients, but also as a result of a healthier microbiome, where protective bacterial species are more abundant while pathogens involved in the development of cancer (e.g., *Fusobacterium nucleatum*) are scant (Yang & Yu, 2021; McCarty, 2021; Nath & Singh, 2017; Glick-Bauer & Yeh, 2014).

People suffering from acne, premenstrual syndrome, menopause symptoms, endometriosis, polycystic ovary syndrome, and fertility problems can significantly benefit from a WFPB diet. A vegan diet also helps reduce the risks of sex hormone–related cancer (e.g., breast and prostate), as animal products negatively affect our oestrogen and testosterone levels (Barnard et al., 2020).

Balanced vegan diets can benefit our brain and our psychological health too. A body of research associates the consumption of animal products, primarily red meat, with vascular dementia, Alzheimer, depression, anxiety, suicidal behaviour, and other mental disorders (Zhang et al., 2021; Mofrad et al., 2021; Dickerson et al., 2019; Haghighatdoost et al., 2019; Borg, 2019; Beezhold et al., 2015; Khoury et al., 2014).

Thirty-seven per cent of respondents in a 2014 survey incorrectly believed that Alzheimer's disease only affects people with a family history. The survey was conducted by the US-based Alzheimer's Association with 6,307 adults aged 18 years and older from Australia, Brazil, Canada, China, Denmark, Germany, Japan, India, Mexico, Nigeria, Saudi Arabia, and the UK (Alzheimer's Association, 2014). Nonetheless, neurologists state that lifestyle changes, including diet, sleep, and exercise, can help prevent 90% of Alzheimer cases, regardless of genetics (Sherzai & Sherzai, 2017). The diet's role in dementia and cognitive decline is mainly ascribed to inflammatory pathways modulated by the gut microbiota (Qian et al., 2021; Nagpal et al., 2020; Sherzai & Sherzai, 2017).

Some studies, however, associate meat abstention (although not specifically vegan diets) with depression, making it hard to derive consistent conclusions (Dobersek et al., 2021). Important to highlight, though, is that mental health is influenced by nutrient intake and abstention, *but also* by worldviews, lifestyle, socioeconomic variables, surrounding culture, life events, and other factors.

For example, deficiencies in B_{12} and omega-3 fatty acids can raise the risk for depression either in vegans, vegetarians, or omnivores. Furthermore, depression can reflect our values of compassion in a fundamentally violent and unjust society. Stigma and stereotyping can also be detrimental to one's mental welfare, especially if a transition to a vegan diet was not caused by ethical beliefs. Therefore, ethical vegans are not likely to become depressed *because of* the diet, but as a result of being aware of the horror we inflict on others due to convenience and taste preferences.

In short, veganism is a blossoming field of study in behavioural and physiological research, with proven mental health benefits (Borg, 2019; Beezhold et al., 2015).

As with *any other diet*, proper food choices and sun exposure are advised to meet the recommendations for overall energy, fibre, zinc, calcium, omega-3 fatty acids, B-complex vitamins, iodine, vitamin D, and other macro and micronutrients. But for the reasons outlined in previous chapters, vegans *must* supplement with B_{12} to avoid the risk of deficiency.

Plant-based diets have also been demonstrated to result in better systolic and diastolic function, higher maximal oxygen consumption (VO_2max), lower relative wall thickness, and reduced muscle inflammation (Król et al., 2020). However, these advantages have not been translated into strength or aerobic and anaerobic exercise capacity (Vitale & Hueglin, 2021). To the best of my knowledge, no significant differences in athletic performance have been reported among omnivores, vegetarians, and vegans in the scientific literature (Vitale & Hueglin, 2021; Lynch et al., 2018; Craddock et al., 2016). The sparse scientific reports on veganism in the field of sports nutrition warrants further research. Nonetheless, anecdotal reports exist about the benefits of vegan diets among athletes, as shown in the 2018 documentary *Game Changers.*

Scientists and health professionals carry the prejudices and biases of their surrounding culture, where diets that exclude animal products are deemed nutrient-deficient by default, even when widespread evidence indicates otherwise. Besides, the terms 'plant-based diet' and 'vegetarian diet' are not always clearly defined in scientific papers and may comprise many dietary variations, some of them including animal products. This prevents unequivocal conclusions from being drawn from the scientific literature.

Even so, more and more nutrition and health professionals recommend a WFPB diet. For example, the US Academy of Nutrition and Dietetics (which represents over 100,000 credentialed practitioners) states that 'appropriately planned vegetarian, including vegan, diets are healthful, nutritionally adequate, and may provide health benefits for the prevention and treatment of certain diseases. These diets are appropriate for all stages of the life cycle, including pregnancy, lactation, infancy, childhood, adolescence, older adulthood, and for athletes' (Melina et al., 2016).

Evidently, being vegan does not mean that we are bulletproof against disease. Cancer, for example, develops over decades, so what we used to eat before we transitioned to a plant-based diet does make a difference. Besides, our lifestyle, genetics, and exposure to pathogens are important factors in contracting diseases or developing health conditions. In other words, the risk of getting sick is *substantially* lowered, but not zeroed. What is guaranteed when we go vegan is that we are not abusing animals anymore.

Also, the beneficial effects of a vegan diet radically outweigh the adverse ones, whether we consider our physical and mental health alone or wider dimensions of health (e.g., environmental, spiritual, and cultural). But most importantly: it does not need to be the best diet ever. This is not a competition to see which is the best way of eating. The vegan diet is cruelty-free, cheap, sustainable, diverse, delicious, and healthy, so what more can we ask for?

The future of food

We have seen that animal agriculture does not operate under the standard supply-and-demand system. Governmental subsidies artificially lower the price of animal products to the extent that a Big Mac can be cheaper than native fruit. However, this does not mean that market demand does not play a crucial role in shaping the food system. As production costs approach the selling prices of animal products, not even public subsidies are saving smaller-scale farmers from bankruptcy. Moreover, a significant fraction of farmers works for major companies on a contractual basis, taking on loans to raise animals and produce meat, milk, and eggs to precise specifications. Compliance with environmental legislation, especially in terms of animal waste generation and on-farm GHG emissions, is commonly transferred to the farmers.

Therefore, many animal farmers are diversifying or fully transitioning into growing crops. Mercy For Animals' Transfarmation Project, for example, supports farmers in shifting away from animal agriculture into growing plants for food and non-food uses. Dairy industries are increasingly migrating to or investing in plant-based milk and derivatives as more consumers switch to dairy-free substitutes. Meat companies are expanding into plant-based and cultivated meat. Animal agriculture has reached a point of economic artificiality and environmental impact that is forcing the food chain back to supply and demand, but now within more sustainable standards.

In my view, and that of many food experts and investors worldwide, the confluence of economic, environmental, public health, and ethical factors will create a future where animal-based foods (at least those obtained from raising and killing animals) are an extravagance rather than the centre of our diet.

I am not naïve, though—animal liberation would take *forever* if based solely on kindness. In our culture of violence and separateness, ethics hardly drives our food choices, but rather taste, convenience, and cost. But as more people awaken to values of justice and compassion, even if they comprise a small percentage of the world's population, they shape the economy by stimulating the production of cruelty-free foods.

Will the rest of the consumers follow suit? More and more of them will (although not necessarily for ethics), especially when we achieve price parity between animal-sourced and animal-free foods. We are already seeing many non-vegan consumers switching to plant-based products for environmental and health reasons, even paying a premium for this. For these reasons combined, I believe a complete value reassessment regarding animal use will happen over a few generations.

The good news is that the food system regeneration is already underway. Science and industry are advancing rapidly towards the use of alternative proteins, especially because the production of animal foods is wasteful and too impactful. Despite the broad spectrum that plants, algae, and fungi have to offer on their own from a nutritional, functional, and sensory perspective, the food industry is investing in plant-based, algae-based, and fungi-based analogues to replace meat, dairy, seafood, and eggs. Shear-cell technology for structuring plant and fungal proteins; fermentation and encapsulation technologies to improve flavour, texture, and functionality of plant ingredients; fortification to enhance the nutritional profile of plant-based products; denaturation, hydrolysis, and cross-linking techniques to alter protein functionality and make non-animal proteins behave like animal proteins; recombinant protein technology to produce casein, ovoalbumin, and other animal-based proteins via microorganisms; and real dairy techniques applied to plant matrices using fermentation by lactic acid bacteria are all being used to mimic traditional animal products.

The food sector is also investing in cellular agriculture. Cultivated meat (also called cultured, lab-grown, or clean meat) is real meat produced by culturing animal cells instead of slaughtering the animal. Muscle cells are extracted from animals and grown into a medium (generally foetal bovine serum, but also animal-free subtracts) to produce muscle tissue. Then, 3D printers can use lab-grown cells to produce meat in the desired shape, typically using animal-free edible scaffolds (Guan et al., 2021).

The global plant-based meat sector was valued at USD 8.6 billion in 2020 and is projected to reach USD 21.9 billion by 2026 (IMARC, 2021). And the cultivated meat sector is growing too. According to the Good Food Institute, 55 cultivated meat and seafood start-ups announced themselves by the end of 2019, up from 35 in the previous year. In the first quarter of 2020, USD 189 million were raised by cultivated meat companies, which represented more than the amount invested in the sector since it was born around 15 years ago. The range of products under development includes beef, pork, chicken, duck, tuna, foie gras, lamb, horse, kangaroo, salmon, and others. And things are moving fast: cultivated chicken by JUST Foods is in the market in Singapore since 2020 (Good Food Institute, 2021).

Governments will not be able to finance businesses that are fundamentally unviable for much longer, and most consumers are not willing or able to pay the actual price of meat, fish, dairy, and eggs. Safety regulations of cultivated meat, for example, are being issued or complemented all over the globe, including the US, the EU, Israel, Singapore, Australia, and New Zealand (Guan et al., 2021).

Considering that we kill trillions of animals every year for food, alternative protein technologies and cell agriculture will undeniably spare many lives. Cultivated meat can also replace the flesh of slaughtered animals in pet food.

Nevertheless, I cannot ignore the fact that cultivated meat for human consumption is still speciesist. Not only because animals will still be raised to serve as cell donors, but also because we will keep believing that we are superior to them. We would not eat human lab meat, would we? The supremacy paradigm will continue creating problems.

Criticisms aside, I am sure of one thing: our food system will radically change in the coming decade, and competitive prices will accelerate the transition.

Changing our outlook on life

By altering our relationship with food, we notice a change in our beliefs and attitudes—for the better. And part of this stems from the very fact that our physical health is improving, since we are not intoxicating our body with harmful substances anymore. People widely report that acne, bloating, heartburn, constipation, and joint pain gradually disappear, and that blood pressure, insulin levels, and cholesterol levels improve. Consequently, we feel calmer, happier, and more energised. We not only save money on groceries, but with pharmaceutical drugs too.

When we go vegan, we become more mindful of our lifestyle choices and our impact on society and the planet. We do not treat animals as commodities anymore. Instead, we engage with others in a deeper way, beyond the shell of form. We see others for their inherent worth and dignity. Therefore, we become less shallow. And as a consequence, our relationships with humans change too.

We realise how much we have been blinded by cultural indoctrination and understand that people surrender to indifference because they are suffering. Happy people do not hurt anyone. And by guiding their attitudes by apathy and cynicism, they reap more problems. It does not need to be like this, though. On the contrary, the key to higher welfare states is questioning our society's narrow outlook on life. By awakening to empathy and self-empowerment, we stop being manipulated by external forces that want us to buy more, eat more, compete more, and numb ourselves with anything that distances us from our true selves. As we grow into more compassionate beings, we become kinder to everyone, including ourselves.

Emancipating ourselves from deleterious conditioning can help us achieve our true potential, as we are no longer held back by obsolete and limiting worldviews (at least not the ones pertaining to animal use). We are now freer to follow our path and live more meaningfully. Plus, the feeling of doing the right thing is priceless.

You may be wondering, but what about ex-vegans? If veganism were indeed so transformative, why would people quit? It is of foremost importance to emphasise that veganism is not just about food. Vegans are individuals who oppose animal use, and they should not be confused with people who simply follow a plant-based way of eating, generally for their own benefit. I have met a few people who are plant-based for health reasons, but to which animal rights are just a secondary concern, if at all. An *ethical vegan* is a vegan for life.

Other aspects of living vegan

If we understand that animals are not inanimate things that we can use as we please, then we cannot limit ourselves to our diet. Eating an entirely plant-based diet does not make us vegans, but abstaining from animal exploitation, as far as possible, does. But one thing at a time, start with your diet and then naturally progress to other aspects of your life, such as clothing, beauty products, and household items.

Unfortunately, animals are exploited in multiple ways. Once we go vegan, we realise how animal products are everywhere—even in the glue used in our phones' electric circuits. Vaccines frequently contain animal-derived materials or are tested on animals as a legal requirement. This is one of the reasons why we might not be able to be 100% vegan in today's society, although we can be as near as that.

Here is a brief outlook on animal-derived products that you might want to avoid:

- **Clothing**: Stay away from fur, leather, suede, and other types of animal skin, wool, angora, cashmere, feathers (used as fillings in winter jackets), and satin (made from silkworms). Instead, wear synthetic or, even better, natural materials, such as cotton, linen, denim, Tencel® (made from wood cellulose), hemp fabrics, bamboo fabrics, and plant-based leather.
- **Shoes:** Leather and suede can be replaced with plant-based leather or recycled plastic.
- **Household furniture, bedding, and accessories**: Leather, wool, and feathers can be replaced with the abovementioned materials.
- **Skincare and make-up:** Chitosan and shellac (both made from the shell of a bug), gelatine, lanolin, propolis, beeswax, and caprylic acid are all animal-derived. Other compounds have both animal-based and plant-based versions, such as biotin, collagen, glycerine, retinol (i.e., vitamin A), cholecalciferol (i.e., vitamin D3), RNA, and oestradiol (i.e., oestrogen). Eyelashes and make-up brushes can be made of animal hair, so look for synthetic options.

Notably, household cleaning products and skincare products labelled as 'vegan' or 'vegan-friendly' may still be tested on animals. Simultaneously, a 'cruelty-free' label means the product was not tested on animals but can still contain animal-derived compounds. I know it can be confusing. But do not stress out—just make sure your products are both 'cruelty-free' and 'vegan-friendly,' or simply 'suitable for vegans.' Websites such as the Ethical Elephant (www.ethicalelephant.com) and Leaping Bunny (www.leapingbunny.org) and apps such as PETAS' Bunny Free can help you on your journey.

It goes without saying that vegans do not support sports and entertainment involving animals, for instance, recreational fishing, bullfight, hunting, circuses, zoos, and aquatic parks that use dolphins and other animals.

As I said, it can seem overwhelming at the beginning, but is a natural journey you will learn as you go along. You will quickly notice that living a life oriented towards *ahimsa* is transformative and fulfilling.

Common mistakes when going vegan

Going vegan is simple, but we can make it complicated. Here are some common mistakes people make that you will want to avoid:

- **Lack of planning:** When it comes to diet transition, eating vegan can backfire if you do not plan it properly—as would happen with any other diet. Exaggerating on processed foods resembling animal products, disregarding fruits and veggies, not eating enough calories, avoiding such staples as potatoes and whole grains and, thus, not feeling sated, and not supplementing with B_{12} are common mistakes. Do your research, maybe sign up for a cooking class, and advance one step at a time.

- **Trusting everything you hear and read:** Nowadays, everyone thinks they are an expert—and even experts can be misinformed or make mistakes. From YouTubers who mindlessly repeat what others say to medical doctors with little to no training in nutrition, there is a lot of controversial advice out there. Also, do not rely entirely on what reporters, book authors, and bloggers say about a research paper, as their claims are not always substantiated by the very work they cite. Do not even trust what I am saying—consult the numerous peer-reviewed references provided and form your opinion. To sum up: do your homework, and as boring as it may sound, this means *reading the research.*

- **Eagerness and anxiety:** I fully understand you no longer want to participate in animal exploitation, but be easy on yourself. Master food first, and then gradually progress to other aspects of living vegan.

- **Guilt:** Do not beat yourself up if you *accidentally* eat or buy something that involved animal use. Accidents can happen. With time, you will understand which products are okay to consume and which are not. Every year we grow as a person and as a vegan—we continuously discover other forms of animal use we were unaware of before. You will adapt as you come across new information, do not worry.

- **Succumbing to (negative) social pressure:** We can still eat out, hang out with friends and family, and do the things we used to do in our pre-vegan times. But I understand our loved ones can make us feel guilty or awkward about our new lifestyle for various reasons. Some well-meaningful but misinformed people may be legitimately concerned about our health; others may worry that we might distance ourselves from them for sharing different principles (at least regarding animal use); and, of course, some will see us as a threat, since eating animals is becoming obsolete. But stay calm and keep growing. Know why you are doing it and be kind to others and yourself.

Useful tips

When I became a vegetarian for the animals, in 1999, I did not know any vegetarians, let alone vegans. In my ignorance, I did not even know the term existed. When I became a vegan, almost twenty years later, I also did not know any ethical vegans. And this delayed my awakening. I blame myself for staying in my bubble and not educating myself on veganism—it was like I was hypnotised. How I wish I had come across a post on social media or met someone who would explain to me that being a vegetarian was not enough But things are what they are. Each of us has a different journey, and for some reason, I had to awaken by myself while there was so much information available out there. So, here are some valuable tips to make your journey easier:

- **Educate yourself:** You will find hundreds of scientific papers and reports on the environment, nutrition, agriculture, food science, veterinary medicine, animal rights, farming practices, and all things related to veganism in this book's references list. You can also read books written by serious professionals, watch documentaries and interviews, and follow social media pages. The information is out there, so keep learning.

- **Get vegan food inspiration:** Follow cooking pages on the Internet, sign up for a cooking class, go to restaurants and see what is possible to recreate at home, ask your vegan friend for the recipe of that dish you liked. There are several free vegan online cooking courses—try www.nutriciously.com, for example.

- **Self-development and connection with Nature:** To deal with the world's resistance to kindness, it is crucial to work on self-development to avoid falling into frustration and grief. Vegan or not, *everyone* should work on personal development. And of course, opting out of violence is already a major step towards a better self. Meditation, yoga, counselling, and psychotherapy are all great ways to start or deepen your journey into self-development—find what works for you. And of course, the good and old connection with Nature is the ultimate healing for anyone. Hiking, surfing, swimming, gardening, contemplating, sunbathing, or simply earthing—follow that which fits your preferences.

- **Acknowledge that vegans are not all the same:** Ethical vegans share the same principle of not using animals. Other than that, we are all different in our political views, lifestyle, spiritual beliefs, and life purpose. We even differ regarding our diet: junk food vegan, super healthy vegan, gluten-free vegan, fruitarianism, and raw foodism are all types of diets that vegans follow. Vegans are not perfect—we are just trying to do our best to end animal exploitation. Becoming vegan is just the beginning of a long journey of self-development and spiritual awakening. *Just the beginning.*

- **Forgive your past:** Many of us regret all the things we have done and bought in the past that involved some form of animal use. But the important thing is that we have woken up. You may have clothes, shoes, and other products that are not vegan and, thus, find yourself in a situation of whether you should get rid of them or not. Some argue that wearing a leather jacket can reinforce normalised speciesism and promote purchases of animal-sourced clothing—in the case when someone wants to buy a similar jacket because they liked your outfit, for example. Others believe it is wasteful to buy new items and prefer to keep their shoes and clothes until they are worn out. Other people will simply gift these items to a charity or someone who is not vegan. This is a touchy subject that you should reflect upon, as there is no ultimate answer. The crucial thing, though, is not to buy items like these from now on.

- **See the world's resistance as a good sign:** It is easy to feel sad and tired as people mock you or bully you because you no longer want to participate in animal abuse. But instead of being dragged by the world's negativity, view the resistance as a sign that people know deep down that you are right. It can be easier for them to attack people promoting justice and kindness than change. But be compassionate (which includes humans too) and remember that you are part of a healthy shift in a bigoted society.

- **Visit farmed animal sanctuaries:** Visiting and supporting animal sanctuaries have the power of bringing us down to Earth again. When we see the lovely work people do for ex-farmed animals and the joy, relief, and gratitude in the animals' eyes, we remember why we are vegan and how rewarding it is on multiple levels.

- **Share vegan food:** Enthusiastically share your food in family and friend gatherings. Delicious vegan food can speak more than a thousand words eloquently put. If cooking is not your thing, then order some nice meal or dessert from that place you like and bring it with you. I am sure people will be much more inclined to learn about veganism after tasting yummy plant-based meals.

- **Seek a vegan community:** Finding support and sharing information with like-minded people helps us thrive in our new lifestyle. Look for vegan discussion forums and vegan professional communities, participate in animal rights groups, volunteer in an animal sanctuary, and so forth. It is comforting to know that other people may be facing the same problems that we are facing. We can make one another's life easier. We may also want to expand our vegan network, make new friends, widen our understanding of the vegan movement, or just laugh at vegan memes.

- **Invite someone to go vegan with you:** Sharing the transition with someone you love is fun and makes things easier. You can share feelings, experiences, recipes, and anything related to veganism. Plus, we are social beings—it feels good not to be the only vegan you know.

- **Keep your transition to yourself:** If you are taking your first steps into veganism and feel that your family and friends might not be supportive, give yourself some time to make your transition before you go out telling this to everyone. Do this for yourself and for the animals.

- **Speak out:** As you grow as a vegan, you may want to enlighten other people about the role of veganism in animal liberation and other forms of social justice, environmental sustainability, climate crisis mitigation, and human health. People are often resistant to worldviews that challenge their habits, so you might want to be as skilful as possible when talking about veganism. If you can, read books on animal rights advocacy. Otherwise, simply imagine how you would have wanted someone to approach you in your pre-vegan times. What arguments would have convinced you? What made you awaken back then? Use the answers to these questions as engagement tools.

- **Encourage others:** Post pictures of beautiful vegan meals on social media and share recipes with your friends and family. When the opportunity arises, explain why you became vegan and how this changed your life for the better. For instance, you can tell people how a plant-based diet helped you get in shape or overcome a health condition, if that was the case.

Always be firm and knowledgeable but also respectful and polite when talking about veganism. Sometimes, people are legitimately interested in learning more about veganism but are afraid of being judged. If they find you are open to addressing their questions in a friendly way, you will be helping them make their breakthrough while gaining an ally in our planet's healing process.

- **Be proud of yourself:** You are making a difference towards a just and kinder world for everyone. Keep going!

I understand that this chapter may leave you with many unanswered questions. But now that you know why and how to be vegan, I trust you will know where to find additional answers.

We are here to *alleviate* the suffering of other beings, not the opposite. And I kindly invite you to adopt this core value.

Thank you for caring.

A note to all animal advocates

I can only imagine all the battles you went through
to help all beings be free
Being the light entails facing the darkness
and I know your wounds are deep

I just want to say
Thank you
You are important
And your work is needed

They forgive your past
and we are grateful for all you do
You are part of something grand and beautiful
So keep shining

REFERENCES

Introduction References

Anthis, J. (2019). US factory farming estimate. *Sentience Institute*, www.sentienceinstitute.org/us-factory-farming-estimates.

Azevedo, D. (2020). Brazil: Cow breaks world record for milk production, *Dairy Global*, www.dairyglobal.net/Milking/Articles/2020/3/Brazil-Cow-breaks-world-record-for-milk-production-553736E/.

Bateman, S. (2020). World's biggest waterfall completely dries out during devastating drought, *Daily Star*, www.dailystar.co.uk/news/world-news/worlds-biggest-waterfall-completely-dries-21807200.

Bekiempis, V. (2021). Record-breaking US Pacific north-west heatwave killed almost 200 people, *The Guardian*, www.theguardian.com/us-news/2021/jul/08/pacific-northwest-heatwave-deaths.

Berlinger, J. & Yee, I. (2020). 66 people now killed by flooding in Jakarta, and more rain appears to be on the way, *CNN*, edition.cnn.com/2020/01/06/asia/jakarta-floods-intl-hnk/index.html.

Bonhommeau, S., Dubroca, L., Le Pape, O., et al. (2013). Eating up the world's food web and the human trophic level. *Proceedings of the National Academy of Sciences*, 110(51):20617-20620.

Bowers, S. (2020). Covid variants 'a real concern' for meat plant workers, *The Irish Times*, www.irishtimes.com/news/ireland/irish-news/covid-variants-a-real-concern-for-meat-plant-workers-1.4491090.

Carne e Osso (2011). Um retrato do trabalho nos frigoríficos brasileiros, carneosso.reporterbrasil.org.br/o-filme/index.html. In Portuguese.

Cassidy, E., West, P., Gerber, J., et al. (2013). Redefining agricultural yields: from tonnes to people nourished per hectare. *Environmental Research Letter*, 8:034015.

Celestial, J. (2020). Summer snowstorm kills nearly 500 livestock animals, strands 400 tourists in Xinjiang, China, *The Watchers*, watchers.news/2020/07/03/summer-snowstorm-kills-nearly-500-livestock-animals-strands-400-tourists-in-xinjiang-china/?utm_source=dlvr.it&utm_medium=facebook.

Clark, M., Domingo, N., Colgan, K., et al. (2020). Global food system emissions could preclude achieving the 1.5° and 2°C climate change targets. *Science*, 370(6517):705-708.

Constable, H. (2021a). The other virus that worries Asia, *The BBC*, www.bbc.com/future/article/20210106-nipah-virus-how-bats-could-cause-the-next-pandemic.

Constable, H. (2021b). The reasons swine flu could return, *The BBC*, www.bbc.com/future/article/20210202-Swine-flu-why-influenza-in-pigs-could-cause-another-pandemic.

Cook, R. (2021). Ranking of countries with the most cattle. *Beef2Live*, beef2live.com/story-world-cattle-inventory-ranking-countries-0-106905.

DAFM – Irish Department of Agriculture, Food and the Marine (2019). Meat, www.marketaccess.agriculture.gov.ie/meat/.

Devnath, P. & Masud, H. (2021). Nipah virus: a potential pandemic agent in the context of the current severe acute respiratory syndrome coronavirus 2 pandemic. *New Microbes and New Infections*, 41:100873.

Ellis-Petersen, H. & Ratcliffe, R. (2020). Super-cyclone Amphan hits coast of India and Bangladesh, *The Guardian*, www.theguardian.com/world/2020/may/20/super-cyclone-amphan-evacuations-in-india-and-bangladesh-slowed-by-virus.

FAO – Food and Agriculture Organization of the United Nations (2002). Intergovernmental group on meat and dairy products. *Work Programme for Meat and Dairy*, www.fao.org/3/Y7029E/Y7029E.htm.

FAO – Food and Agriculture Organization of the United Nations (2012). The state of world fisheries and aquaculture. FAO Fisheries and Aquaculture Department.

FAO – Food and Agriculture Organization of the United Nations (2013). Tackling climate change through livestock. A global assessment of emissions and mitigation opportunities, www.fao.org/3/i3437e/i3437e.pdf.

FAO – Food and Agriculture Organization of the United Nations (2021). Dairy market review: Overview of global dairy market developments in 2020, www.fao.org/3/cb4230en/cb4230en.pdf.

Foer, J. (2019). We are the weather. *Saving the planet begins at breakfast* (London: Hamish Hamilton).

Fremstad, S., Rho, H., & Brown, H. (2020). Meatpacking workers are a diverse group who need better protections, cepr.net/meatpacking-workers-are-a-diverse-group-who-need -better-protections/.

Greening, G. & Cannon, J. (2016). Human and animal viruses in food (including taxonomy of enteric viruses). In Goyal, S. & Cannon, J. (Eds.), *Viruses in Foods: Food Microbiology and Food Safety* (Switzerland: Springer).

Goodland, R. & Anhang, J. (2009). Livestock and climate change: what if the key actors in climate change are... cows, pigs, and chickens? *World Watch Magazine*, pp 10-19 (Washington, D.C: Worldwatch Institute).

Greger, M. (2020). *How to Survive a Pandemic* (New York, US: Macmillan Publishers).

Haque, T. & Haque, M. (2018). The swine flu and its impacts on tourism in Brunei. *Journal of Hospitality and Tourism Management*, 36:92-101.

Hassan, M., El Zowalaty, M., Lundkvist, Å., et al. (2021). Residual antimicrobial agents in food originating from animals. *Trends in food Science & Technology*, 111:141-150.

Hollan, M. (2020). Tyson publicly reveals COVID-19 test results after plant-wide testing at North Carolina facility, *Fox News*, www.foxnews.com/food-drink/tyson-north-carolina-coronavirus-testing-results.

HHS & USDA – US Department of Health and Human Services & US Department of Agriculture (2015). 2015–2020 Dietary guidelines for Americans, www.dietaryguidelines.gov/sites/default/files/2019-05/2015-2020_Dietary_Guidelines.pdf.

Hubbard, B. (2015). *Conscious Evolution: Awakening Our Social Potential*. Revised ed (Novato, CA: New World Library).

Huber, B. (2020). How did Europe avoid the COVID-19 catastrophe ravaging US meatpacking plants? *Mother Jones*, www.motherjones.com/food/2020/06/meatpacking-plants-covid-hotspots-europe-regulations-line-speed/.

Kendall, L. (2020). Revealed: Covid-19 outbreaks at meat-processing plants in US being kept quiet, *The Guardian*, www.theguardian.com/environment/2020/jul/01/revealed-covid-19-outbreaks-meat-processing-plants-north-carolina.

Koneswaran, G. & Nierenberg, D (2008). Global farm animal production and global warming: impacting and mitigating climate change. *Environmental Health Perspectives*, 116(5):578-582.

Kraaijenbrink, P., Bierkens, M., Lutz, A., et al. (2017). Impact of a global temperature rise of 1.5° Celsius on Asia's glaciers. *Nature*, 549(7671):257-260.

Kushner, J. (2021a). The new mosquito bringing disease to North America, *The BBC*, www.bbc.com/future/article/20210115-aedes-vittatus-a-mosquito-that-carries-zika-and-dengue.

Kushner, J. (2021b). Why camels are worrying coronavirus hunters, *The BBC*, www.bbc.com/future/article/20210122-the-coronavirus-10-times-more-deadly-than-covid.

Kushner, J. (2021c). How vaccinating monkeys could stop a pandemic, *The BBC*, www.bbc.com/future/article/20210208-yellow-fever-this-virus-could-be-the-next-epidemic.

IATP – Institute for Agriculture and Trade Policy & GRAIN (2018). Emissions impossible. How big meat and dairy are heating up the planet. Global Agriculture, www.globalagriculture.org/fileadmin/files/weltagrarbericht/Weltagrarbericht/04Fleisch/2018GRAINIATPEmissionsimpossible.pdf.

IATP – Institute for Agriculture and Trade Policy (2020). Milking the planet, www.iatp.org/milking-planet.

India Today (2020). Climate change 'fuelling deadly India lightning strikes', www.indiatoday.in/india/story/climate-change-india-lightning-strikes-1697455-2020-07-06.

IPCC – United Nations Intergovernmental Panel on Climate Change (2015). Special Report: Global Warming of 1.5 °C, www.ipcc.ch/sr15.

Johns Hopkins University & Medicine (2021). Coronavirus Research Centre, coronavirus.jhu.edu/map.html.

Johnson, D. (2021). Reports: WHO analysis indicates animal farms were likely origin of COVID-19, *UPI,* www.upi.com/Top_News/World-News/2021/03/30/china-who-report-covid-origins/5101617098543/.

Kindy, K. (2020). More than 200 meat plant workers in the U.S. have died of covid-19. Federal regulators just issued two modest fines, *The Washington Post,* www.washingtonpost.com/national/osha-covid-meat-plant-fines/2020/09/13/1dca3e14-f395-11ea-bc45-e5d48ab44b9f_story.html.

Laughland, O. & Holpuch, A. (2020). 'We're modern slaves': How meat plant workers became the new frontline in Covid-19 war, *The Guardian,* www.theguardian.com/world/2020/may/02/meat-plant-workers-us-coronavirus-war.

Loh, J., Zhao, G., Presti, R., et al. (2009). Detection of novel sequences related to African Swine Fever virus in human serum and sewage. *Journal of Virology,* 83(24):13019-13025.

Machovina, B., Feeley, K., & Ripple, W. (2015). Biodiversity conservation: The key is reducing meat consumption. *Science of the Total Environment,* 536:419-431.

Magome, M. (2019). Southern Africa's deadly drought leaving millions hungry. *ABC News,* abcnews.go.com/International/wireStory/southern-africas-deadly-drought-leaving-millions-hungry-67189786.

McCauley, D., Pinsky, M., Palumbi, S., et al. (2015). Marine defaunation: Animal loss in the global ocean, *Science,* 347(6219): 1255641.

McDonald, C. (2015). How many Earths do we need? *The BBC,* www.bbc.com/news/magazine-33133712.

Mood, A. (2010). Worse things happen at sea: the welfare of wild caught-fish. *fishcount,* fishcount.org.uk/publications.

MRCI – Migrant Rights Centre Ireland (2020). Working to the bone: New research on the experiences of migrant workers in the meat sector in Ireland, www.mrci.ie/2020/11/25/working-to-the-bone-new-research-on-the-experiences-of-migrant-workers-in-the-meat-sector-in-ireland/.

National Geographic (2020). A plague of locusts has descended on East Africa. Climate change may be to blame, www.nationalgeographic.com/science/article/locust-plague-climate-science-east-africa.

Nellemann, C., MacDevette, M., Mandes, T., et al. (2009). The Environmental Food Crisis: The Environment's Role in Averting Future Food Crises. GRID-Arendal, www.grida.no/publications/154.

Nielsen, J. & Vigh, H. (2012). Adaptive lives. Navigating the global food crisis in a changing climate. *Global Environmental Change,* 22(3):659-669.

NIFC – US National Interagency Fire Center (2021). National fire news, www.nifc.gov/fire-information/nfn.

NOAA – US National Oceanic and Atmospheric Administration & NWS – US National Weather Service (2021). Austin/San Antonio Weather Forecast Office. February 2021 Historical winter storm event South-Central Texas, www.weather.gov/media/ewx/wxevents/ewx-20210218.pdf.

Poore, J. & Nemecek, T. (2018). Reducing food's environmental impacts through producers and consumers. *Science,* 360:987-992.

Phillips, D. (2020). Climate crisis blamed for rains and floods that have killed 150 in Brazil, *The Guardian,* www.theguardian.com/environment/2020/mar/13/climate-crisis-blamed-for-rains-and-floods-that-have-killed-150-in-brazil.

Phillips, C., Caldas, A., Cleetus, R., et al. (2020). Compound climate risks in the COVID-19 pandemic. *Nature Climate Change,* 10:586-588.

Pilkington, E. & Canon, G. (2021). Western US and Canada brace for another heatwave as wildfires spread, *The Guardian,* www.theguardian.com/world/2021/jul/15/heatwave-us-west-canada-wildfires.

Raiten, D. & Aimone, A. (2017). The intersection of climate/environment, food, nutrition and health: crisis and opportunity. *Current Opinion in Biotechnology,* 44:52-62.

Ranganathan, J., Vennard, D., Waite, R., et al. (2016). Shifting Diets for a Sustainable Food Future. *Creating a Sustainable Food Future, 11* (Washington, D.C.: World Resources Institute).

Rao, S. (2021). Animal agriculture is the leading cause of climate change – A position paper. *Journal of the Ecological Society,* 32-33:155-167.

Readfearn, G. (2020). Tropical cyclones have become more destructive over past 40 years, data shows, *The Guardian,* www.theguardian.com/environment/2020/may/20/tropical-cyclones-have-become-more-destructive-over-past-40-years-data-shows.

Reisinger, A. & Clark, H. (2018). How much do direct livestock emissions actually contribute to global warming? *Global Change Biology,* 24(4):1749-1761.

Reperant, L. & Osterhaus, A. (2017). AIDS, Avian flu, SARS, MERS, Ebola, Zika... what next? *Vaccine,* 35(35Pt A):4470-4474.

Ritchie, H. & Roser, M. (2019). Meat and dairy production. *Our World in Data,* ourworldindata.org/meat-production.

Ritchie, H. (2020). Environmental impacts of food production, *Our World in Data,* ourworldindata.org/environmental-impacts-of-food.

Saenz, R., Hethcote, H., & Gray, G. (2006). Confined animal feeding operations as amplifiers of influenza. *Vector-Borne and Zoonotic Diseases,* 6(4):338-346.

Simon, D. (2013). *Meatonomics: How the Rigged Economics of Meat and Dairy Make You Consume Too Much—and How to Eat Better, Live Longer, and Spend Smarter* (Newburyport, US: Conari Press).

Simpson, H., Deribe, K., Tabah, E., et al. (2019). Mapping the global distribution of Buruli ulcer: a systematic review with evidence consensus. *The Lancet Global Health,* 7(7):e912-e922.

Soric, M. (2020). Opinion: Putting the foxes at Tönnies in charge of the hen house? *Deutsche Welle,* www.dw.com/en/opinion-putting-the-foxes-at-t%C3%B6nnies-in-charge-of-the-hen-house/a-53929565.

Sun, H., Xiao, Y., Liu, J., et al. (2020). Prevalent Eurasian avian-like H1N1 swine influenza virus with 2009 pandemic viral genes facilitating human infection. *Proceedings of the National Academy of Sciences,* 117(29):17204-17210.

The Local (2021). Death toll rises to 133 after flood disaster in western Germany, www.thelocal.de/20210716/latest-death-toll-in-western-german-floods-rises-to-at-least-81/.

The University of Sydney (2020). A statement about the 480 million animals killed in NSW bushfires since September, www.sydney.edu.au/news-opinion/news/2020/01/03/a-statement-about-the-480-million-animals-killed-in-nsw-bushfire.html.

Theurl, M., Lauk, C., Kalt, G., et al. (2020). Food systems in a zero-deforestation world: Dietary change is more important than intensification for climate targets in 2050. *Science of the Total Environment,* 735:139353.

Tiseo, K., Huber, L., Gilbert, M., et al. (2020). Global trends in antimicrobial use in food animals from 2017 to 2030. *Antibiotics,* 9(12):918.

UN – United Nations (2015). Paris Agreement, unfccc.int/files/essential_background/convention/application/pdf/english_paris_agreement.pdf.

UNEP – United Nations Environmental Programme (2008). Fisheries subsidies: A critical issue for trade and sustainable development at the WTO: An introductory guide (Geneva, Switzerland: UNEP).

UNHCR – The United Nations Refugee Agency (2018). Rohingya refugee crisis: Learn the facts, www.unrefugees.org/news/rohingya-refugee-crisis-learn-the-facts/.

Van Boeckel, T., Glennon, E., Chen, D., et al. (2017). Reducing antimicrobial use in food animals. *Science*, 357:1350-1352.

Wagler, R. (2018). 6th mass extinction. In Dellasala, D. & Goldstein, M. (Eds). *Encyclopedia of the Anthropocene*, p. 9-12 (Amsterdam, Netherlands: Elsevier).

Wester, P., Mishra, A., Mukherji, A., et al. (2019). *The Hindu Kush Himalaya Assessment—Mountains, Climate Change, Sustainability and People* (Switzerland: Springer).

WHO – World Health Organization (2017). Stop using antibiotics in healthy animals to prevent the spread of antibiotic resistance, www.who.int/news-room/detail/07-11-2017-stop-using-antibiotics-in-healthy-animals-to-prevent-the-spread-of-antibiotic-resistance.

WHO – World Health Organization (2020a). Global and regional food consumption patterns and trends, www.who.int/nutrition/topics/3_foodconsumption/en/index4.html.

WHO – World Health Organization (2020b). Origin of SARS-CoV-2, apps.who.int/iris/bitstream/handle/10665/332197/WHO-2019-nCoV-FAQ-Virus_origin-2020.1-eng.pdf.

Williams, M., Zalasiewicz, J., Haff, P., et al. (2015). The Anthropocene biosphere. *The Anthropocene Review*, 2(3):196-219.

World Population Review (2021). CO_2 emissions by country, worldpopulationreview.com/country-rankings/co2-emissions-by-country.

Yang, Y., Huisman, W., Hettinga, K., et al. (2019). Fraud vulnerability in the Dutch milk supply chain: Assessments of farmers, processors and retailers, *Food Control*, 95:308-317.

Zeder, M. (2011). The origins of agriculture in the Near East. *Current Anthropology*, 52(4):S221-S335.

Chapter 1 References

Aghasi, M., Golzarand, M., Shab-Bidar, S., et al. (2019). Dairy intake and acne development: A meta-analysis of observational studies. *Clinical Nutrition*, 38(3):1067-1075.

Agri Benchmark (2013). Country information – Sweden, www.agribenchmark.org/dairy/country-information/sweden.html.

AHDB – Agriculture and Horticulture Development Board (2020). Mastitis in dairy cows, ahdb.org.uk/knowledge-library/mastitis-in-dairy-cows.

Alwarith, J., Kahleova, H., Rembert, E., et al. (2019). Nutrition interventions in rheumatoid arthritis: the potential use of plant-based diets. A review. *Frontiers in Nutrition*, 10(6):141.

Alwarith, J., Kahleova, H., Crosby, L., et al. (2020). The role of nutrition in asthma prevention and treatment. *Nutrition Reviews*, 78(11):928-938.

Animal Equality (2019). BabyBell? More like BabyHell, animalequality.org/action/babyhell.

Animal Equality (2020). Dairy's dark secrets, animalequality.org.uk/act/dairys-dark-secrets.

Archila, L., Khan, F., Bhatnagar, N., et al. (2017). αS1-casein elucidate major T-cell responses in cow's milk allergy. *The Journal of Allergy and Clinical Immunology*, 140(3):854-857.e6.

Australian Department of Agriculture, Water and the Environment (2020). Dairy industry overview, www.agriculture.gov.au/abares/research-topics/surveys/dairy.

Bandaw, T. & Herago, T. (2017). Review on abattoir waste management. *Global Veterinaria*, 19(2): 517-524.

Barker, Z., Leach, K., Whay, H., et al. (2010). Assessment of lameness prevalence and associated risk factors in dairy herds in England and Wales. *Journal of Dairy Science*, 93(3), 932-941.

Barnard, N. (2020). *Your Body in Balance: The New Science of Food, Hormones, and Health* (New York: Hachette Book Group).

BBC News (2021). COP26: US and EU announce global pledge to slash methane, www.bbc.com/news/world-59137828.

Beggs, D., Jongman, E., Hemsworth, P., et al. (2019). The effects of herd size on the welfare of dairy cows in a pasture-based system using animal- and resource-based indicator. *Journal of Dairy Science*, 102(4):3406-3420.

Benkerroum, N. (2016). Mycotoxins in dairy products: A review. *International Dairy Journal*, 62:63-75.

Bischoff-Ferrari, H., Dawson-Hughes, B., Baron, J., et al. (2011). Milk intake and risk of hip fracture in men and women: a meta-analysis of prospective cohort studies. *Journal of Bone and Mineral Research*, 26:833-839.

Bolaños, C., de Paula, C., Guerra, S., et al. (2017). Diagnosis of mycobacteria in bovine milk: an overview. *Revista do Instituto de Medicina Tropical de São Paulo*, 59:e40.

Borges, A., Silva, D., Gonçalves, R., et al. (2003). Fraturas vertebrais em grandes animais: estudo retrospectivo de 39 casos (1987-2002). *Arquivo Brasileiro de Medicina Veterinária e Zootecnia*, 55(2). In Portuguese.

Brouwer-Brolsma, E., Sluik, D., Singh-Povel, C., et al. (2018). Dairy product consumption is associated with pre-diabetes and newly diagnosed type 2 diabetes in the Lifelines Cohort Study. *British Journal of Nutrition*, 119(4):442-455.

Burger, J., Kirchner, M., Bramanti, B., et al. (2007). Absence of the lactase-persistence-associated allele in early Neolithic Europeans. *Proceedings of the National Academy of Sciences*, 104(10):3736-3741.

Bustillo-Lecompte, C. & Mehrvar, M. (2017). Slaughterhouse wastewater: Treatment, management and resource recovery. In Farooq, R & Ahmad, Z. (Eds.), *Physico-chemical wastewater treatment and resource recovery*. IntechOpen.

Byberg, L. & Lemming, E. (2020). Milk consumption for the prevention of fragility fractures. *Nutrients*, 12(9):2720.

Campbell, T. C. & Campbell II, T. M. (2005). *The China Study* (New York, US: BenBella).

Canal Rural (2015). Produtores gaúchos desistem do leite por causa da crise do setor, www.canalrural.com.br/programas/produtores-gauchos-desistem-leite-por-causa-crise-setor-54871/. In Portuguese.

Canal Rural (2017). Indústria paga 60% a mais para produção de leite orgânico, www.canalrural.com.br/programas/industria-paga-mais-para-producao-leite-organico-67993/. In Portuguese.

Canal Rural (2018). A cada 24 horas cerca de 45 produtores de leite abandonam a atividade, www.canalrural.com.br/noticias/a-cada-24-horas-cerca-de-45-produtores-de-leite-abandonam-a-atividade/. In Portuguese.

Carrington, D. (2020). Emissions from 13 dairy firms match those of entire UK, says report, *The Guardian*, www.theguardian.com/environment/2020/jun/15/emissions-from-13-dairy-firms-match-those-of-entire-uk-says-report.

Chia, J., McRae, J., Kukuljan, S., et al. (2017). A1 beta-casein milk protein and other environmental pre-disposing factors for type 1 diabetes. *Nutrition & Diabetes*, 7:e274.

CIWF – Compassion in World Farming (2012). Statistics: Dairy cows, www.ciwf.org.uk/media/5235182/Statistics-Dairy-cows.pdf.

CIWF – Compassion in World Farming (2020). About dairy cows, www.ciwf.com/farmed-animals/cows/dairy-cows/.

Chan, J., Giovannucci, E., Andersson, S., et al. (1998). Dairy products, calcium, phosphorous, vitamin D, and risk of prostate cancer. *Cancer Causes & Control*, 9:559-566.

Cremonesi, P., Monistero, V., Moroni, P., et al. (2020). Main pathogens detected in milk. *Reference Module in Food Science* (Cambridge, MA: Academic Press).

da Silva, A., da Silva, F., & Bett, V. (2017). A prevalência de mastites em vacas leiteiras do município de Carlinda (MT), no ano de 2016. *PUBVET*, 11(8):761-766. In Portuguese.

Dairy Australia (2019). International market overview,www.dairyaustralia.com.au/industry/exports-and-trade/international-market-overview.

Dairy Global (2019). India's demand for dairy products increasing, www.dairyglobal.net/Market-trends/Articles/2019/10/Indias-demand-for-dairy-products-increasing-479014E/.

Dalton, J. (2019). Cows sexually abused, hit and punched at company owned by NFU deputy president, footage shows. *The Independent*, www.independent.co.uk/climate-change/news/cow-sexual-abuse-violence-dairy-farm-punch-kick-hit-essex-nfu-a9215306.html.

de Passillé, A. (2001). Sucking motivation and related problems in calves. *Applied Animal Behaviour Science*, 72(3):75-187.

Dervilly-Pinel, G., Prévost, S., Monteau, F., et al. (2014). Analytical strategies to detect use of recombinant bovine somatotropin in food-producing animals. *Trends in Analytical Chemistry*, 53:1-10.

Embrapa – Empresa Brasileira de Pesquisa Agropecuária (2020). Cadeia produtiva do leite no Brasil: produção primária. *Circular Técnica*, 123. In Portuguese.

European Commission (2013). Organic versus conventional farming, which performs better financially? An overview of organic field crop and milk production in selected Member States, ec.europa.eu/agriculture/rica/pdf/FEB4_Organic_farming_final_web.pdf.

European Commission (2017). Mastitis control in organic dairy, ec.europa.eu/eip/agriculture/en/content/mastitis-control-organic-dairy.

FAO – Food and Agriculture Organization of the United Nations (2019). Overview of global dairy market developments in 2018, www.fao.org/3/ca3879en/ca3879en.pdf.

FAO – Food and Agricultural Organization of the United Nations (2020). Mitigating the impacts of COVID-19 on the livestock sector, reliefweb.int/sites/reliefweb.int/files/resources/CA8799EN.pdf.

FAO – Food and Agriculture Organization of the United Nations (2021).Dairy market review: Overview of global dairy market developments in 2020, www.fao.org/3/cb4230en/cb4230en.pdf.

FAWEC – Farm Animal Welfare Education Centre (2014). Cow's welfare during milking, www.fawec.org/en/practical-notes/63-cattle/297-cow-s-welfare-during-milking.

FAWEC – Farm Animal Welfare Education Centre (2015). Welfare issues during the dry period in dairy cattle, https://www.fawec.org/en/fact-sheets/31-cattle/132-welfare-issues-during-the-dry-period-in-dairy-cattle.

Fazenda Colorado (2020). Globo Rural – 70 mil litros de leite por dia com sistema carrossel, www.fazendacolorado.com.br. In Portuguese.

Fox, M. (1983). Animal welfare and the dairy industry. *Journal of Dairy Sciences*, 66:2221-2225.

García-Fernández, N. (2019). Mastitis treatment in dairy farms. *Dairy Knowledge Center, DKC quarterly 2019 Vol. 1*, www.researchgate.net/publication/332401241_DKC_Quarterly_2019_Issue_1.

Gaskins, A., Pereira, A., Quintiliano, D., et al. (2017). Dairy intake in relation to breast and pubertal development in Chilean girls. *The American Journal of Clinical Nutrition*, 105(5):1166-1175.

Geary, B. (2015). Shocking video footage of New Zealand dairy industry shows animals bashed and kicked, with four-day old calves regarded as 'trash' thrown onto bloodied concrete floor, *Daily Mail*, www.dailymail.co.uk/news/article-3340565/Shocking-video-shows-New-Zealand-dairy-farmers-bashing-kicking-throwing-four-day-old-calves-regarded-trash-bloodied-concrete-floors.html.

Gibson, M., Rogers, C., Dittmer, K., et al. (2019). Can bone measures of the bovine metacarpus predict humeral bone structure? *New Zealand Journal of Animal Science and Production*, 79:8-12.

González, L., Mantecab, X., Calsamiglia, S., et al. (2012). Ruminal acidosis in feedlot cattle: Interplay between feed ingredients, rumen function and feeding behavior (a review). *Animal Feed Science Technology*, 172:66-79.

Guibourg, C. & Briggs, H. (2019). Climate change: Which vegan milk is best? *The BBC,* www.bbc.com/news/science-environment-46654042.

Gulliksen, S., Jor, E., Lie, K., et al. (2009). Respiratory infections in Norwegian dairy calves. *Journal of Dairy Science,* 92(10):5139-5146.

Harrison, S., Lennon, R., Holly, J., et al. (2017). Does milk intake promote prostate cancer initiation or progression via effects on insulin-like growth factors (IGFs)? A systematic review and meta-analysis. *Cancer Causes Control,* 28(6):497-528.

Hodges, C. (2008). The Link: Cruelty to animals and violence towards people. Animal Legal and Historical Center, www.animallaw.info/article/link-cruelty-animals-and-violence-towards-people.

Hristov, A., Oh, J., Firkins, J., et al. (2013). Special topics—Mitigation of methane and nitrous oxide emissions from animal operations: I. A review of enteric methane mitigation options. *Journal of Animal Sciences,* 91(11):5045-5069.

Hsu, E. (2020). Plant-based diets and bone health: sorting through the evidence. *Current Opinion in Endocrinology, Diabetes and Obesity,* 27(4):248-252.

Hublin, J., Ben-Ncer, A., Bailey, S., et al. (2017). New fossils from Jebel Irhoud, Morocco and the pan-African origin of *Homo sapiens. Nature,* 546:289-292.

The Irish Independent (2020). It's not dairy versus beef, it's integration versus isolation, www.independent.ie/business/farming/beef/its-not-dairy-versus-beef-its-integration-versus-isolation-39763642.html.

IATP – Institute for Agriculture and Trade Policy & GRAIN (2018). Emissions impossible. How big meat and dairy are heating up the planet, www.globalagriculture.org/fileadmin/files/weltagrarbericht/Weltagrarbericht/04Fleisch/2018GRAINIATPEmissionsimpossible.pdf.

IATP – Institute for Agriculture and Trade Policy (2020). Milking the planet, www.iatp.org/milking-planet.

IMARC (2021). Top companies in the plant-based meat industry, www.imarcgroup.com/plant-based-meat-companies.

IPCC – Intergovernmental Panel on Climate Change (2019). Summary for policymakers. In Climate change and land: an IPCC special report on climate change, desertification, land degradation, sustainable land management, food security, and greenhouse gas fluxes in terrestrial ecosystems.

Johnston, B., Zeraatkar, D., C Han, M., et al. (2019). Unprocessed red meat and processed meat consumption: Dietary guideline recommendations from the Nutritional Recommendations (NutriRECS) Consortium. *Annals of Internal Medicine,* 171(10):756-764.

Juhl, C., Bergholdt, H., Miller, I., et al. (2018). Dairy intake and acne vulgaris: A systematic review and meta-analysis of 78,529 children, adolescents, and young adults. *Nutrients,* 10:1049.

Kanis, J., Johansson, H., Oden, A., et al. (2005). A meta-analysis of milk intake and fracture risk: low utility for case finding. *Osteoporosis International,* 16:799-804.

Kavanagh, S. (2016). Feeding the dairy cow. Section 6. Chapter 34. Teagasc – Irish Agriculture and Food Development Authority, www.teagasc.ie/media/website/publications/2016/Dairy-Manual-Section6.pdf.

Keaveny, M. (2020). Meath farmer on expanding from 60 to 260 cows based on zero-grazing approach, *The Irish Independent,* www.independent.ie/business/farming/dairy/dairy-farm-profiles/meath-farmer-on-expanding-from-60-to-260-cows-based-on-zero-grazing-approach-38985922.html.

Kevany, S. & Busby, M. (2020). 'It would be kinder to shoot them': Ireland's calves set for live export, *The Guardian,* www.theguardian.com/environment/2020/jan/20/it-would-be-kinder-to-shoot-them-irelands-calves-set-for-live-export.

Knapp, J., Laur, G., Vadas, P., et al. (2014). Invited review: Enteric methane in dairy cattle production: quantifying the opportunities and impact of reducing emissions. *Journal of Dairy Science*, 97(6):3231-3261.

Koneswaran, G. & Nierenberg, D. (2008). Global farm animal production and global warming: Impacting and mitigating climate change. *Environmental Health Perspectives*, 116(5):578-582.

Kuebler, M. & Schauenberg, T. (2020). Record heat wave in Siberia: What happens when climate change goes extreme? *Deutsche Welle*, www.dw.com/en/siberia-heatwave-climate-change/a-54120019.

Lean, I., Degaris, P., Little, S. (2010). Transition cow management: A review for nutritional professionals, veterinarians and farm advisers (Victoria, Australia: Dairy Australia).

Leung, A.& Sauve, R. (2003). Whole cow's milk in infancy. *Paediatrics & Child Health*, 8(7):419-421.

Lima, S., Bicalho, M., de S., Bicalho, R. (2018) Evaluation of milk sample fractions for characterization of milk microbiota from healthy and clinical mastitis cows. *PLoS ONE*, 13(3):e0193671.

Lindahl, C., Pinzke, S., Herlin, A., et al. (2016). Human-animal interactions and safety during dairy cattle handling—Comparing moving cows to milking and hoof trimming. *Journal of Dairy Science*, 99(3):2131-2141.

MAPA – Ministério da Agricultura Pecuária e Abastecimento do Brasil (2016). Projeções do agronegócio Brasil 205/2016 a 2025/2016. In Portuguese.

Maurer, P., Lücker, E., Riehn, K. (2016). Slaughter of pregnant cattle in German abattoirs—current situation and prevalence: a cross-sectional study. *BMC Veterinary Research*, 12(91):1-9.

Meagher, R., Beaver, A., Weary, D. M., von Keyserlingk, M. A. G. (2019). Invited review: A systematic review of the effects of prolonged cow–calf contact on behavior, welfare, and productivity. *Journal of Dairy Science*, 102(7), 5765-5783.

Melendez, P., Bartolome, J., Archbald, L. F., Donovan, A. (2003). The association between lameness, ovarian cysts and fertility in lactating dairy cows. *Theriogenology*, 59(3-4):927-937.

Michaëlsson, K., Wolk, A., Langenskiöld, S., et al. (2014). Milk intake and risk of mortality and fractures in women and men: cohort studies, *BMJ*, 349:g6015.

Moore, J., Spink, J., & Lipp, M. (2012). Development and application of a database of food ingredient fraud and economically motivated adulteration from 1980 to 2010. *Journal of Food Science*, 77(4):R118-R126.

Motamedi, N., Mohamadnia, A., Khoramian, B., et al. (2018). Evaluation of mastitis impact on lameness and digital lesions in dairy cows. *Iranian Journal of Veterinary Surgery*, 13-1(28):39-46.

Nagraik, R., Sharma, A., Kumar, D., et al. (2021). Milk adulterant detection: Conventional and biosensor based approaches: A review. *Sensing and Bio-Sensing Research*, 33:100433.

Nero, L., de Mattos, M., Beloti, V., et al. (2007). Resíduos de antibióticos em leite cru de quatro regiões leiteiras no Brasil. *Ciência e Tecnologia de Alimentos*, 27(2): 391-393. In Portuguese.

NIH – US National Library of Medicine (2020). Lactose intolerance, medlineplus.gov/genetics/condition/lactose-intolerance.

Oltenacu, P. & Broom, D. (2010). The impact of genetic selection for increased milk yield on the welfare of dairy cows. *Animal Welfare*, 19(S):39-49.

Outwater, J., Nicholson, A., & Barnard, N. (1997). Dairy products and breast cancer: the IGF-I, estrogen, and bGH hypothesis. *Medical Hypotheses*, 48(6):453-461.

Picinin, L. (2010). Quantidade e qualidade da água na produção de bovinos de leite. *Annals of the Symposium on Animal Production and Water Resources*. Embrapa, www.cnpsa.embrapa.br/sgc/sgc_publicacoes/publicacao_e1u76v6p.pdf#page=65. In Portuguese.

Poore, J. & Nemecek, T. (2018). Reducing food's environmental impacts through producers and consumers. *Science*, 360:987-992.

ProVeg (2019). Plant Milk Report 2019, proveg.com/plant-based-food-and-lifestyle/vegan-alternatives/plant-milk-report/.

Qin, L., Wang, P., Kaneko, T., et al. (2004). Estrogen: one of the risk factors in milk for prostate cancer. *Medical Hypotheses*, 62(1):133-142.

Raghunath, B., Punnagaiarasi, A., Rajarajan, Get al. (2016). Impact of dairy effluent on environment – A review. In Prashanthi, M. & Sundaram, R. (Eds.), *Integrated waste management in India, Environmental Science and Engineering* (Denmark: SpringerCham).

Rao, S. (2021). Animal agriculture is the leading cause of climate change – A position paper. *Journal of the Ecological Society*, 32-33:155-167.

Rich-Edwards, J., Ganmaa, D., Pollak, M., et al. (2007). Milk consumption and the prepubertal somatotropic axis. *Nutrition Journal*, 6(28).

Riehn, K., Domel, G., Einspanier, A., et al. (2010). Slaughter of pregnant cattle – ethical and legal aspects. *Fleischwirtschaft*, 90(8):100-106.

Robbins, J., Von Keyserlingk, M., Fraser, D., et al. (2016). Invited review: Farm size and animal welfare. *Journal of Animal Sciences*, 94:5439-5455.

Röös, E. & Nylinder, J. (2013). Uncertainties and variations in the carbon footprint of livestock products. *Report 063. Swedish University of Agricultural Sciences.* ISSN 1654-9406.

Sachi, S., Ferdous, J., Sikder, M., et al. (2019). Antibiotic residues in milk: Past, present, and future. *Journal of Advanced Veterinary and Animal Research*, 6(3):315-332.

Sartori, R. (2007). Manejo reprodutivo da fêmea leiteira. *XVII Congresso Brasileiro de Reprodução Animal* (Belo Horizonte, Brazil: CBRA). In Portuguese.

Shahbandeh, M (2021a). Number of milk cows worldwide by country 2020. Statista, www.statista.com/statistics/869885/global-number-milk-cows-by-country/.

Shahbandeh, M. (2021b). Estimated dairy market value worldwide in 2019 and 2024. Statista., https://www.statista.com/topics/4649/dairy-industry/#dossierKeyfigures.

Sharda, S. (2015). Three out of four Indians have no milk tolerance: Study. *Times of India*, timesofindia.indiatimes.com/city/lucknow/Three-out-of-four-Indians-have-no-milk-tolerance-Study/articleshow/46522488.cms

Shindell, D., Faluvegi, G., Koch, D., et al. (2009). Improved attribution of climate forcing to emissions. *Science*, 326: 716-718.

Shwe, T., Pratchayasakul, W., Chattipakorn, N., et al. (2018). Role of D-galactose-induced brain aging and its potential used for therapeutic interventions. *Experimental Gerontology*: 101:13-36.

Sieber, R., Rehberger, B., & Walther, B. (2011). Milk lipids | Removal of cholesterol from dairy products. In Fuquay, J. (Ed.), *Encyclopedia of Dairy Sciences*, pp. 734-740 (Cambridge, US: Academic Press).

Sigsgaard, T., Balmes, J. (2017). Environmental effects of intensive livestock farming. *American Journal of Respiratory and Critical Care Medicine*, 196(9):1092-1093.

Simon, D. R. (2013). *Meatonomics: How the Rigged Economics of Meat and Dairy Make You Consume Too Much—and How to Eat Better, Live Longer, and Spend Smarter* (Newburyport, US: Conari Press).

Smart Protein Project (2021). Plant-based foods in Europe: How big is the market? Smart protein plant-based food sector report by Smart Protein Project, European Union's Horizon 2020 research and innovation programme (No 862957), smartproteinproject.eu/plant-based-food-sector-report.

Summit Calf Ranch (2022). Welcome to Summit Calf Ranch, tulsdairies.com/index.php/our-dairies/summit-calf-ranch.

Tibola, C., da Silva, S., Dossa, A., et al. (2018). Economically motivated food fraud and adulteration in Brazil: Incidents and alternatives to minimize occurrence. *Journal of Food Science*, 83(8):2028-2038.

The Courtyard Dairy (2017). Grass fed cheese, meat and dairy – What is it and who does it? www.thecourtyarddairy.co.uk/blog/cheese-musings-and-tips/grass-fed-pasture-milk-cheese/.

The Good Food Institute (2021). U.S. retail market data for the plant-based industry, gfi.org/marketresearch/.

Theurl, M., Lauk, C., Kalt, G., et al. (2020). Food systems in a zero-deforestation world: Dietary change is more important than intensification for climate targets in 2050. *Science of The Total Environment*, 735:139353.

Tiseo, K., Huber, L., Gilbert, M., et al. (2020). Global trends in antimicrobial use in food animals from 2017 to 2030. *Antibiotics*, 9(12):918.

USDA – United States Department of Agriculture (2014). Pesticide use in US agriculture: 21 selected crops, 1960-2008, www.ers.usda.gov/webdocs/publications/43854/46734_eib124.pdf.

USDA – United States Department of Agriculture (2015). Livestock slaughter – 2014 summary, web.archive.org/web/20150912085407/http://usda.mannlib.cornell.edu/usda/current/LiveSlauSu/LiveSlauSu-04-27-2015.pdf.

USDA – United States Department of Agriculture (2016). Organic production and handling standards. National Organic Program, www.ams.usda.gov/sites/default/files/media/OrganicProductionandHandlingStandards.pdf.

USDA – United States Department of Agriculture (2019). Dairy: World markets and trade. December 2019, apps.fas.usda.gov/psdonline/circulars/dairy.pdf.

University of Massachusetts (2002). Crops, Dairy, Livestock, & Equine Program. Manure inventory, ag.umass.edu/crops-dairy-livestock-equine/fact-sheets/manure-inventory.

Van Boeckel, T., Brower, C., Gilbert, M., et al. (2015). Global trends in antimicrobial use in food animals. *Proceedings of the National Academy of Sciences*, 112(18):5649-5654.

Van Middelaar, C., Berentsen, P., Dijkstra, J., et al. (2015). Effect of feed-related farm characteristics on relative values of genetic traits in dairy cows to reduce greenhouse gas emissions along the chain. *Journal of Dairy Science*, 98(7):4889-4903.

Vilar, M., Hovinen, M., Simojoki, P., et al. (2018). Short communication: Drying-off practices and use of dry cow therapy in Finnish dairy herds. *Journal of Dairy Science*, 101(8):7487-7493.

von Keyserlingk, M., Barrientos, A., Ito, K., et al. (2012). Benchmarking cow comfort on North American freestall dairies: Lameness, leg injuries, lying time, facility design, and management for high-producing Holstein dairy cows. *Journal of Dairy Sciences*, 95:7399-7408.

von Keyserlingk, M. & Weary, D. (2017). A 100-year review: Animal welfare in the Journal of Dairy Science—The first 100 years. *Journal of Dairy Science*, 100(12):10432-10444.

Weaver, C. (2014). How sound is the science behind the dietary recommendations for dairy? *The American Journal of Clinical Nutrition*, 99(5 Suppl):1217S-22S.

Webster, J. (1993). *Understanding the Dairy Cow* (Hoboken, US: Wiley).

Wilkerson, J., Dobosy, R., Sayres, D., et al. (2019). Permafrost nitrous oxide emissions observed on a landscape scale using the airborne eddy-covariance method, *Atmospheric Chemistry and Physics*, 19:4257-4268.

Willett, W. & Ludwig, D. S. (2020). Milk and health. *The New England Journal of Medicine*, 382:644-654.

WHO – World Health Organization & IARC – International Agency for Research on Cancer (2015). IARC monographs evaluate consumption of red meat and processed meat. Press release n. 240 of 26 October 2015, www.iarc.who.int/wp-content/uploads/2018/07/pr240_E.pdf.

WHO – World Health Organization (2017). Stop using antibiotics in healthy animals to prevent the spread of antibiotic resistance, www.who.int/news-room/detail/07-11-2017-stop-using-antibiotics-in-healthy-animals-to-prevent-the-spread-of-antibiotic-resistance.

WHO – World Health Organization (2020a). Origin of SARS-CoV-2, apps.who.int/iris/bitstream/handle/10665/332197/WHO-2019-nCoV-FAQ-Virus_origin-2020.1-eng.pdf.

WHO – World Health Organization (2020b). Tuberculosis, www.who.int/news-room/fact-sheets/detail/tuberculosis.

Zehetmeier, M., Baudracco, J., Hoffmann, H., et al. (2012). Does increasing milk yield per cow reduce greenhouse gas emissions? A system approach. *Animal*, 6(1):154-166.

Chapter 2 References

Agriculture Fairness Alliance (2020). Fiscal conservatism, www.agriculturefairness-alliance.org/mission.

Animal Equality (2019). INVESTIGATION: Animal Equality reveals truth behind Italy's pig farms, animalequality.org/news/animal_equality_investigation_reveals_truth_behind_italian_pig_farms/.

Animal Equality (2020). Dozens of pigs killed in slaughterhouse transport accident in Spain, animalequality.org/blog/2020/02/25/dozens-of-pigs-killed-in-slaughterhouse-transport-accident-in-spain/.

Anthis, J. (2019). US factory farming estimate. *Sentience Institute,* www.sentienceinstitute.org/us-factory-farming-estimates.

Anthropocene (2013). Climate change. www.anthropocene.info/short-films.php.

Azevedo, D. (2019). Small steps forward for Brazil's pig industry, www.pigprogress.net/World-of-Pigs1/Articles/2019/5/Small-steps-forward-for-Brazils-pig-industry-422679E/.

Badcock, J. (2019). Madrid Charity races to save 'starving' cattle stuck on board transport ship, www.telegraph.co.uk/news/2019/08/26/charity-races-save-starving-cattle-stuck-board-transport-ship/.

Barth, A. (2000). *Bull breeding soundness evaluation.* Western Canadian Association of Bovine Practitioners.

Baxter, E., Andersen, I., & Edwards, S. (2018). Chapter 2 – Sow welfare in the farrowing crate and alternatives. In Špinka, M. (Ed.), *Advances in Pig Welfare* (Cambridge, UK: Woodhead Publishing).

Bedford-Guaus, S. (2014). Breeding soundness examination of bulls. Merck Sharp & Dohme (MSD), www.msdvetmanual.com/management-and-nutrition/breeding-soundness-examination-of-the-male/breeding-soundness-examination-of-bulls.

BeefPoint (2006). Métodos auxiliares na detecção de cios: rufiões, www.beefpoint.com.br/metodos-auxiliares-na-deteccao-de-cios-rufioes-30310/. In Portuguese.

Berendes, D., Yang, P., Lai, A., et al. (2018). Estimation of global recoverable human and animal faecal biomass. *Nature Sustainability,* 1:679-685.

Bergeron R., Meunier-Salaun, M., Robert, S. (2008). The welfare of pregnant and lactating sows. In Faucitano, L. & Schaefer A. (Eds.), *Welfare of Pigs From Birth to Slaughter* (Wageningen, Netherlands: Wageningen Academic Publishers).

Brasil Notícia (2020). Após Covid-19 em funcionários, justiça determina que JBS teste empregados, www.brasilnoticia.com.br/agronegocio/apos-covid-19-em-funcionarios-justica-determina-que-jbs-teste-empregados/14173. In Portuguese.

Bray, F., Ferlay, J., Soerjomataram, I., et al. (2018). Global cancer statistics 2018: GLOBOCAN estimates of incidence and mortality worldwide for 36 cancers in 185 countries. *CA: A Cancer Journal for Clinicians,* 68(6):394-424.

Broom, D. M., Sena, H., Moynihan, K. L. (2009). Pigs learn what a mirror image represents and use it to obtain information. *Animal Behaviour,* 78(5):1037-1041.

Buchholz, K. (2020). The biggest producers of beef in the world. Statista, www.statista.com/chart/19127/biggest-producers-of-beef/.

CDIAC – Carbon Dioxide Information Analysis Center (2017). ESS-DIVE CDIAC data transition. Lawrence Berkeley Laboratory, cdiac.ess-dive.lbl.gov/.

Certified Humane (2020). Fact sheets, certifiedhumane.org/fact-sheet/.

Certified Humane Brasil (2021). Bem-estar dos suínos: 9 cuidados para a criação dos animais, certifiedhumanebrasil.org/9-cuidados-para-o-bem-estar-dos-suinos/. In Portuguese.

CIWF – Compassion in World Farming (2020a). Germany bans sow stalls, www.ciwf.org.uk/news/2020/07/germany-bans-sow-stalls.

CIWF – Compassion in World Farming (2020b). Pig welfare, www.ciwf.org.uk/farm-animals/pigs/pig-welfare/.

Claffey, N. (2018). Analysis: The Irish live cattle export market in 2018. Agriland, www.agriland.ie/farming-news/analysis-the-irish-live-cattle-export-market-in-2018/.

Clark, M. & Tilman, D. (2017). Comparative analysis of environmental impacts of agricultural production systems, agricultural input efficiency, and food choice. *Environmental Research Letters*, 12:064016.

Clarke, S. (2021). US bracing for record temperatures, www.nationnews.com/2021/07/11/us-bracing-record-temperatures/.

Cook, R. (2020). World cattle inventory: Ranking of countries. *Beef2Live,* beef2live.com/story-world-cattle-inventory-ranking-countries-0-106905.

Cox, D. (209), The planet's prodigious poo problem, *The Guardian,* www.theguardian.com/news/2019/mar/25/animal-waste-excrement-four-billion-tonnes-dung-poo-faecebook.

Christen, C. (2021). Investigation: How the meat industry is climate-washing its polluting business model. DeSmog UK, www.desmog.com/2021/07/18/investigation-meat-industry-greenwash-climatewash/.

Dalla Costa, O., Dalla Costa, F., Feddern, V., et al. (2019). Risk factors associated with pig pre-slaughtering losses. *Meat Science,* 155:61-68.

Daragahi, B. (2018). 'This one has heat stress': the shocking reality of live animal exports, *The Guardian,* www.theguardian.com/environment/2018/jul/30/this-one-has-heat-stress-the-shocking-reality-of-live-animal-exports.

Davis, W. (2018). Overflowing hog lagoons raise environmental concerns In North Carolina. *NPR,* www.npr.org/2018/09/22/650698240/hurricane-s-aftermath-floods-hog-lagoons-in-north-carolina?t=1596539841586.

Donnelly, M. (2020a). Meat factory workers 'petrifed' of catching Covid-19. *The Irish Independent,* www.independent.ie/business/farming/news/farming-news/meat-factory-workers-petrifed-of-catching-covid-19-39446346.html.

Donnelly, M. (2020b). Elderly meat factory vets have signed waivers to not sue if they contract Covid-19, *The Irish Independent,* www.independent.ie/business/farming/news/farming-news/elderly-meat-factory-vets-have-signed-waivers-to-not-sue-if-they-contract-covid-19-39447307.html.

Dorning, M. (2020). U.S. to buy milk and meat as part of $15.5 billion aid. *Bloomberg,* www.bloomberg.com/news/articles/2020-04-15/usda-wants-to-buy-milk-meat-to-shield-farms-from-virus-logjams.

EFSA – European Food Safety Authority (2017). The animal welfare aspects in respect of the slaughter or killing of pregnant livestock animals (cattle, pigs, sheep, goats, horses). *EFSA Journal,* 15(5):4782.

EFSA – European Food Safety Authority (2018). The European Union summary report on trends and sources of zoonoses, zoonotic agents and food-borne outbreaks in 2017. *EFSA Journal,* 16(12):5500.

EFSA – European Food Safety Authority (2020). Welfare of pigs at slaughter. *EFSA Journal,* 18(6):6148.

Embrapa – Empresa Brasileira de Pesquisa Agropecuária & ABCS – Associação Brasileira de Criadores de Suínos (2011). *Manual Brasileiro de Boas Práticas Agropecuárias na Produção de Suínos,* p. 140 (Brasília, Brasil: ABCS). In Portuguese.

Enríquez, D., Ungerfeld, R., Quintans, G., et al. (2010). The effects of alternative weaning methods on behaviour in beef calves. *Livestock Science,* 128:20-27.

Enríquez, D., Hötzel, M., & Ungerfeld, R. (2011). Minimising the stress of weaning of beef calves: a review. *Acta Veterinaria Scandinavica,* 53(28).

EPRS–European Parliamentary Research Service (2021). The EU pig meat sector, www.europarl.europa.eu/RegData/etudes/BRIE/2020/652044/EPRS_BRI(2020)652044_EN.pdf.

Extinction Rebellion (2020). 'We're looking at billions of people not being able to survive' | Peter Carter, Expert IPCC Reviewer. Online interview available at: www.youtube.com/watch?v=6VSE5ubpKhg.

Fangman, T. & Amass, S. (2007). Chapter 104 – Postpartum care of the sow and neonates. In Youngquist R. & Threlfall W. (Eds.), *Current therapy in large animal theriogenology* (Amsterdam, Netherlands: Elsevier).

FAO – Food and Agriculture Organization (2006). Livestock's long shadow – Environmental issues and options, www.fao.org/3/a0701e/a0701e00.htm.

Fiebrig, I., Bombardi, L., & Nepomuceno, P. (2020). Sars-CoV-2, suinocultura intensiva e a agricultura industrializada. *Le Monde Diplomatique Brasil*, diplomatique.org.br/sars-cov-2-suinocultura-intensiva-e-a-agricultura-industrializada/. In Portuguese.

Foer, J. S. (2019). *We Are the Weather: Saving the Planet Begins at Breakfast* (London, UK: Hamish Hamilton).

Forster, P., Forster, H., Evans, M., et al. (2020). Current and future global climate impacts resulting from COVID-19. *Nature Climate Change*, 10:913-919.

Fraga, P., Gerencsér, L., Lovas, M., et al. (2021). Who turns to the human? Companion pigs' and dogs' behaviour in the unsolvable task paradigm. *Animal Cognition*, 24:33-40.

Fretwell, P., Pritchard, H., Vaughan, D., et al. (2013). Bedmap2: improved ice bed, surface and thickness datasets for Antarctica. *Cryosphere*, 7:375-393.

Garza, E. (2014). Meat vs veg: An energy perspective. *The University of Vermont*, learn.uvm.edu/foodsystemsblog/2014/07/10/meat-vs-veg-an-energy-perspective/.

Gerber, P., Steinfeld, H., Henderson, B., et al. (2013). *Tackling climate change through livestock – A global assessment of emissions and mitigation opportunities.* (Rome, Italy: FAO).

Global Forest Atlas (2020). Soy agriculture in the Amazon Basin. *Yale University*, globalforestatlas.yale.edu/amazon/land-use/soy.

Greene, T. (2021). Poo overload: Northern Ireland could be forced to export a third of its animal waste, *The Guardian*, www.theguardian.com/environment/2021/jun/23/poo-overload-northern-ireland-could-be-forced-to-export-a-third-of-its-animal-waste.

Guarino, M., Sime, L., Schröeder, D., et al. (2020). Sea-ice-free Arctic during the Last Interglacial supports fast future loss. *Nature Climate Change*, 10:928-932.

Guimarães, D., Amaral, G., Maia, G., et al. (2017). Suinocultura: Estrutura da cadeia produtiva, panorama do setor no Brasil e no mundo e o apoio do BNDES. *Agroindústria|BNDES Setorial*, 45, p. 85-136. In Portuguese.

Haley, D., Bailey, D., & Stookey, J. (2005). The effects of weaning beef calves in two stages on their behavior and growth rate. *Journal of Animal Science*, 83(9):2205-2214.

Hamilton, T. (2007). Maximizing beef bull fertility and reproduction. *The Cattle Site*, www.thecattlesite.com/articles/838/maximizing-beef-bull-fertility-and-reproduction/.

Held, L. (2020). Industrial meat 101: Could large livestock operations cause the next pandemic?, civileats.com/2020/05/29/industrial-meat-101-could-large-livestock-operations-cause-the-next-pandemic/.

Horrell, R., A'Ness, P., Edwards, S., et al. (2001). The use of nose-rings in pigs: Consequences for rooting, other functional activities, and welfare. *Animal Welfare*, 10(1):3-22(20).

Humane Choice (2020). Unravelling Coles' cage & sow stall free advertising, www.humanechoice.com.au/coles.

IATP – Institute for Agriculture and Trade Policy & GRAIN (2018). Emissions impossible. How big meat and dairy are heating up the planet. Global Agriculture, www.globalagriculture.org/fileadmin/files/weltagrarbericht/Weltagrarbericht/04Fleisch/2018GRAINIATPEmissionsimpossible.pdf.

Imfeld-Mueller, S., Wezemael, L., Stauffacher, M., et al. (2011). Do pigs distinguish between situations of different emotional valences during anticipation? *Applied Animal Behaviour Science*, 131(3-4):86-93.

IPCC – Intergovernmental Panel on Climate Change (2013). Climate change 2013: The physical science basis. Contribution of Working Group I to the Fifth Assessment Report of the Intergovernmental Panel on Climate Change [Stocker, T., Qin D., Plattner, K., et al. (Eds.)], pp. 15-35 (Cambridge, UK: Cambridge University Press).

Janjevic, D. (2020). Coronavirus: Gütersloh mayor Henning Schulz slams Tönnies meat producer after massive outbreak forces lockdown. *Deutsche Welle*, www.dw.com/en/coronavirus-g%C3%BCtersloh-mayor-henning-schulz-slams-t%C3%B6nnies-meat-producer-after-massive-outbreak-forces-lockdown/a-53932101.

Kauffman, M. (2001). In pig farming, growing concern raising sows in crates is questioned. *Washington Post*, web.archive.org/web/20110724013229/http://www.pmac.net/AM/pigs_in_crates.html.

Keaveny, M. (2020). Covid outbreaks at meat plants for discussion at Oireachtas committee, *The Irish Independent*, www.independent.ie/business/farming/news/farming-news/covid-outbreaks-at-meat-plants-for-discussion-at-oireachtas-committee-39443941.html.

Kevany, S. (2020). 'Sending animals to war zones': Irish cattle export to Libya may breach laws, *The Guardian*, www.theguardian.com/environment/2020/jul/24/sending-animals-to-war-zones-irish-cattle-export-to-libya-may-breach-laws.

Kiernan, A. (2020). Union officials portray a day in the life of a meat worker. *The Irish Examiner*, www.irishexaminer.com/farming/arid-40034282.html.

KilBride, A., Mendl, M., Statham, P., et al. (2012). A cohort study of preweaning piglet mortality and farrowing accommodation on 112 commercial pig farms in England. *Preventive Veterinary Medicine*, 104(3-4):281-291.

Knox, R. (2016). Artificial insemination in pigs today. *Theriogenology*, 85(1):83-93.

Koneswaran, G., Nierenberg, D (2008). Global farm animal production and global warming: impacting and mitigating climate change. *Environmental Health Perspectives*, 116(5):578-582.

Kość, W. & Standaert, M. (2020). African swine fever outbreak reported in western Poland, *The Guardian*, www.theguardian.com/environment/2020/apr/08/african-swine-fever-outbreak-reported-in-western-poland.

Lazarus, O., McDermid, S., & Jacquet, J. (2021). The climate responsibilities of industrial meat and dairy producers. *Climatic Change*, 165(30).

Lorenz, I., Fagan, J., & More, S. (2011a). Calf health from birth to weaning. II. Management of diarrhoea in pre-weaned calves. *Irish Veterinary Journal*, 64(1):9.

Lorenz, I., Earley, B., Gilmore, J., et al. (2011b). Calf health from birth to weaning. III. Housing and management of calf pneumonia. *Irish Veterinary Journal*, 64(14).

Loh, J., Zhao, G., Presti, R., et al. (2009). Detection of novel sequences related to African Swine Fever virus in human serum and sewage. *Journal of Virology*, 83(24):13019-13025.

Matarneh, S., England, E., Scheffler, T. L., et al. (2017). Chapter 5 – The conversion of muscle to meat. In Toldra, F. (Ed.), *Lawrie's Meat Science* (Cambridge, UK: Woodhead Publishing).

May, J. & Bates, R. (2007). Managing the Sow and Gilt Estrous Cycle. *The Pig Site*, www.thepigsite.com/articles/managing-the-sow-and-gilt-estrous-cycle.

McLennan, K. (2012). Farmyard friends. *The Biologist*, 59(4):18-22.

Miranda-de la Lama, M., Genaro, C., Sepulveda, W., et al. (2011). Livestock vehicle accidents in Spain: Causes, consequences, and effects on animal welfare. *Journal of Applied Animal Welfare*, 14(2):109-123.

Molony, S. (2020a). Health and Safety Authority 'not notified' of Covid-19 cases in meat plants, Dáil committee hears, *The Irish Independent*, www.independent.ie/irish-news/politics/health-and-safety-authority-not-notified-of-covid-19-cases-in-meat-plants-dail-committee-hears-39447011.html.

Molony, S. (2020b). Meat plants will no longer be warned in advance of health-and-safety inspections, *The Irish Independent*, www.independent.ie/world-news/coronavirus/meat-plants-will-no-longer-be-warned-in-advance-of-health-and-safety-inspections-39448867.html.

Mood, A. (2010). Worse things happen at sea: the welfare of wild caught-fish. *fishcount*, fishcount.org.uk/publications.

Morrison, R. (2020). Britain's obsession with timber, leather and beef 'is having a heavy impact' on the Amazon rainforest and contributing to wildfires, charities say, *Daily Mail*, www.dailymail.co.uk/sciencetech/article-8507103/Britains-obsession-beef-impacting-Amazon-rainforest-wildfires.html.

NAWAC – New Zealand National Animal Welfare Advisory Committee (2016). NAWAC review of the use of farrowing crates for pigs in New Zealand. Ministry for Primary Industries, www.mpi.govt.nz/dmsdocument/11959.

Notz, D. & SIMIP Community (2020). Arctic sea ice in CMIP6. *Geophysical Research Letters*, 47(10):e2019GL086749.

OECD – Organisation for Economic Co-operation and Development (2020). Meat consumption, data.oecd.org/agroutput/meat-consumption.htm.

OECD-FAO – Organisation for Economic Co-operation and Development & Food and Agriculture Organization (2020). OECD-FAO Agricultural Outlook 2020-2029, www.agri-outlook.org/.

O'Brien, T. (2017). Irish cattle slaughtered in conditions 'breaching EU law, *The Irish Times*, www.irishtimes.com/news/ireland/irish-news/irish-cattle-slaughtered-in-conditions-breaching-eu-law-1.3033583.

Open Secrets (2019). Meat processing & products: Lobbying, 2019, www.opensecrets.org/industries/lobbying.php?cycle=2018&ind=G2300.

Patton, D. (2018). China supersizes pig farms to cut costs in world's top pork market. *Reuters*, www.reuters.com/article/us-china-pigs-production-idUSKBN1FQ39W.

Pedersen, L. (2018). Chapter 1 – Overview of commercial pig production systems and their main welfare challenges. In Špinka, M. (Ed.), *Advances in Pig Welfare* (Cambridge, UK: Woodhead Publishing).

Perry, G. & Walker, J. (2008). Beef reproduction and management specialist. Reproductive fertility in herd bulls. *The Beef Site*, www.thebeefsite.com/articles/1491/reproductive-fertility-in-herd-bulls/.

Petersen, S., Olsen, A., Elsgaard, L., et al. (2016). Estimation of methane emissions from slurry pits below pig and cattle confinements. *PLoS ONE*, 11(8): e0160968.

Poore, J. & Nemecek, T. (2018). Reducing food's environmental impacts through producers and consumers. *Science*, 360:987-992.

Portal Rondônia (2020). 60% dos casos de Covid-19 em São Miguel do Guaporé são de funcionários da empresa JBS, www.portalrondonia.com/2020/05/27/60-dos-casos-de-covid-19-em-sao-miguel-do-guapore-sao-de-funcionarios-da-empresa-jbs/. In Portuguese.

Potenteau, D. (2019). 'We tried saving as many as we could': Truck driver hauling pigs in B.C. highway crash defends efforts. *Global News*, globalnews.ca/news/4819361/bc-pigs-crash-truck-driver/.

Price, E., Harris, J., Borgwardt, R., et al. (2003). Fenceline contact of beef calves with their dams at weaning reduces the negative effects of separation on behavior and growth rate. *Journal of Animal Science*, 81:116-121.

Pulz, R. (2020). Transporte marítimo de bovinos por longas distâncias. *VEG VETS*, vegvets.com/transporte-maritimo-de-bovinos-por-longas-distancias/.

Ranganathan, J., Vennard, D., Waite, R., et al. (2016). Shifting diets for a sustainable food future. *Creating a Sustainable Food Future, 11* (Washington, US: World Resources Institute).

Rao, S., Jain, A., & Shu, S. (2015). The lifestyle carbon dividend: Assessment of the carbon sequestration potential of grasslands and pasturelands reverted to native forests. AGU Fall Meeting, 14-18 December 2015.

Rao, S. (2021). Animal agriculture is the leading cause of climate change – A position paper. *Journal of the Ecological Society*, 32-33:155-167.

Red Tractor (2017). Pigs standards, /assurance.redtractor.org.uk/standard-categories/pigs/.

Reinhardt, V. & Reinhardt, A. (1981). Natural sucking performance and age of weaning in zebu cattle (*Bos indicus*). *Journal of Agricultural Science*, 96:309-312.

Reinhardt, V. (1983a). Reproduktionsdaten einer halb-wilden Rinderherde. Verhandlungsbericht der 7. *Veterinärmedizinischen Gemeinschaftstagung*: 124-126. In German.

Reinhardt, V. (1983b). Flehmen, mounting and copulation among members of a semi-wild cattle herd. *Animal Behaviour*, 31:641-650.

Reuters (2020). China's 2019 pork output plunges to 16-year low as disease culls herd. *Reuters*, www.reuters.com/article/us-china-economy-output-pork-idUSKBN1ZG08H.

Riggs, B., Mueller, C., & Cooke, R. (2019). Weaning management of beef calves. *Oregon State University*, extension.oregonstate.edu/animals-livestock/beef/weaning-management-beef-calves.

Ritchie, H. & Roser, M. (2019). Meat and dairy production. *Our World in Data*, ourworldindata.org/meat-production.

Ritchie, H. (2020). Environmental impacts of food production. *Our World in Data*, ourworldindata.org/environmental-impacts-of-food.

Rivera, I. (2017). The truth behind the pork we eat, *The Independent*, www.independent.co.uk/life-style/food-and-drink/pork-production-truth-pig-farming-uk-factory-hughfearnley-whittingstall-sienna-miller-mick-jagger-a7813746.html.

RSPCA Assured – Royal Society for the Prevention of Cruelty to Animals (2020). Key pig welfare problems and how RSPCA Assured help, www.rspcaassured.org.uk/.

RSPCA Australia – Royal Society for the Prevention of Cruelty to Animals (2018). RSPCA approved farming scheme standards, rspcaapproved.org.au/wp-content/uploads/2018/11/2018-11_PIGS_Standards.pdf.

RSPCA Australia – Royal Society for the Prevention of Cruelty to Animals (2020). What is the difference between, free range, outdoor bred, organic, sow-stall free, RSPCA Approved? kb.rspca.org.au/knowledge-base/what-is-the-difference-between-free-range-outdoor-bred-organic-sow-stall-free-rspca-approved/#outdoor-bred-pork.

Samuel, S. (2020). A Utah meat plant is staying open even after 287 workers got coronavirus. *Vox*, www.vox.com/future-perfect/2020/6/11/21286840/meat-plant-covid-19-utah-coronavirus.

Santé-Lhoutellier, V. & Monin, G. (2014). Conversion of muscle to meat | Slaughter-line operation and pig meat quality. In Dikeman, M. & Devine, C. (Eds.), *Encyclopedia of Meat Sciences* (Cambridge, US: Academic Press).

Scown, M., Brazy, M., & Nicholas, K. (2020). Billions in misspent EU agricultural subsidies could support the Sustainable Development Goals. *One Earth*, 3:237-250.

Searchinger, T., Wirsenius, S., Beringer, T., et al. (2018). Assessing the efficiency of changes in land use for mitigating climate change. *Nature*, 564:249-253.

Selistre, A. (2018). A verdade sobre a exportação de gado vivo. *BeefPoint*, www.beefpoint.com.br/a-verdade-sobre-a-exportacao-de-gado-vivo-por-alexandre-valente-selistre/. In Portuguese.

Simon, D. (2013). *Meatonomics: How the Rigged Economics of Meat and Dairy Make You Consume Too Much—And How to Eat Better, Live Longer, and Spend Smarter* (Newburyport, US: Conari Press).

Shahbandeh, M. (2021). Global number of pigs 2012-2021. Statista, www.statista.com/statistics/263963/number-of-pigs-worldwide-since-1990/.

Shepherd, A., Ivins, E., Rignot, E., et al. (The IMBIE team) (2018). Mass balance of the Antarctic ice sheet from 1992 to 2017. *Nature*, 558:219-222.

Shindell, D., Faluvegi, G., Koch, D., et al. (2009). Improved attribution of climate forcing to emissions. *Science*, 326: 716-718.

Soares, J., Neves, D., & De Carvalho, J. (2014). Produção de carne bovina em sistema orgânico: desafios e tecnologias para um mercado em expansão. *Embrapa Cerrados*, /www.embrapa.br/busca-de-publicacoes/-/publicacao/1002261/producao-de-carne-bovina-em-sistema-organico-desafios-e-tecnologias-para-um-mercado-em-expansao. In Portuguese.

Soil Association (2020). Farming & Growing Standards, www.soilassociation.org/our-standards/read-our-organic-standards/farming-growing-standards/.

Springmann, M, Wiebe, K., Mason-D'Croz, D., et al. (2018a). Health and nutritional aspects of sustainable diet strategies and their association with environmental impacts: a global modelling analysis with country-level detail. *Lancet Planet Health*, 2:e451-61.

Springmann, M., Clark, M., Mason-D'Croz, D., et al. (2018b). Options for keeping the food system within environmental limits. *Nature*, 562:519-525.

Standaert, M. (2020). African swine fever destroying small pig farms, as factory farming booms – report, *The Guardian*, www.theguardian.com/environment/2020/mar/11/african-swine-fever-destroying-small-pig-farms-as-factory-farming-booms-report.

Statham, J. (2015). Breeding in cattle reproduction. Merck Sharp & Dohme (MSD), www.msdvetmanual.com/management-and-nutrition/management-of-reproduction-cattle/breeding-in-cattle-reproduction.

Strokal, M., Ma, L., Bai, Z., et al. (2016). Alarming nutrient pollution of Chinese rivers as a result of agricultural transitions. *Environmental Research Letters*, 11(2):024014.

Teagasc – Irish Agriculture and Food Development Authority (2020). Castration – Best practice, www.teagasc.ie/publications/2020/castration—best-practice.php.

Terra Brasilis (2021). Taxas de desmatamento na Amazônia Legal, terrabrasilis.dpi.inpe.br/app/dashboard/deforestation/biomes/legal_amazon/rates. In Portuguese.

Terras Indígenas no Brasil (2021a). Terras indígenas por jurisdição legal, terrasindigenas.org.br/pt-br/brasil. In Portuguese.

Terras Indígenas no Brasil (2021b). Brasil é citado na ONU como caso de 'risco de genocídio' de índios, terrasindigenas.org.br/pt-br/noticia/212343. In Portuguese.

The Courtyard Dairy (2017). Grass fed cheese, meat and dairy – What is it and who does it? www.thecourtyarddairy.co.uk/blog/cheese-musings-and-tips/grass-fed-pasture-milk-cheese/.

The Local (2021). Death toll rises to 133 after flood disaster in western Germany, www.thelocal.de/20210716/latest-death-toll-in-western-german-floods-rises-to-at-least-81/.

The Netherland's Environmental Data Compendium (2021). Manure production at application standards: farms with overproduction, 2000-2020, www.clo.nl/indicatoren/nl0528-mestproductie-bij-gebruiksnormen-bedrijven-met-overproductie.

Thring, O. (2012). Is Red Tractor pork really 'high welfare'? *The Guardian*, www.theguardian.com/lifeandstyle/wordofmouth/2012/jan/26/is-red-tractor-pork-high-welfare.

USDA – United States Department of Agriculture (2016a). Swine 2012. Part II: Reference of swine health and health management in the United States, 2012. Revised Feb 2020, www.aphis.usda.gov/animal_health/nahms/swine/downloads/swine2012/Swine2012_dr_PartII_revised.pdf/.

USDA – United States Department of Agriculture (2016b). Grades and standards – Organic, www.ams.usda.gov/grades-standards/organic-standards.

USDA – United States Department of Agriculture (2020). Agriculture and food sectors and the economy, www.ers.usda.gov/data-products/ag-and-food-statistics-charting-the-essentials/ag-and-food-sectors-and-the-economy.aspx.

Velarde, A. & Dalmau, A. (2018a). Chapter 10 – Slaughter of pigs. In Špinka, M. (Ed.), *Advances in Pig Welfare* (Cambridge, UK: Woodhead Publishing).

Velarde, A. & Dalmau, A. (2018b). Chapter 12 – Slaughter without stunning. In Mench, J. (Ed.), *Advances in Pig Welfare* (Cambridge, UK: Woodhead Publishing).

Viva! (2016). Pig farming: The inside story. A Viva! Report by Juliet Gellatley, BSc Zoology, Founder & Director, cdn.viva.org.uk/wp-content/uploads/2020/03/pig-report.pdf.

Viva! (2020). Pig meat standards schemes, www.viva.org.uk/pig-farming-report/pig-meat-standard-schemes.

Wahlquist, C. (2019). Live export footage shows Australian cattle dragged by ropes before slaughter in Indonesia, *The Guardian,* www.theguardian.com/australia-news/2019/sep/11/live-exports-footage-australian-cattle-dragged-ropes-slaughter-indonesia.

Wasley, A., Harvey, F., Davies, M., et al. (2017). UK has nearly 800 livestock mega farms, investigation reveals, *The Guardian,* www.theguardian.com/environment/2017/jul/17/uk-has-nearly-800-livestock-mega-farms-investigation-reveals.

WHO – World Health Organization & IARC – International Agency for Research on Cancer (2015). IARC monographs evaluate consumption of red meat and processed meat. Press release n. 240 of 26 October 2015, www.iarc.who.int/wp-content/uploads/2018/07/pr240_E.pdf.

Willett, W., Rockström, J., Loken, B., et al. (2019). Food in the Anthropocene: the EAT–Lancet Commission on healthy diets from sustainable food systems. *The Lancet Commissions,* 393.

Willett, W. & Ludwig, D. (2020). Milk and health. *The New England Journal of Medicine,* 382:644-654.

World Animal Protection (2020). Quit stalling: Are companies making good on promises to end sow confinement? dkt6rvnu67rqj.cloudfront.net/cdn/ff/TG_6t-y4ZXJSt50Dsyq2RdZ-zzXOWFn61xw4j05_EYE/1600450508/public/media/quit-stalling-2020.pdf.

World Bank (2019). Why the Amazon's biodiversity is critical for the globe: An interview with Thomas Lovejoy, www.worldbank.org/en/news/feature/2019/05/22/why-the-amazons-biodiversity-is-critical-for-the-globe.

WWF – World Wildlife Fund (2016). The story of soy, www.worldwildlife.org/stories/the-story-of-soy.

Zhang, M., Li, X., Zhang, X., et al. (2017). Effects of confinement duration and parity on stereotypic behavioral and physiological responses of pregnant sows. *Physiology & Behavior,* 179:369-376.

Chapter 3 References

ABNT – Brazilian Association of Technical Standards (2015). Avicultura - Produção, abate, processamento e identificação do frango caipira, colonial ou capoeira, www.abnt.org.br/noticias/4402-avicultura-producao-abate-processamento-e-identificacao-do-frango-caipira-colonial-ou-capoeira. In Portuguese.

ABNT – Associação Brasileira de Normas Técnicas (2019). Dossiê técnico. Ovo caipira, www.abnt.org.br/paginampe/biblioteca/files/upload/anexos/pdf/6004589acb595833bbccc9f09fa09f18.pdf. In Portuguese.

Aboubakr, H., Sharafeldin, T., & Goyal, S. (2020). Stability of SARS-CoV-2 and other coronaviruses in the environment and on common touch surfaces and the influence of climatic conditions: A review. *Transboundary and Emerging Diseases,* 68(2):296-312.

Abramovay, R. (2020). Um setor infectado: propostas para as doenças do sistema agroalimentar. *UOL,* tab.uol.com.br/colunas/ricardo-abramovay/2020/06/05/trabalhadores-infectados-para-que-a-carne-chegue-a-sua-mesa.htm. In Portuguese.

Aderibigbe, A., Cowieson, A., Sorbara, J., et al. (2020). Growth performance and amino acid digestibility responses of broiler chickens fed diets containing purified soybean trypsin inhibitor and supplemented with a monocomponent protease. *Poultry Science,* 99(10):5007-5017.

AEB – American Egg Board (2020). About us, www.incredibleegg.org/about-us/.

Agriculture Fairness Alliance (2020). Fiscal conservatism, www.agriculturefairness-alliance.org/mission.

Ahammed, M., Chae, B., Lohakare, J., et al. (2014). Comparison of aviary, barn and conventional cage raising of chickens on laying performance and egg quality. *Asian-Australasian Journal of Animal Sciences*, 27(8):1196-1203.

Akiba, Y., Takahashi, K., Kimura, M., et al. (1983). The influence of environmental temperature, thyroid status and a synthetic oestrogen on the induction of fatty livers in chicks. *British Poultry Science*, 24:71-80.

Aljazeera (2020). COVID-19 in US poultry plants, interactive.aljazeera.com/aje/2020/covid-19-in-us-poultry-plants/index.html.

Almuhanna, E., Ahmed, A., & Al-Yousif, Y. (2011). Effect of air contaminants on poultry immunological and production performance. *International Journal of Poultry Science*, 10(6):461-470.

Andreatti Filho, R., Milbradt, E., Okamoto, A., et al. (2019). *Salmonella* Enteritidis infection, corticosterone levels, performance and egg quality in laying hens submitted to different methods of molting. *Poultry Science*, 98(10):4416-4425.

Animal Equality (2020a). Spain working toward ban on live chick shredding in egg industry, animalequality.org/blog/2020/03/06/spain-egg-industry-chick-culling/.

Animal Equality (2020b). Investigation: India's egg industry, animalequality.org/news/investigation-indias-egg-industry/.

Animal Equality (2020c). Chicks routinely killed with worker's bare hands at major chicken supplier, animalequality.org/news/moy-park-chicken-cruelty-investigation/.

Animal Equality (2020d). Organic egg farms exposed!, animalequality.org/news/organic-egg-farms-exposed/.

Armitage, J. (2017). Jim Armitage on Tesco: Willow Farms' branding just makes us customers wary. *The Evening Standard*, www.standard.co.uk/business/jim-armitage-on-tesco-willow-farms-branding-just-makes-us-customers-wary-a3650436.html.

Aury, K., Le Bouquin, S., Toquin, M., et al. (2011). Risk factors for *Listeria monocytogenes* contamination in French laying hens and broiler flocks. *Preventive Veterinary Medicine*, 98(4):271-278.

Averós, X., Balderas, B., Cameno, E., et al. (2020). The value of a retrospective analysis of slaughter records for the welfare of broiler chickens. *Poultry Science*, 99(11):5222-5232.

AVMA – American Veterinary Animal Association (2020). AVMA guidelines for the euthanasia of animals.

AWC – Animal Welfare Committee of the Australian Government (2002). *Model Code of Practice for the Welfare of Animals: Domestic Poultry* (Clayton, Australia: CSIRO Publishing).

Barnard, N., Long, M., Ferguson, J., et al. (2019). Industry funding and cholesterol research: A systematic review. *American Journal of Lifestyle Medicine*, 15(2):165-172.

Batkowska, J., Drabik, K., Brodacki, A., et al. (2021). Fatty acids profile, cholesterol level and quality of table eggs from hens fed with the addition of linseed and soybean oil. *Food Chemistry*, 334(1):127612.

BBC News (2020). Coronavirus: 200 cases at 2 Sisters meat plant outbreak in Llangefni, www.bbc.com/news/uk-wales-politics-53152362.

Bedanova, I., Voslarova, E., Chloupek, P., et al. (2007). Stress in broilers resulting from shackling. *Poultry Science*, 86(6):1065-1069.

Bellows, L. & Moore, R. (2012). Dietary fat and cholesterol. *Fact Sheet No. 9.319. Colorado State University*, mountainscholar.org/bitstream/handle/10217/195108/AEXT_093192012.pdf.

Belz, A. (2020). Egg demand shifted, and 61,000 Minnesota chickens were euthanized. *The Star Tribune*, www.startribune.com/egg-demand-shifted-and-61-000-minnesota-chickens-were-euthanized/569817312/.

Benson, E., Malone, G., Alphin, R., et al. (2007). Foam-based mass emergency depopulation of floor-reared meat-type poultry operations. *Poultry Science*, 86:219-224.

Berger, S., Raman, G., Vishwanathan, R., et al. (2015). Dietary cholesterol and cardiovascular disease: a systematic review and meta-analysis. *American Journal of Clinical Nutrition*, 102:276-294.

Bertran, K., Susta, l., & Miller, P. (2017). Chapter 51 – Avian Influenza virus and Newcastle Disease virus. In Hester, P. (Ed.), *Egg Innovations and Strategies for Improvements* (Cambridge, US: Academic Press).

Bessei, W. (2006). Welfare of broilers: A review. *World's Poultry Science Journal*, 62(3):455-466.

Blezinger, S. (2008). When feeding vitamins, a little can go a long way. *Cattle Today*, www.cattletoday.com/archive/2008/October/CT1772.shtml.

Bourassa, D., Bowker, B., Zhuang, H., et al. (2017). Impact of alternative electrical stunning parameters on the ability of broilers to recover consciousness and meat quality. *Poultry Science*, 96(9):3495-3501.

Bracke, M., Vermeer, H., & van Emous, R. (2019). Animal welfare regulations and practices in 7 (potential) trade-agreement partners of the EU with a focus on laying hens, broilers and pigs: Mexico, Chile, Indonesia, Australia, New Zealand, Turkey and the Philippines. Wageningen Livestock Research.

Bryden, W., Cumming, R., & Balnave, D. (1979). The influence of vitamin A status on the response of chickens to aflatoxin B1 and changes in the liver metabolism associated with aflatoxicosis. *British Journal of Nutrition*, 41:529-540.

Bureau of Labour Statistics of the United States of America (2018). Injuries, illnesses and fatalities, www.bls.gov/iif/oshwc/osh/os/summ1_00_2018.htm.

Carne e Osso (2011). Um retrato do trabalho nos frigoríficos brasileiros, carneosso.reporterbrasil.org.br/o-filme/index.html. In Portuguese.

Casey-Trott, T., Guerin, M., Sandilands, V., et al. (2017). Rearing system affects prevalence of keel-bone damage in laying hens: a longitudinal study of four consecutive flocks. *Poultry Science*, 96(7):2029-2039.

CDC – Centers for Disease Control and Prevention of the United States of America (2019). Salmonella, www.cdc.gov/salmonella/index.html.

CDC – Centers for Disease Control and Prevention of the United States of America (2020a). Outbreak of *Listeria* infections linked to hard-boiled eggs, www.cdc.gov/listeria/outbreaks/eggs-12-19/index.html.

CDC – Centers for Disease Control and Prevention of the United States of America (2020b). Update: COVID-19 among workers in meat and poultry processing facilities. Morbidity and Mortality Weekly Report (MMWR), 69(27):887-892, www.cdc.gov/mmwr/volumes/69/wr/mm6927e2.htm.

CDFA – California Department of Food and Agriculture (2020). Virulent Newcastle disease, www.cdfa.ca.gov/ahfss/Animal_Health/newcastle_disease_info.html.

Certified Humane Brasil (2017). Ovo caipira ou orgânico? Quais as diferenças e o que isso tem a ver com o bem-estar animal? certifiedhumanebrasil.org/ovo-caipira-ou-organico-quais-as-diferencas/. In Portuguese.

Certified Humane (2020). Regras de uso do selo Certified Humane®, certifiedhumanebrasil.org/quem-pode-usar-a-marca-de-certificacao-de-bem-estar-animal-certified-humane/. In Portuguese.

Chemaly, M., Toquin, M., Le Nôtre, Y., et al. (2008). Prevalence of *Listeria monocytogenes* in poultry production in France. *Journal of Food Protection*, 71(10):1996-2000.

Chin, A., Chu, J., Perera, M., et al. (2020). Stability of SARS-CoV-2 in different environmental conditions. *The Lancet Microbe*, 1(1):e10.

CIWF – Compassion in World Farming (2020). Chickens farmed for meat, www.ciwf.org.uk/farm-animals/chickens/meat-chickens/.

Cordero, T. (2020). Philippines bans chicken imports from Brazil amid reports of COVID-19 contamination, *GMA News Online*, www.gmanetwork.com/news/news/nation/751399/philippines-bans-chicken-imports-from-brazil-amid-reports-of-covid-19-contamination/story/.

Curry, L. (2017). Certified "organic" livestock are supposed to have outdoor access. In practice, they don't. *The Counter*, thecounter.org/usda-organic-animal-welfare-rule-livestock-poultry-practices/.

da Silva, I., Broch, J., Wachholz, L., et al. (2019). Dry residue of cassava associated with carbohydrases in diets for broiler chickens. *Journal of Applied Poultry Research*, 28(4):1189-1201.

Dawkins, M. & Hardie, S. (1989). Space needs of laying hens. *British Poultry Science*, 30:413-416.

De Carvalho, S., Barros, M., & Bastos, F. (2013). Resíduos da produção de frangos de corte: incubatório. *III Simpósio Internacional sobre Gerenciamento de Resíduos Agropecuários e Agroindustriais*. www.sbera.org.br/3sigera/obras/ag_tec_04_SabrinaCarvalho.PDF (São Pedro, Brasil: SIGRA). In Portuguese.

De Jong, I., van Voorst, A., & Blokhuis, H. (2003). Parameters for quantification of hunger in broiler breeders. *Physiology & Behavior*, 78(4-5):773-783.

De Jong, I. & Guémené, D. (2011). Major welfare issues in broiler breeders. *World's Poultry Science Journal*, 67(1):73-82.

De Lima, V., Ceballos, M., Gregory, N., et al. (2019). Effect of different catching practices during manual upright handling on broiler welfare and behavior. *Poultry Science*, 98(10):4282-4289.

Deutsche Welle (2020). Germany to delay ban on slaughter of male chicks. *Deutsche Welle*, www.dw.com/en/germany-to-delay-ban-on-slaughter-of-male-chicks/a-54870048.

Dickie, D. (2020). Investigation into Coupar Angus 2 Sisters coronavirus outbreak is wound down. *The Daily Record*, www.dailyrecord.co.uk/news/local-news/investigation-coupar-angus-2-sisters-22621324.

Douglas, L. (2020). Mapping Covid-19 outbreaks in the food system. *The Food & Environment Reporting Network (FERN)*, thefern.org/2020/04/mapping-covid-19-in-meat-and-food-processing-plants/.

DSM (2011). DSM vitamin supplementation guidelines 2011 for domestic animals, www.dsm.com/content/dam/dsm/anh/en_US/documents/OVN_supplementation_guidelines.pdf.

Duizer, E. & Koopmans, M. (2008). Chapter 3 – Emerging food-borne viral diseases. In Koopmans, M., Cliver, D., Bosch, A., et al (Eds.), *Food-borne viruses: Progress and challenges* (Hoboken, US: Wiley).

EFSA – European Food Safety Authority (2019a). Killing for purposes other than slaughter: poultry. *EFSA Journal*, 17(11):5850.

EFSA – European Food Safety Authority (2019b). The European Union One Health 2018 zoonoses report. *EFSA Journal*, 17(12):5926.

Egg Info (2020a). Organic egg production, www.egginfo.co.uk/egg-facts-and-figures/production/organic-egg.

Egg Info (2020b). Lion Code of Practice, www.egginfo.co.uk/british-lion-eggs/lion-code-practice.

El-Katcha, M., Soltan, M., El-Naggar, K., et al. (2019). Laying performance, fat digestibility and liver condition of laying hens supplemented with vitamin B12 or biotin and or/bile acids in diet. *Slovenian Veterinary Research*, 56(Suppl 22):341-352.

Ennaji, Y., Khataby, K., & Ennaji, M. (2020). Chapter 3 – Infectious bronchitis virus in poultry: Molecular epidemiology and factors leading to the emergence and reemergence of novel strains of infectious bronchitis virus. In Ennaji, M. (Ed.), *Emerging and Reemerging Viral Pathogens* (Cambridge, US: Academic Press).

Erickson, A. (2019). Cobalt deficiency in sheep and cattle. Department of Primary Industries and Regional Development, www.agric.wa.gov.au/livestock-biosecurity/cobalt-deficiency-sheep-and-cattle.

European Commission (1999). European Union Council Directive 1999/74/EC. Official Journal of the European Communities. European Union.

European Commission (2017). New rules extend marketing standards for free range eggs hit by avian flu restrictions, ec.europa.eu/info/news/new-rules-extend-marketing-standards-free-range-eggs-hit-avian-flu-restrictions_en.

European Commission (2021). EU market situation for eggs, circabc.europa.eu/sd/a/18f7766e-e9a9-46a4-bbec-94d4c181183f/23.03.2017_eggs_Europa.pdf.

FAO – Food and Agriculture Organization of the United Nations (2003). Egg marketing – A guide for the production and sale of eggs. ISSN 1010-1365. FAO Agricultural Services Bulletin 150.

FAO – Food and Agriculture Organization of the United Nations & and WHO – World Health Organization (2020). COVID-19 and food safety: guidance for food businesses. Interim guidance, 7 April 2020.

Farm Animal Welfare Council (2010). Opinion on osteoporosis and bone fractures in laying hens, www.gov.uk/government/publications/fawc-opinion-on-osteoporosis-and-bone-fractures-in-laying-hens.

FDA – Food and Drug Administration of the United States of America (2018). Trans fats, www.fda.gov/food/food-additives-petitions/trans-fat.

FDA – Food and Drug Administration of the United States of America (2019a). *Escherichia coli* (*E. coli*), www.fda.gov/food/foodborne-pathogens/escherichia-coli-e-coli.

FDA – Food and Drug Administration of the United States of America (2019b). *Salmonella* (Salmonellosis), www.fda.gov/food/foodborne-pathogens/salmonella-salmonellosis.

Fisher, D., Reilly, A., Zheng, A., et al. (2020). Seeding of outbreaks of COVID-19 by contaminated fresh and frozen food. *BioRxiv*, https://doi.org/10.1101/2020.08.17.255166.

Flock, D. & Anderson, K. (2016). Molting of laying hens: test results from North Carolina and implications for US and German egg producers. *Lohmann Information*, 50:12-17.

FSAI – Food Safety Authority of Ireland (2018). Marketing standards for poultrymeat, www.fsai.ie/legislation/food_legislation/poultrymeat/marketing_standards.html

Fulton, R. (2017). Causes of normal mortality in commercial egg-laying chickens. *Avian Diseases*, 61(3):289-295.

Garcia, E., Mendes, A., Pizzolante, C., et al. (2001). Alterações morfológicas de codornas poedeiras submetidas a muda forçada. *Brazilian Journal of Poultry Science*, 3(3):265-273. In Portuguese.

Geske, D. (2019). Egg recall 2019: 88 egg products recalled from Kroger, Giant Eagle, Walmart, ShopRite, and more. *International Business Times*, www.ibtimes.com/egg-recall-2019-88-egg-products-recalled-kroger-giant-eagle-walmart-shoprite-more-2891685.

GGN (2015). Friboi é campeã em acidentes de trabalho em frigoríficos, jornalggn.com.br/trabalho/friboi-e-campea-em-acidentes-de-trabalho-em-frigorificos/. In Portuguese.

Gingerich, E. (2016). *E. coli* infections and peritonitis in egg layers. Dutchland Farms, www.dutchlandfarms.com/e-coli-infections-and-peritonitis-in-egg-layers/.

Gongruttananun, N. & Saengkudrua, K. (2016). Responses of laying hens to induce molting procedures using cassava meal of variable length with or without recovery period. *Agriculture and Natural Resources*, 50(5):400-407.

Glatz, P. & Bourke, M. (2006). Beak Trimming Handbook for Egg Producers. Best Practice for Minimising Cannibalism in Poultry. ISBN: 9780643093539 (Collingwood, Australia: Landlinks Press).

Glatz P. & Tilbrook A. (2020). Welfare issues associated with moulting of laying hens. *Animal Production Science*, 61(10):1006-1012.

Greene, J. & Cowan, T. (2014). Table egg production and hen welfare: Agreement and legislative proposals. Congressional Research Service, fas.org/sgp/crs/misc/R42534.pdf.

Greening, G. & Cannon, J. (2016). Human and animal viruses in food (including taxonomy of enteric viruses). In Goyal, S. & Cannon, J. (Eds.), *Viruses in Foods: Food Microbiology and Food Safety* (Switzerland: Springer).

Griffin, J. & Lichtenstein, A. (2013). Dietary cholesterol and plasma lipoprotein profiles: randomized-controlled trials. *Current Nutrition Reports*, 2:274-282.

Häne, M., Huber-Eicher, B., & Fröhlich, E. (2000). Survey of laying hen husbandry in Switzerland. *World's Poultry Science Journal*, 56(1):21-31.

Hernández, A. (2014). Poultry and avian diseases. In Van Alfen, N. (Ed.), *Encyclopedia of Agriculture and Food Systems* (Cambridge, US: Academic Press).

Hezarjaribi, A., Rezaeipour, V., & Abdollahpour, R. (2016). Effects of intramuscular injections of vitamin E-selenium and a gonadotropin releasing hormone analogue (GnRHa) on reproductive performance and blood metabolites of post-molt male broiler breeders. *Asian Pacific Journal of Reproduction*, 5(2):156-160.

Hindle, V., Lambooij, E., Reimert, H., et al. (2010). Animal welfare concerns during the use of the water bath for stunning broilers, hens, and ducks. *Poultry Science*, 89(3):401-412.

HSUS – The Humane Society of the United States of America (n.d.). An HSUS report: The welfare of animals in the egg industry, www.humanesociety.org/sites/default/files/docs/hsus-report-welfare-egg-industry.pdf.

Humane Society International (2012). Bhutan bans extreme confinement cages for egg-laying hens, www.hsi.org/news-media/bhutan_cage_free_080212/.

IBD (2019). Selo orgânico IBD, www.ibd.com.br/selo-organico-ibd/. In Portuguese.

In Ovo (2020). In Ovo awarded €2.5 million EU grant, inovo.nl/news/in-ovo-awarded-e2-5-million-eu-grant/.

Jiang, X. & Groen, A. (1999). Combined crossbred and purebred selection for reproduction traits in a broiler dam line. *Journal of Animal Breeding and Genetics*, 116(2):111-125.

Jones, E., Wathes, C., & Webster, A. (2005). Avoidance of atmospheric ammonia by domestic fowl and the effect of early experience. *Applied Animal Behaviour Science*, 90(3-4):293-308.

Kannan, G. & Mench, J. (1996). Influence of different handling methods and crating periods on plasma corticosterone concentrations in broilers. *British Poultry Science*, 37(1):21-31.

Kasaeinasab, A., Jahangiri, M., Karimi, A., et al. (2017). Respiratory disorders among workers in slaughterhouses. *Safety and Health at Work*, 8(1):84-88.

Keaveny, M. (2020). Covid outbreaks at meat plants for discussion at Oireachtas committee, *The Irish Independent*, www.independent.ie/business/farming/news/farming-news/covid-outbreaks-at-meat-plants-for-discussion-at-oireachtas-committee-39443941.html.

Kevany, S. (2020a). Hundreds of thousands of chickens to be culled after Covid disruption, *The Guardian*, www.theguardian.com/environment/2020/aug/31/hundreds-of-thousands-of-chickens-to-be-culled-after-covid-disruption.

Kevany, S. (2020b). Millions of US farm animals to be culled by suffocation, drowning and shooting, *The Guardian*, www.theguardian.com/environment/2020/may/19/millions-of-us-farm-animals-to-be-culled-by-suffocation-drowning-and-shooting-coronavirus.

Kijlstra, A. & Eijck, I. (2006). Animal health in organic livestock production systems: a review. *NJAS - Wageningen Journal of Life Sciences*, 54(1):77-94.

Kim, J. & Campbell, W. (2018). Dietary cholesterol contained in whole eggs is not well absorbed and does not acutely affect plasma total cholesterol concentration in men and women: Results from 2 randomized controlled crossover studies. *Nutrients*, 10(9):1272.

Knierim, U. & Gocke, A. (2003). Effect of catching broilers by hand or machine on rates of injuries and dead-on-arrivals. *Animal Welfare*, 12:63-73.

Krautwald-Junghanns, M., Cramer, K., Fischer, B., e. (2018). Current approaches to avoid the culling of day-old male chicks in the layer industry, with special reference to spectroscopic methods. *Poultry Science*, 97(3):749-757.

La Mora, Z., Macías-Rodríguez, M., Arratia-Quijada, L., et al. (2020). *Clostridium perfringens* as foodborne pathogen in broiler production: Pathophysiology and potential strategies for controlling necrotic enteritis. *Animals*, 10(9):1718.

Le Monde (2020). La France veut interdire la castration à vif des porcelets et le broyage des poussins à la fin de 2021, www.lemonde.fr/planete/article/2020/01/28/la-france-veut-interdire-le-broyage-des-poussins-et-la-castration-a-vif-des-porcelets-a-la-fin-de-2021_6027528_3244.html. In French.

Lee, M. (2019). Avian campylobacter infection. Merck Sharp & Dohme (MSD), www.msdvet-manual.com/poultry/avian-campylobacter-infection/avian-campylobacter-infection.

Leeson, S. (2015). Vitamin deficiencies in poultry. Merck Sharp & Dohme (MSD), www.merckvetmanual.com/poultry.

Lesiów, T. & Kijowski, J. (2003). Impact of PSE and DFD meat on poultry processing – A review. *Polish Journal of Food and Nutrition Sciences*, 12(2):3-8.

Li, Y., Yang, X., Zhang, H., et al. (2020). Prevalence and antimicrobial susceptibility of *Salmonella* in the commercial eggs in China. *International Journal of Food Microbiology*, 325:108623.

Livetec Systems (2021). Nitrogen Foam Delivery System, www.livetecsystems.co.uk/nitrogen-foam-delivery-system/.

Madaan, N. (2017). Poultry industry uses battery cages despite government ban. *The Times of India*, timesofindia.indiatimes.com/city/pune/poultry-industry-uses-battery-cages-despite-government-ban/articleshow/57354968.cms.

Mann, T. (2016). Tesco's new range of meat is from another made-up farm. *Metro*, metro.co.uk/2016/03/25/tescos-new-range-of-meat-is-from-another-made-up-farm-5772514/.

Marshall, S. & Unger, C. (2020). Treating workers like meat: what we've learnt from COVID-19 outbreaks in abattoirs. *The Conversation*, theconversation.com/treating-workers-like-meat-what-weve-learnt-from-covid-19-outbreaks-in-abattoirs-145444.

Matthews, D. & Pinkerton, B. (2020). How chicken plants became more dangerous places to work than coal mines. US chicken plants process 140 birds a minute. The Trump administration thinks that's too slow. *Vox*, www.vox.com/future-perfect/21502225/chicken-meatpacking-plant-future-perfect-podcast.

Mattioli, S., Dal Bosco, A., Martino, M., et al. (2016). Alfalfa and flax sprouts supplementation enriches the content of bioactive compounds and lowers the cholesterol in hen egg. *Journal of Functional Foods*, 22:454-462.

Maurice, D., Jensen, L., & Tojo, H. (1979). Comparison of fish meal and soybean meal in the prevention of Fatty Liver-Hemorrhagic Syndrome in caged layers. *Poultry Science*, 58(4): 864-870.

McClane, B. (2014). *Clostridium perfringens*. In Wexler, P. (Ed.), *Encyclopedia of Toxicology* (Cambridge, US: Academic Press).

McDougall, J. (2010). Five major poisons inherently found in animal foods. *McDougall Newsletter*, 9(1):1-5.

Mitrović, S., Pandurević, T., Dimitrijević, M., et al. (2016). Impact on productivity of forced molt and duration of production cycle of commercial laying hens held in cage systems. *Acta Agriculturae Serbica*, XXI(42):145-154.

Molino, A., Garcia, E., Berto, D., et al. (2009). The effects of alternative forced-molting methods on the performance and egg quality of commercial layers. *Brazilian Journal of Poultry Science*, 11 (2):109-113.

Mönch, J., Rauch, E., Hartmannsgruber, S., et al. (2020). The welfare impacts of mechanical and manual broiler catching and of circumstances at loading under field conditions. *Poultry Science*, 99(11):5233-5251.

Muthulakshmi, M., Rajkumar, S., Rajkumar, R., et al. (2015). Incidence of egg bound syndrome in culled commercial layers. *International Journal of Science, Environment and Technology*, 4(3):583-587.

Naseem, S. & King, A. (2018). Ammonia production in poultry houses can affect health of humans, birds, and the environment – techniques for its reduction during poultry production. *Environmental Science and Pollution Research*, 25:15269-15293.

NCC – National Chicken Council of the United States of America (2013). National Chicken Council brief on stunning of chickens, www.nationalchickencouncil.org/national-chicken-council-brief-on-stunning-of-chickens/.

NCC – US National Chicken Council of the United States of America (2019). Broiler chicken industry key facts 2019, www.nationalchickencouncil.org/about-the-industry/statistics/broiler-chicken-industry-key-facts/.

NCC – US National Chicken Council of the United States of America (2020). U.S. broiler performance, www.nationalchickencouncil.org/about-the-industry/statistics/u-s-broiler-performance/.

Nevin, R., Bernt, J., & Hodgson, M. (2017). Association of poultry processing industry exposures with reports of occupational finger amputations: Results of an analysis of OSHA Severe Injury Report (SIR) Data. *Journal of Occupation and Environmental Medicine*, 59(10):e159-e163.

NHS – National Health System of the United Kingdom (2018). Bird flu. World Health Organization, www.who.int/influenza/human_animal_interface/avian_influenza/h5n1_research/faqs/en/.

Nijdam, E., Delezie, E., Lambooij, E., et al. (2005). Comparison of bruises and mortality, stress parameters, and meat quality in manually and mechanically caught broilers. *Poultry Science*, 84:467-474.

Nunes, F. (2018). Apanha das aves: Manejo fundamental para uma empresa avícola. *aviNews*, avicultura.info/pt-br/apanha-empresa-avicola/.

Oakenfull, R. & Wilson, A. (2020). Qualitative Risk Assessment: What is the risk of food or food contact materials being a source or transmission route of SARS-CoV-2 for UK consumers? UK: Food Standards Agency, www.food.gov.uk/sites/default/files/media/document/web-version-qualitative-risk-assessment-risk-of-food-or-food-contact-materials-as-transmission-route-of-sars-cov-2-002.pdf.

Ohier, F. (2020). Carrefour et Loué s'associent pour réduire le broyage des poussins males. *France Inter*, www.franceinter.fr/economie/carrefour-et-loue-s-associent-pour-reduire-le-broyage-des-poussins-males. In French.

O'Kane, C. (2020). Chicken wings imported to China from Brazil test positive for COVID-19, Chinese officials say. *CBS News*, www.cbsnews.com/news/china-covid-19-chicken-wings-brazil/.

Olgun, O. & Aygun, A. (2016). Nutritional factors affecting the breaking strength of bone in laying hens. *World's Poultry Science Journal*, 72(4):821-832.

Ong, S., Tan, Y., Chia, P., et al. (2020). Air, surface environmental, and personal protective equipment contamination by severe acute respiratory syndrome coronavirus 2 (SARS-CoV-2) from a symptomatic patient. *The Journal of the American Medical Association*, 323(16):1610-1612.

Open Secrets (2020). Poultry & eggs: Lobbying, 2020, www.opensecrets.org/industries/lobbying.php?ind=A05++.

Pattison, B. (2020a). 80 coronavirus cases confirmed at O'Brien Fine Foods plant in Timahoe, Co Kildare. *The Irish Mirror*, www.irishmirror.ie/news/irish-news/health-news/80-coronavirus-cases-confirmed-obrien-22482404.

Pattison, B. (2020b). Another six cases of coronavirus confirmed at O'Brien Fine Foods plant in Kildare. *The Irish Mirror*, www.irishmirror.ie/news/irish-news/another-six-cases-coronavirus-confirmed-22487384.

Patwardhan, D. & King, A. (2011). Review: feed withdrawal and non feed withdrawal moult. *World's Poultry Science Journal*, 67(2):253-268.

Peccia, J., Zulli, A., Brackney, D., et al. (2020). SARS-CoV-2 RNA concentrations in primary municipal sewage sludge as a leading indicator of COVID-19 outbreak dynamics. *MedRxiv.* doi.org/10.1101/2020.05.19.20105999.

Perrett, M. (2020). 2 Sisters and Fyffes close sites hit by COVID-19. *Food Manufacture,* www.foodmanufacture.co.uk/Article/2020/08/18/2-Sisters-and-Fyffes-close-sites-hit-by-COVID-19.

Petrović, T. & D'Agostino, M. (2016). Chapter 5 – Viral contamination of food. In Barros-Velázquez, J. (Ed.), *Antimicrobial Food Packaging* (Cambridge, US: Academic Press).

Phillips, K. (2018). 200 million eggs recalled after nearly two dozen were sickened with salmonella, officials say. *The Washington Post,* www.washingtonpost.com/news/business/wp/2018/04/15/200-million-eggs-recalled-after-nearly-two-dozen-were-sickened-with-salmonella-officials-say/.

Poole, T. (2015). Introduction to developing a free-range poultry enterprise. *University of Maryland,* extension.umd.edu/sites/default/files/2021-02/Developing%20a%20Free-Range%20Poultry%20Enterprise.pdf.

Poultry DVM (2020a). Egg binding, www.poultrydvm.com/condition/egg-binding#:~:text=%20%20%20%20Name%20%20%20,is%20%20...%20%206%20more%20rows%20.

Poultry DVM (2020b). Vitamin B12 deficiency, www.poultrydvm.com/condition/vitamin-B12-deficiency.

Poultry Hub (2020a). Beak trimming, www.poultryhub.org/all-about-poultry/health-management/beak-trimming.

Poultry Hub (2020b). Infectious bursal disease (or Gumboro), www.poultryhub.org/health/disease/types-of-disease/infectious-bursal-disease-or-gumboro/.

Rangel, J., Sparling, P., Crowe, C., et al. (2005). Epidemiology of *Escherichia coli* O157:H7 outbreaks, United States, 1982–2002. *Emerging Infectious Diseases,* 11(4):603-609.

Rise, H. (2018). Nortura kjenner seg pressa vekk frå buregg. *Nationen,* www.nationen.no/landbruk/nortura-kjenner-seg-pressa-vekk-fra-buregg/. In Norwegian.

Ritchie, H. & Roser, M. (2019). Meat and dairy production. *Our World in Data,* ourworldindata.org/meat-production.

Rivera, L. (2017). Unmasking the truth behind food labelling in the chicken industry, *The Independent,* www.independent.co.uk/life-style/food-and-drink/supermarket-chicken-labels-truth-free-range-battery-treatment-organic-a7751536.html.

Rivoal, K., Fablet, A., Courtillon, C., et al. (2013). Detection of *Listeria* spp. in liquid egg products and in the egg breaking plants environment and tracking of *Listeria monocytogenes* by PFGE. *International Journal of Food Microbiology,* 166(1):109-16.

Rodrigues, D., Café, M., Jardim Filho, R., et al. (2017). Metabolism of broilers subjected to different lairage times at the abattoir and its relationship with broiler meat quality. *Arquivo Brasileiro de Medicina Veterinária e Zootecnia,* 69(3):733-741. In Portuguese.

Rodriguez-Navarro, A., McCormack, H., Fleming, R., et al. (2018). Influence of physical activity on tibial bone material properties in laying hens. *Journal of Structural Biology,* 201(1):36-45.

Rouhani, M., Rashidi-Pourfard, N., Salehi-Abargouei, A., et al. (2018). Effects of egg consumption on blood lipids: a systematic review and meta-analysis of randomized clinical trials. *Journal of the American College of Nutrition,* 37:99-110.

RSCPA – Royal Society for the Prevention of Cruelty to Animals (2017a). RSPCA welfare standards for meat chickens, science.rspca.org.uk/documents/1494935/9042554/RSPCA+welfare+standards+for+meat+chickens+(8.48+MB).pdf/e7f9830d-aa9e-0908-aebd-2b8fbc6262ea.

RSPCA – Royal Society for the Prevention of Cruelty to Animals (2017b). RSPCA welfare standards for hatcheries (chicks, poults and ducklings). Revised edition, science.rspca.org.uk/sciencegroup/farmanimals/standards/hatcheries.

RSPCA UK – Royal Society for the Prevention of Cruelty to Animals of the United Kingdom (2020). RSPCA Assured chicken welfare. Revised edition, www.rspcaassured.org.uk/farm-animal-welfare/meat-chickens/.

Rueda, J. & Khosla, P. (2013). Impact of breakfasts (with or without eggs) on body weight regulation and blood lipids in university students over a 14-week semester. *Nutrients*, 5:5097-5113.

Rufener C. & Makagon, M. (2020). Keel bone fractures in laying hens: a systematic review of prevalence across age, housing systems, and strains. *Journal of Animal Sciences*, 98(1):S36-S51.

Ruiz, K. (2019). BREAKING: Urgent recall is issued on one of Australia's most popular egg brands over contamination fears in the FOURTH salmonella scare in just one month, *Daily Mail*, www.dailymail.co.uk/news/article-6927029/NSW-Food-Authority-issues-egg-recall-salmonella-fears.html.

Russo, J. (2019). Tudo que você precisa saber sobre os sistemas de produção de ovos. Agroceres Multimix, agroceresmultimix.com.br/blog/tudo-que-voce-precisa-saber-sobre-os-sistemas-de-producao-de-ovos/. In Portuguese.

Ryan, C. (2019). Tesco to launch 'higher welfare' indoor chicken range, in line with Better Chicken Commitment. *Poultry News*, www.poultrynews.co.uk/news/tesco-to-launch-higher-welfare-indoor-chicken-range-in-line-with-better-chicken-commitment.html.

Sander, J. (2019a). Sudden Death Syndrome of broiler chickens. Merck & Co., www.merckvetmanual.com/poultry/sudden-death-syndrome-of-broiler-chickens/sudden-death-syndrome-of-broiler-chickens.

Sander, J. (2019b). Fowl Cholera. Merck Sharp & Dohme (MSD), www.msdvetmanual.com/poultry/fowl-cholera/fowl-cholera.

Sandilands, V., Moinard, C., & Sparks, N. (2009). Providing laying hens with perches: fulfilling behavioural needs but causing injury? *British Poultry Science*, 50:395-406.

Saraiva, S., Esteves, A., Oliveira, I., et al. (2020). Impact of pre-slaughter factors on welfare of broilers. *Veterinary and Animal Science*, 10:100146.

Saul, H. (2015). Hatched, discarded, gassed: what happens to male chicks in the UK, *The Independent*, www.independent.co.uk/life-style/food-and-drink/hatched-discarded-gassed-what-happens-male-chicks-uk-10088509.html.

Saunders, W. (2013). Dystocia. In Mayer, J. & Donnelly, T. (Eds.), *Clinical veterinary advisor – Birds and exotic pets* (Amsterdam, Netherlands: Elsevier).

Savenije, B., Lambooij, E., Pieterse, C., et al. (2000). Electrical stunning and exsanguination decrease the extracellular volume in the broiler brain as studied with brain impedance recordings. *Poultry Science*, 79(7):1062-1066.

Savory, C., Maros, K., & Rutter, S. (1993). Assessment of hunger in growing broiler breeders in relation to a commercial restricted feeding programme. *Animal Welfare*, 2(2):131-152.

Shahbandeh, M. (2020). Number of chickens worldwide from 1990 to 2018. Statista, www.statista.com/statistics/263962/number-of-chickens-worldwide-since-1990/.

Shahbandeh, M. (2021). Global chicken meat production 2012-2020. Statista, www.statista.com/statistics/237637/production-of-poultry-meat-worldwide-since-1990/.

Sherwin, C., Richards, G., & Nicol, C. (2010). Comparison of the welfare of layer hens in 4 housing systems in the UK. *British Poultry Science*, 51(4):488-499.

Shetty, J. (2019). Antonio Singh Boparan. *WaliKali*, www.walikali.com/antonio-singh-boparan/.

Shini, A., Shini, S., & Bryden, W. (2019). Fatty liver haemorrhagic syndrome occurrence in laying hens: impact of production system. *Avian Pathology*, 48(1):25-34.

Shini, S., Shini, A., & Bryden, W. (2020). Unravelling fatty liver haemorrhagic syndrome: 1. Oestrogen and inflammation. *Avian Pathology*, 49(1):87-98.

Silva, D., Maciel, M., Arouca, C., et al. (2017). Alternative forced-molting methods in japanese quails. *Revista Brasileira de Saúde e Produção Animal*, 18(2):359-368.

Soil Association (2020). Soil Association standards: Farming and growing, www.soilassociation.org/media/15931/farming-and-growing-standards.pdf.

Spence, J., Jenkins, D., & Davignon, J. (2010). Dietary cholesterol and egg yolks: Not for patients at risk of vascular disease. *Canadian Journal of Cardiology*, 26(9):e336-e339.

Squires, E. & Leeson, S. (1988). Aetiology of fatty liver syndrome in laying hens. *British Veterinary Journal*, 144(6): 602-609.

Squires, M. & Naber, E. (1992). Vitamin profiles of eggs as indicators of nutritional status in the laying hen: Vitamin B_{12} study. *Poultry Science*, 71(12):2075-2082.

Srinivasan, P., Balasubramaniam, G., Murthy, T., et al. (2013). Bacteriological and pathological studies of egg peritonitis in commercial layer chicken in Namakkal area. *Asian Pacific Journal of Tropical Biomedicine*, 3(12):988-994.

Starkey, L. (2020). Australia faces meat shortages as abattoirs are hit with coronavirus outbreaks - while KFCs are forced to CLOSE after running out of chicken, *Daily Mail*, www.dailymail.co.uk/news/article-8583015/Australia-faces-meat-shortages-abattoirs-hit-coronavirus-outbreaks.html.

Stratmann, A., Fröhlich, E., Gebhardt-Henrich, S., et al. (2015). Modification of aviary design reduces incidence of falls, collisions and keel bone damage in laying hens. *Applied Animal Behaviour Science*, 165:112-123.

Sun, Y., Yang, H., Zhong, X., et al. (2011). Ultrasonic-assisted enzymatic degradation of cholesterol in egg yolk. *Innovative Food Science & Emerging Technologies*, 12(4):505-508.

Sun, P., Lu, Y., Cheng, H., et al. (2018). The effect of grape seed extract and yeast culture on both cholesterol content of egg yolk and performance of laying hens. *Journal of Applied Poultry Research*, 27(4):564-569.

Taylor, C. (2020a). Covid-19 outbreak at 2 Sisters poultry plant leads to temporary closure. *The Scottish Farmer*,www.thescottishfarmer.co.uk/news/18660654.covid-19-outbreak-2-sisters-poultry-plant-leads-temporary-closure/.

Taylor, C. (2020b). Scottish food and drink businesses awarded £5 million to strengthen local food supply chains. *The Scottish Farmer*, www.thescottishfarmer.co.uk/news/18648710.scottish-food-drink-businesses-awarded-5-million-strengthen-local-food-supply-chains/.

Teixeira, R., Santos, I., Sampaio, F., et al. (2014). Muda forçada a partir do jejum: importância, aspectos relacionados ao bem estar animal e visão do consumidor. *PUBVET*, 8(11), ed. 260, art. 1729. In Portuguese.

The Poultry Site (2018). Marek's disease control in broiler breeds, www.thepoultrysite.com/articles/mareks-disease-control-in-broiler-breeds.

The Poultry Site (2019). Five steps to going cage-free in China, www.thepoultrysite.com/articles/five-steps-to-going-cage-free-in-china.

Thornton, G. (2012). New Zealand's Tegel Poultry achieves world's best feed conversion. *Feed Strategy*, www.feedstrategy.com/poultry/new-zealands-tegel-poultry-achieves-worlds-best-feed-conversion/.

Tiseo, K., Huber, L., Gilbert, M., et al. (2020). Global trends in antimicrobial use in food animals from 2017 to 2030. *Antibiotics*, 9(12):918.

Togoh, I. (2020). China suspends Tyson Foods poultry imports from Arkansas facility following outbreak. *Forbes*, www.forbes.com/sites/isabeltogoh/2020/06/21/china-suspends-tyson-foods-poultry-imports-from-arkansas-facility-following-outbreak.

Tsiouris, V., Georgopoulou, I., Batzios, C., et al. (2015). High stocking density as a predisposing factor for necrotic enteritis in broiler chicks. *Avian Pathology*, 44(2):59-66.

Tufts (2020). Jean Mayer USDA Human Nutrition Research Center on Aging, hnrca.tufts.edu/mission/.

United Egg Producers (2017). Animal husbandry guidelines for U.S. egg laying flocks, unitedegg.com/wp-content/uploads/2017/11/2017-UEP-Animal-Welfare-Complete-Guidelines-11.01.17-FINAL.pdf.

United Egg Producers (2021). Facts & stats, unitedegg.com/facts-stats/.

United Poultry Concerns (2019). "Free-range" poultry and eggs, www.upc-online.org/freerange.html.

USDA – United States Department of Agriculture (2015a). Dietary guidelines for Americans 2015-2020,www.fns.usda.gov/2015-2020-dietary-guidelines-americans.

USDA – United States Department of Agriculture (2015b). Meat and poultry labelling terms,www.fsis.usda.gov/food-safety/safe-food-handling-and-preparation/food-safety-basics/meat-and-poultry-labeling-terms. .

USDA – United States Department of Agriculture (2017). USDA graded cage-free eggs: All they're cracked up to be, www.usda.gov/media/blog/2016/09/13/usda-graded-cage-free-eggs-all-theyre-cracked-be.

USDA – United States Department of Agriculture (2020). Chicken meat production – Top countries summary, apps.fas.usda.gov/psdonline/app/index.html#/app/downloads.

van der Made, S., Kelly, E., Berendschot, T., et al. (2014). Consuming a buttermilk drink containing lutein-enriched egg yolk daily for 1 year increased plasma lutein but did not affect serum lipid or lipoprotein concentrations in adults with early signs of age-related macular degeneration. *The Journal of Nutrition*, 144:1370-1377.

Van Doremalen, N., Bushmaker, T., & Munster, V. (2013). Stability of Middle East respiratory syndrome coronavirus (MERS-CoV) under different environmental conditions. *Euro Surveillance*, 18(38):20590.

Van Doremalen, N., Bushmaker, T., Morris, D., et al. (2020). Aerosol and surface stability of SARS-CoV-2 as compared with SARS-CoV-1. *The New England Journal of Medicine*, 382:1564-1567.

Vega, G. & Grundy, S. (2019). Current trends in non–HDL cholesterol and LDL cholesterol levels in adults with atherosclerotic cardiovascular disease. *Journal of Clinical Lipidology*, 13(4):563-567.

Vincent, M., Allen, B., Palacios, O., et al. (2019). Meta-regression analysis of the effects of dietary cholesterol intake on LDL and HDL cholesterol. *American Journal of Clinical Nutrition*, 109:7-16.

Vizzier-Thaxton, Y., Christensen, K., Schilling, M., et al. (2010). A new humane method of stunning broilers using low atmospheric pressure. *Journal of Applied Poultry Research*, 19(4):341-348.

Wallop, H. (2017). FOWL PLAY. Lidl chicken 'repackaged and sold as Tesco Willow Farm chicken. *The Sun*, www.thesun.co.uk/money/4577402/lidl-chicken-repackaged-and-sold-as-tesco-willow-farm-chicken/.

Webster, A. (2004). Welfare implications of avian osteoporosis. *Poultry Science*, 83(2):184-192.

Webster, A. (2007). The commercial egg industry should consider controlled atmosphere stunning for spent hens. *The Poultry Site*, www.thepoultrysite.com/articles/the-commercial-egg-industry-should-consider-controlled-atmosphere-stunning-for-spent-hens.

Webster, B. (2016). Supermarket poultry firm boiled halal chickens alive. *The Times*, www.thetimes.co.uk/article/supermarket-poultry-firm-boiled-halal-chickens-alive-wkx2xh329.

Weggemans, R., Zock, P., & Katan, M. (2001). Dietary cholesterol from eggs increases the ratio of total cholesterol to high-density lipoprotein cholesterol in humans: a metaanalysis. *American Journal of Clinical Nutrition*, 73:885-891.

WHO – World Health Organization (2011). Influenza, www.who.int/influenza/human_animal_interface/avian_influenza/h5n1_research/faqs/en/.

WHO – World Health Organization (2018). The top 10 causes of death, www.who.int/news-room/fact-sheets/detail/the-top-10-causes-of-death.

Wood, Z. (2017). Tesco vows to keep Willow Farms brand despite chicken scandal, *The Guardian*, www.theguardian.com/business/2017/oct/04/tesco-willow-farms-chicken-scandal-2-sisters.

Wright, P. (2016). Halal slaughterhouse boils 81 chickens alive after equipment failure at Suffolk factory. *International Business Times*, www.ibtimes.co.uk/halal-slaughterhouse-boils-81-chickens-alive-after-equipment-failure-suffolk-factory-1598808.

Yeakel, S. (2019). Pullorum Disease in poultry. Merck Sharp & Dohme (MSD), www.msdvet-manual.com/poultry/salmonelloses/pullorum-disease-in-poultry?query=pullorum.

Yi, B., Chen, L., Sa, R., et al. (2016). High concentrations of atmospheric ammonia induce alterations of gene expression in the breast muscle of broilers (*Gallus gallus*) based on RNA-Seq. *BMC Genomics*, 17(1):598.

Chapter 4 references

Abdulqader, E., Abdurahiman, P., Mansour, L., et al. (2020). Bycatch and discards of shrimp trawling in the Saudi waters of the Arabian Gulf: ecosystem impact assessment and implications for a sustainable fishery management. *Fisheries Research*, 229:105596.

Acerete, L., Balasch, J., Espinosa, E., et al. (2004). Physiological responses in Eurasian perch (*Perca fluviatilis*, L.) subjected to stress by transport and handling. *Aquaculture*, 237(1-4):167-178.

Agnew, D., Pearce, J., Pramod, G., et al. (2009). Estimating the worldwide extent of illegal fishing. *PLoS ONE*, 4(2):e4570.

Anderson, O., Small, C., Croxall, J., et al. (2011). Global seabird bycatch in longline fisheries. *Endangered Species Research*, 14:91-106.

Animal Equality (2012). The killing of tuna in Carloforte, animalequality.org/investigation/the-killing-of-tuna-in-carloforte/.

Animal Equality (2021a). INVESTIGATION: Fish killed while fully conscious in Scottish salmon slaughterhouse, animalequality.org/news/investigation-fish-killed-while-fully-conscious-in-scottish-salmon-slaughterhouse/.

Animal Equality (2021b). INVESTIGATION: The cruel and illegal practices of India's fishing industry, animalequality.org/news/investigation-the-cruel-and-illegal-practices-of-indias-fishing-industry/.

Arends, R., Mancera, J., Munoz, J., et al. (1999). The stress response of the gilthead sea bream (*Sparus aurata* L.) to air exposure and confinement. *Journal of Endocrinology*, 163(1):149-157.

Arrifano, G., Martín-Doimeadios, R., Jiménez-Moreno, M., et al. (2018). Large-scale projects in the Amazon and human exposure to mercury: The case-study of the Tucuruí Dam. *Ecotoxicology and Environmental Safety*, 147:299-305.

Australian Associated Press (2018). Court rejects challenge to salmon farm near Tasmanian world heritage area, *The Guardian*, www.theguardian.com/australia-news/2018/apr/12/court-rejects-challenge-to-salmon-farm-near-tasmanian-world-heritage-area.

Avery, J., Aagaard, K., Burkhalter, J., et al. (2017). Seabird longline bycatch reduction devices increase target catch while reducing bycatch: A meta-analysis. *Journal for Nature Conservation*, 38:37-45.

Basu, N., Horvat, M., Evers, D., et al. (2018). A state-of-the-science review of mercury biomarkers in human populations worldwide between 2000 and 2018. *Environmental Health Perspectives*, 126(10):1-14.

Belhabib, D., Le Billon, P., & Wrathall, D. (2020). Narco-fish: Global fisheries and drug trafficking. *Fish and Fisheries*, 21(5):992-1007.

Bellows, L. & Moore, R. (2012). Dietary fat and cholesterol. *Fact Sheet No. 9.319. Colorado State University*, mountainscholar.org/bitstream/handle/10217/195108/AEXT_093192012.pdf.

Bourret, V., O'Reilly, P., Carr, J., et al. (2011). Temporal change in genetic integrity suggests loss of local adaptation in a wild Atlantic salmon (*Salmo salar*) population following introgression by farmed escapees. *Heredity*, 106:500-510.

Boyland, N. & Brooke, P. (2017). Farmed fish welfare during slaughter. EU Aquaculture Advisory Council (AAC) Working Group I (Finfish), www.aac-europe.org/images/Slaughter_report__AAC_report.pdf.

Bradbury, I., Duffy, S., Lehnert, S., et al. (2020). Model-based evaluation of the genetic impacts of farm-escaped Atlantic salmon on wild populations. *Aquaculture Environment Interactions*, 12:45-59.

Bradford, M. R. (1995). Comparative aspects of forebrain organization in the ray-finned fishes – touchstones or not. *Brain, Behavior and Evolution*, 46(4-5):259-274.

Carey, E. (2018). Is there cholesterol in fish? *Healthline*, www.healthline.com/health/high-cholesterol/-is-there-cholesterol-in-fish.

Carver, A. & Gallicchio, V. (2017). Heavy metals and cancer. In Atroshi, F. (Ed.), *Cancer Causing Substances*. IntechOpen. DOI: 10.5772/intechopen.70348

Chandroo, K., Duncan, I., & Moccia, R. (2004). Can fish suffer?: perspectives on sentience, pain, fear and stress. *Applied Animal Behaviour Science*, 86(3-4):225-250.

Chervova, L. (1997). Pain sensitivity and behavior of fishes. *Journal of Ichthyology (Voprosy Ikhtiologii)*, 37:98-102.

Chinnakali, P., Thekkur, P., Manoj, K., et al. (2016). Alarmingly high level of alcohol use among fishermen: A community based survey from a coastal area of south India. *Journal of Forensic and Legal Medicine*, 42:41-44.

Chiocchetti, G., Jadán-Piedra, C., Vélez, D., et al. (2017). Metal(loid) contamination in seafood products. *Critical Reviews in Food Science and Nutrition*, 57(17):3715-3728.

Council of Europe (2006). Standing Committee of the European Convention for the Protection of Animals kept for farming purposes. Recommendation concerning farmed fish, www.coe.int/t/e/legal_affairs/legal_co-operation/biological_safety_and_use_of_animals/farming/rec%20fish%20e.asp.

Coulthard, S. (2012). Can we be both resilient and well, and what choices to people have? Incorporating agency into the resilience debate from a fisheries perspective. *Ecology and Society*, 17(1):4.

Crespo-Lopez, M., Augusto-Oliveira, M., Lopes-Araújo, A., et al. (2021). Mercury: What can we learn from the Amazon? *Environment International*, 146:106223.

Croxall, J., Butchart, S., Lascelles, B., et al. (2012). Seabird conservation status, threats and priority actions: A global assessment. *Bird Conservation International*, 22:1-34.

Crozier, W. (1993). Evidence of genetic interaction between escaped farmed salmon and wild Atlantic salmon (*Salmo salar* L.) in a Northern Irish river. *Aquaculture*, 113(1–2):19-29.

Daly, E. & White, M. (2021). Bottom trawling noise: Are fishing vessels polluting to deeper acoustic habitats? *Marine Pollution Bulletin*, 162:111877.

Davies, R.; Cripps, S.; Nickson, A., et al. (2009). Defining and estimating global marine fisheries bycatch. *Marine Policy*, 33(4): 661-672.

De Vries, M. & De Boer, I. (2010). Comparing environmental impacts for livestock products: a review of life cycle assessments. *Livest Science*, 128(1-3):1-11.

DeFilippis, A. & Sperling, L. (2006). Understanding omega-3's. *American Heart Journal*, 151(3):564-570.

Denzer, D. & Laudien, H. (1987). Stress induced biosynthesis of a 31 kd-glycoprotein in goldfish brain. *Comparative biochemistry and Physiology Part B: Comparative Biochemistry*, 86(3):555-559.

Donlan, C., Wilcox, C., Luque, G., et al. (2020). Estimating illegal fishing from enforcement officers. *Scientific Reports*, 10:12478.

Dunlop, R. & Laming, P. (2005). Mechanoreceptive and nociceptive responses in the central nervous system of goldfish (*Carassius auratus*) and trout (*Oncorhynchus mykiss*). *Journal of Pain*, 6(9):561-568.

EFSA – European Food Safety Authority (2004). EFSA provides risk assessment on mercury in fish: precautionary advice given to vulnerable groups, www.efsa.europa.eu/en/press/news/040318.

EFSA – European Food Safety Authority (2009). General approach to fish welfare and to the concept of sentience in fish. *EFSA Journal*, 954:1-27.

EFSA – European Food Safety Authority (2012). Scientific opinion on the tolerable upper intake level of eicosapentaenoic acid (EPA), docosahexaenoic acid (DHA) and docosapentaenoic acid (DPA). EFSA Panel on Dietetic Products, Nutrition and Allergies. *EFSA Journal*, 10(7):2815.

EFSA – European Food Safety Authorities (2018). Risk for animal and human health related to the presence of dioxins and dioxin-like PCBs in feed and food. *EFSA Journal*, 16(11):e05333.

Erkinharju, T., Dalmo, R., Hansen, M., et al. (2020). Cleaner fish in aquaculture: review on diseases and vaccination. *Reviews in Aquaculture*, 13(1):189-237.

FAO – Food and Agriculture Organization (2020). The state of world fisheries and aquaculture 2020. Sustainability in action, www.fao.org/3/ca9229en/CA9229EN.pdf.

FAO – Food and Agriculture Organization (2021). Aquaculture, www.fao.org/aquaculture/en/.

FDA – Food and Drug Administration of the United States (2019). Advice about eating fish – For women who are or might become pregnant, breastfeeding mothers, and young children, www.fda.gov/food/consumers/advice-about-eating-fish.

Ferreira, N., de Araújo, R., & Campos, E. (2018). Good practices in pre-slaughter and slaughter of fish. *PUBVET*, 12(7):1-14. In Portuguese.

fishcount (2019). Slaughter of farmed fish. *fishcount*, fishcount.org.uk/farmed-fish-welfare/farmed-fish-slaughter.

Føre, H. & Thorvaldsen, T. (2021). Causal analysis of escape of Atlantic salmon and rainbow trout from Norwegian fish farms during 2010–2018. *Aquaculture*, 532, 736002.

Fort, E., Massardier-Pilonchéry, A., & Bergeret, A. (2010). Psychoactive substances consumption in French fishermen and merchant seamen. *International Archive of Occupational and Environmental Health*, 83:497-509.

Fort, E., Lassiège, T., & Bergeret, A. (2016). Prevalence of cannabis and cocaine consumption in French fishermen in South Atlantic region in 2012-2013 and its policy consequences. *International Maritime Health*, 67(2):88-96.

Frankowiak, K., Wang, X., Sigman, D., et al. (2016). Photosymbiosis and the expansion of shallow-water corals. *Science Advances*, 2(11):e1601122.

FRDC – Australian Fisheries Research and Development Corporation (2020). Australian aquaculture, www.frdc.com.au/media-publications/fish/FISH-Vol-27-4/australian-aquaculture.

Fry, J., Love, D., MacDonald, G., et al. (2016). Environmental health impacts of feeding crops to farmed fish. *Environment International*, 91:201-14.

Fürst, P. (2010). What did we learn so far from data resulting from dioxin monitoring? Joint AESAN/EFSA Workshop 'Science Supporting Risk Surveillance of Imports'. European Food Safety Authority, www.efsa.europa.eu/sites/default/files/event/documentset/corporate100210-p09.pdf.

Gall, S. & Thompson, R. (2015). The impact of debris on marine life. *Marine Pollution Bulletin*, 92(1-2):170-179.

Ganapathiraju, P., Pitcher, T., & Mantha, G. (2019). Estimates of illegal and unreported seafood imports to Japan. *Marine Policy*, 108:103439.

Garcés, J. (2021). Chile: Covid shutdown 'could make things worse'. *FishFarmingExpert*, www.fishfarmingexpert.com/article/chile-salmon-plant-shutdown-could-make-covid-crisis-worse/.

Gerbens-Leenes, P., Nonhebel, S., & Ivens, W. (2002). A method to determine land requirements relating to food consumption patterns. *Agriculture, Ecosystems and Environment*, 90 (1):47-58.

Gerbens-Leenes, P., Mekonnen, M., & Hoekstra, A. (2013). The water footprint of poultry, pork and beef: a comparative study in different countries and production systems. *Water Resources and Industry*, 1:25-36.

Giacomini, A., Abreu, M., Zanandrea, R., et al. (2016). Environmental and pharmacological manipulations blunt the stress response of zebrafish in a similar manner. *Scientific Reports*, 6:28986.

Gilmour, K., DiBattista, J., & Thomas, J. (2005). Physiological causes and consequences of social status in salmonid fish. *Integrative and Comparative Biology*, 45:263-273.

Glover, K., Solberg, M., McGinnity, P., et al. (2017). Half a century of genetic interaction between farmed and wild Atlantic salmon: status of knowledge and unanswered questions. *Fish and Fisheries*, 18:890-927.

Glover, K., Wennevik, V., Hindar, K., et al. (2020). The future looks like the past: Introgression of domesticated Atlantic salmon escapees in a risk assessment framework. *Fish and Fisheries*, 21(6):1077-1091.

Gray, C. & Kennelly, S. (2018). Bycatches of endangered, threatened and protected species in marine fisheries. *Reviews in Fish Biology and Fisheries*, 28:521-541.

Guzmán-Luna, P., Gerbens-Leenes, P., & Vaca-Jiménez, S. (2021). The water, energy, and land footprint of tilapia aquaculture in Mexico, a comparison of the footprints of fish and meat. *Resources, Conservation and Recycling*, 165:105224.

Haenen, O. (2017). Major bacterial diseases affecting aquaculture. Aquatic AMR Workshop 1: 10-11 April 2017, Mangalore, India. Food and Agriculture Organization (FAO), www.fao.org/fi/static-media/MeetingDocuments/WorkshopAMR/presentations/07_Haenen.pdf.

Hagenbuch, B. (2020a). American Seafoods trawler reports 85 COVID-19 cases. *SeafoodSource*, www.seafoodsource.com/news/supply-trade/american-seafoods-trawler-reports-85-covid-19-cases.

Hagenbuch, B. (2020b). Alaska hit with second COVID-19 trawler outbreak. *SeafoodSource*, www.seafoodsource.com/news/supply-trade/alaska-hit-with-second-covid-19-trawler-outbreak.

Hagenbuch, B. (2020c). 24 of 25 crewmembers aboard US Seafoods trawler test positive for COVID-19. *SeafoodSource*, www.seafoodsource.com/news/supply-trade/all-but-one-crewmembers-aboard-us-seafoods-trawler-test-positive-for-covid-19.

Hansen, L., Jacobsen, J., & Lund, R. (1993). High numbers of farmed Atlantic salmon. *Salmo salar* L., observed in oceanic waters north of the Faroe Islands. *Aquaculture Research*, 24:777-781.

Hansen, L. & Jacobsen, J. (2003). Origin and migration of wild and escaped farmed Atlantic salmon, *Salmo salar* L., in oceanic areas north of the Faroe Islands. *ICES Journal of Marine Science*, 60:110-119.

Henderson, R., Forrest, D., Black, K., et al. (1997). The lipid composition of sealoch sediments underlying salmon cages. *Aquaculture*, 158(1-2):69-83.

Herz, N. (2020). COVID-19 outbreak In Pacific Northwest seafood industry as season ramps up. *NPR*, www.npr.org/sections/coronavirus-live-updates/2020/06/05/870312092/pacific-northwest-seafood-industry-faces-covid-19-outbreak-as-season-ramps-up.

Hightower, J. & Moore, D. (2003). Mercury levels in high-end consumers of fish. *Environmental Health Perspectives*, 111(4):604-608.

Himes-Cornell, A. & Hoelting, K. (2015). Resilience strategies in the face of short- and long-term change: Out-migration and fisheries regulation in Alaskan fishing communities. *Ecology and Society*, 20(2):9.

Howe, K., Clark, M., Torroja, C., et al. (2013). The zebrafish reference genome sequence and its relationship to the human genome. *Nature* 496:498-503.

Hull, E., Barajas, M., Burkart, K., et al. (2021). Human health risk from consumption of aquatic species in arsenic-contaminated shallow urban lakes. *Science of the Total Environment*, 770:145318.

Huntingford, F. & Adams, C. (2005). Behavioural syndromes in farmed fish: implications for production and welfare. *Behaviour,* 142:1207-1221.

Ibidhi, R., Hoekstra, A., Gerbens-Leenes, P., et al. (2017). Water, land and carbon footprints of sheep and chicken meat produced in Tunisia under different farming systems. *Ecological Indicators,* 77:304-313.

International Organization for Migration (2016). Report on human trafficking, forced labour and fisheries crime in the Indonesian fishing industry.

Jambeck, J., Geyer, R., Wilcox, C., et al. (2015). Plastic waste inputs from land into the ocean. *Science,* 347(6223):768-771.

Jensen, A., Karlsson, S., Fiske, P., et al. (2013). Escaped farmed Atlantic salmon grow, migrate and disperse throughout the Arctic Ocean like wild salmon. *Aquaculture Environment Interactions,* 3:223-229.

Jia, Y., Wang, L., Li, S., et al. (2018). Species-specific bioaccumulation and correlated health risk of arsenic compounds in freshwater fish from a typical mine-impacted river. *Science of The Total Environment,* 625:600-607.

Johnson, K. (2020). Shocking facts about fishing that you need to know. Animal Equality, animalequality.org/blog/2020/05/29/shocking-fishing-facts.

Johnsson, C., Sällsten, G., Schütz, A., et al. (2004). Hair mercury levels versus freshwater fish consumption in household members of Swedish angling societies. *Environmental Research,* 96(3):257-63.

Kadfak, A. & Linke, S. (2021). More than just a carding system: Labour implications of the EU's illegal, unreported and unregulated (IUU) fishing policy in Thailand. *Marine Policy,* 127:104445.

Kihslinger, R. & Nevitt, G. (2006). Early rearing environment impacts cerebellar growth in juvenile salmon. *Journal of Experimental Biology,* 209(3):504-509.

Kim, S., Kim, P., Lim, J., et al. (2016). Use of biodegradable driftnets to prevent ghost fishing: physical properties and fishing performance for yellow croaker. *Animal Conservation,* 19(4):309-319.

Kissling, E., Allison, E., Seeley, J., et al. (2005). Fisherfolk are among groups most at risk of HIV: Cross-country analysis of prevalence and numbers infected. *AIDS,* 19(17):19391946.

Kramer, K. & Moll, H. (1995). Energie voedt, nadere analyses van het directe energieverbruik van voedingsmiddelen [Energy feeds, analysis of indirect energy use of food items]. *Report Center for Energy and Environmental studies (IVEM), report 71* (Groningen, The Netherlands: University of Groningen). In Dutch.

Krienitz, L. (2009). Algae. In. Likens, G., *Encyclopedia of Inland Waters,* pp. 103-113 (Cambridge, US: Academic Press).

Kurlansky, M. (2020). Net loss: the high price of salmon farming, *The Guardian,* www.theguardian.com/news/2020/sep/15/net-loss-the-high-price-of-salmon-farming.

Lane, K., Wilson, M., Hellon, T., et al. (2021). Bioavailability and conversion of plant based sources of omega-3 fatty acids – a scoping review to update supplementation options for vegetarians and vegans. *Critical Reviews in Food Science and Nutrition,* 10.1080/10408398.2021.1880364.

Lange, K. (2020). Omega-3 fatty acids and mental health. *Global Health Journal,* 4(1):18-30.

Larsen, M. & Vormedal, I. (2021). The environmental effectiveness of sea lice regulation: Compliance and consequences for farmed and wild salmon. *Aquaculture,* 532, 736000.

Lau, W., Shiran, Y., Bailey, R., et al. (2020). Evaluating scenarios toward zero plastic pollution. *Science,* 369(6510):1455-1461.

Lawrence, F. & McSweeney, E. (2017). UK police rescue nine suspected victims of slavery from British trawlers, *The Guardian,* www.theguardian.com/world/2017/dec/12/uk-police-rescue-nine-suspected-victims-of-slavery-from-british-trawlers.

Lele, U., Klousia-Marquis, M., & Goswami, S. (2013). Good governance for food, water and energy security. *Aquatic Procedia,* 1:44-63.

Lines, J. & Spence, J. (2014). Humane harvesting and slaughter of farmed fish. *Revue Scientifique et Technique*, 33(1):255-264.

Link, R. (2017). The 7 best plant sources of omega-3 fatty acids. *Healthline*, www.healthline.com/nutrition/7-plant-sources-of-omega-3s#TOC_TITLE_HDR_2.

MacPherson, E., Phiri, M., Sadalaki, J., et al. (2020). Sex, power, marginalisation and HIV amongst young fishermen in Malawi: Exploring intersecting inequalities. *Social Science & Medicine*, 266:113429.

Manciocco, A., Calamandrei, G., & Alleva, E. (2014). Global warming and environmental contaminants in aquatic organisms: the need of the etho-toxicology approach. *Chemosphere*, 100:1-7.

Manoj, K., Gomathi, M., Balaji, B., et al. (2018). Alcohol, harmful use and dependence: Assessment using the WHO Alcohol Use Disorder Identification Test tool in a South Indian fishermen community. *Industrial Psychiatry Journal*, 27(2):259-263.

Marschke, M. & Vandergeest, P. (2016). Slavery scandals: Unpacking labour challenges and policy responses within the off-shore fisheries sector. *Marine Policy*, 68:39-46.

Marshall, N. & Marshall, P. (2007). Conceptualizing and operationalizing social resilience within commercial fisheries in Northern Australia. *Ecology and Society*, 12(1):1.

Matthews, C. (2021). COVID-19 outbreak at Trident Seafoods Akutan plant grows to 135. *Alaska's News Source*, www.msn.com/en-us/news/us/covid-19-outbreak-at-trident-seafoods-akutan-plant-grows-to-135/ar-BB1d7BfX.

May, C. (2017). Transnational crime and the developing world. Global Financial Integrity, www.gfintegrity.org/wp-content/uploads/2017/03/Transnational_Crime-final.pdf.

McBey, L. (2018). West Coast beach closed to swimmers until further notice. *The Advocate*, www.theadvocate.com.au/story/5239862/west-coast-beach-closed-to-swimmers/.

McGinnity, P., Stone, C., & Taggart, J. (1997). Genetic impact of escaped farmed Atlantic salmon (*Salmo salar* L.) on native populations: use of DNA profiling to assess freshwater performance of wild, farmed, and hybrid progeny in a natural river environment. *ICES Journal of Marine Science*, 54(6), 998-1008.

McKenney, H. (2020). Unalaska's biggest fish processor sees new COVID-19 cases. *Alaska Public Media*, www.alaskapublic.org/2020/09/15/unalaskas-biggest-fish-processor-sees-new-covid-19-cases/.

McKenney, H. (2021). Unalaska seafood plant remains on lockdown with 20 more COVID-19 cases. *Alaska Public Media*, www.alaskapublic.org/2021/01/19/unalaska-seafood-plant-remains-on-lockdown-with-20-more-covid-19-cases/.

Mekonnen, M. & Hoekstra, A. (2010). The green, blue and grey water footprint of farm animals and animal products. UNESCO-IHE, Delft, the Netherlands. *Value of Water Research Report Series No. 48*.

Mendoza, M. & Mason, M. (2016). Hawaiian seafood caught by foreign crews confined on boats. *The Associated Press*, www.ap.org/explore/seafood-from-slaves/hawaiian-seafood-caught-foreign-crews-confined-boats.html.

Mikolajczyk, S., Warenik-Bany, M., Maszewski, S., et al. (2020). Dioxins and PCBs – Environment impact on freshwater fish contamination and risk to consumers. *Environmental Pollution*, 263(Part B):114611.

Milman, O. (2015). FDA approves genetically modified salmon in agency first, *The Guardian*, www.theguardian.com/environment/2015/nov/19/fda-approves-genetically-modified-salmon.

Mood, A. (2010). Worse things happen at sea: the welfare of wild caught-fish. *fishcount*, fishcount.org.uk/publications.

Moore, K. (2021). NMFS reports fishing revenue crashed 29 percent in pandemic. *National Fisherman*, www.nationalfisherman.com/national-international/nmfs-reports-fishing-revenue-crashed-29-percent-in-pandemic.

Morash, A., Lyle, J., Currie, S., et al. (2020). The endemic and endangered Maugean Skate (*Zearaja maugeana*) exhibits short-term severe hypoxia tolerance. *Conservation Physiology*, 8(1):coz105.

Morris, M., Fraser, D., Heggelin, A., et al. (2008) Prevalence and recurrence of escaped farmed Atlantic salmon (*Salmo salar*) in eastern North American rivers. *Canadian Journal of Fisheries and Aquatic Sciences*, 65:2807-2826.

Morton, A. (2018). The battle over big salmon: industry at a crossroads as Tasmania votes, *The Guardian*, www.theguardian.com/australia-news/2018/feb/26/the-battle-over-big-salmon-industry-at-a-crossroads-as-tasmania-votes.

MRCI – Migrant Research Centre Ireland (2017). Left high and dry. The exploitation of migrant workers in the Irish fishing industry, www.mrci.ie/app/uploads/2020/01/MRCI-FISHER-REPORT-Dec-2017-2KB.pdf.

MSC – Marine Stewardship Council (2021). What is sustainable fishing? www.msc.org/what-we-are-doing/our-approach/what-is-sustainable-fishing.

Murray, P. (2014). Boat slave shame of fishing industry. *Express*, www.express.co.uk/news/uk/456085/Boat-slave-shame-of-fishing-industry.

NCA – UK's National Crime Agency (2020). Three men charged in NCA people smuggling investigation, nationalcrimeagency.gov.uk/news/three-men-charged-in-nca-people-smuggling-investigation?highlight=WyJmaXNoaW5nIiwiZmlzaCIsImZpc2gnIl0=.

Njåstad, M. & Riise, O. (2020). 29 employees infected with COVID-19 at Mowi Poland factory. *IntraFish*, www.intrafish.com/processing/29-employees-infected-with-covid-19-at-mowi-poland-factory/2-1-848608.

NOAA – US National Oceanic and Atmospheric Administration (2017). Chapter 1: Assessment of the Walleye Pollock stock in the Eastern Bering Sea, archive.fisheries.noaa.gov/afsc/REFM/Docs/2017/EBSpollock.pdf.

Norði, G., Glud, R., Gaard, E., et al. (2011). Environmental impacts of coastal fish farming: carbon and nitrogen budgets for trout farming in Kaldbaksfjørður (Faroe Islands). *Marine Ecology Progress Series*, 431:223-241.

NRC – National Research Council (2000). Committee on the Toxicological Effects of Methylmercury. Toxicological effects of methylmercury (Washington, US: National Academies Press).

Oceana (2015). Oceana reveals mislabeling of America's favorite fish: salmon, usa.oceana.org/publications/reports/oceana-reveals-mislabeling-americas-favorite-fish-salmon.

OEHHA – Office of Environmental Health Hazard Assessment (n.d.). Fish advisories, oehha.ca.gov/fish/advisories.

OIE – World Organisation for Animal Health (2019). Aquatic animal health code, www.oie.int/en/standard-setting/aquatic-code/access-online/.

Oliver, S., Braccini, M., Newman, S. J., Harvey, E. S. (2015). Global patterns in the bycatch of sharks and rays. *Marine Policy*, 54:86-97.

Pennisi, E. (2017). Meet the obscure microbe that influences climate, ocean ecosystems, and perhaps even evolution. *Science*, www.sciencemag.org/news/2017/03/meet-obscure-microbe-influences-climate-ocean-ecosystems-and-perhaps-even-evolution.

Pérez Roda, M., (2019) A third assessment of global marine fisheries discards. In Gilman, E., Huntington, T., et al. (Eds.). *FAO Fisheries and Aquaculture Technical Paper No. 633*, p. 78.

Poli, B., Parisi, G., Scappini, F., et al. (2005). Fish welfare and quality as affected by pre-slaughter and slaughter management. *Aquaculture International*, 13(1-2):29-49.

Poore, J. & Nemecek, T. (2018). Reducing food's environmental impacts through producers and consumers. *Science*, 360:987-992.

Pounder, K., Mitchell, J., Thomson, J., et al. (2016). Does environmental enrichment promote recovery from stress in rainbow trout? *Applied Animal Behaviour Science*, 176:136-142.

Pramod, G., Pitcher, T., & Mantha, G. (2019). Estimates of illegal and unreported seafood imports to Japan. *Marine Policy,* 108:103439.

Princeton University (2016). When corals met algae: Symbiotic relationship crucial to reef survival dates to the Triassic. *Phys.org,* phys.org/news/2016-11-corals-met-algae-symbiotic-relationship.html.

Rachwani, M. (2020a). Fears for environment after 50,000 fish escape salmon farm in Tasmania, *The Guardian,* www.theguardian.com/australia-news/2020/nov/24/fears-for-environment-after-50000-fish-escape-salmon-farm-in-tasmania.

Rachwani, M. (2020b). 'A circus': second mass salmon outbreak in Tasmania outrages conservationists, *The Guardian,* www.theguardian.com/australia-news/2020/dec/03/a-circus-second-mass-salmon-outbreak-in-tasmania-outrages-conservationists.

Rahmanifarah, K., Shabanpour, B., & Sattari A. (2011). Effects of clove oil on behavior and flesh quality of common carp (*Cyprinus carpio* L.) in comparison with pre-slaughter CO_2 stunning, chilling and asphyxia. *Turkish Journal of Fisheries and Aquatic Sciences,* 11:139-147.

Rajão, R., Soares-Filho, B., Nunes, F., et al. (2020). The rotten apples of Brazil's agribusiness. *Science,* 369(6501):246-248.

Rice, K., Walker, E., Wu, M., et al. (2014). Environmental mercury and its toxic effects. *Journal of Preventive Medicine and Public Health,* 47(2):74-83.

Riise, O. (2021). Mowi employees test positive for COVID-19 at Norway plant. *IntraFish,* www.intrafish.com/coronavirus/mowi-employees-test-positive-for-covid-19-at-norway-plant/2-1-942980.

Ritchie, H. (2020). Environmental impacts of food production. *Our World in Data,* ourworldindata.org/environmental-impacts-of-food.

Roberts, S., Jaremin, B., & Lloyd, K. (2013). High-risk occupations for suicide. *Psychological Medicine,* 43:1231-1240.

Rochman, C., Browne, M., Underwood, A., et al. (2016). The ecological impacts of marine debris: Unraveling the demonstrated evidence from what is perceived. *Ecology,* 97:302-312.

Roth, B., Imsland A. K., Foss A. (2009). Live chilling of turbot and subsequent effect on behaviour, muscle stiffness, muscle quality, blood gases and chemistry. *Animal Welfare,* 18:33-41.

RSPCA Australia – Royal Society for the Prevention of Cruelty to Animals Australia (2019). What are farmed Atlantic salmon fed?, kb.rspca.org.au/knowledge-base/ what-are-farmed-atlantic-salmon-fed/.

Sakkas, H., Bozidis, P., Touzios, C., et al. (2020). Nutritional status and the influence of the vegan diet on the gut microbiota and human health. *Medicina,* 56(2):88.

Sandström, V., Valin, H., Krisztin, T., et al. (2018). The role of trade in the greenhouse gas footprints of EU diets. *Global Food Security,* 19:48-55.

Santos, L. & Ramos, F. (2018). Antimicrobial resistance in aquaculture: Current knowledge and alternatives to tackle the problem. *International Journal of Antimicrobial Agents,* 52(2):135-143.

Schecter, A. & Päpke, O. (1998). Comparison of blood dioxin, dibenzofuran and coplanar PCB levels in strict vegetarians (vegans) and the general United States population. *Organohalogen Compounds,* 38:179-182.

Schecter, A., Cramer, P., Boggess, K., et al. (2001). Intake of dioxins and related compounds from food in the US population. *Journal of Toxicology and Environmental Health, Part A,* 63:1-18.

Schenone, N., Vackova, L., & Cirelli, A. (2011). Fish-farming water quality and environmental concerns in Argentina: a regional approach. *Aquaculture International,* 19:855-863.

Sequeira, I., Prata, J., da Costa, J., et al. (2020). Worldwide contamination of fish with microplastics: A brief global overview. *Marine Pollution Bulletin,* 160:111681.

Shark Research Institute (2021). Shark finning, www.sharks.org/shark-finning.

Sloth, J. & Julshamn, K. (2008). Survey of total inorganic Arsenic content in blue mussels (*Mytilus edulis* L.) from Norwegian Fiords: revelation of unusual high levels of inorganic Arsenic. *Journal of Agricultural and Food Chemistry*, 56:1269-1273.

Smith, Q. (2020). Pacific Seafood announces that 124 workers at its Newport plants have tested positive for COVID-19, closes all five facilities. *YachatsNews.com*, yachatsnews.com/pacific-seafood-announces-that-124-workers-at-its-newport-plants-have-tested-positive-for-covid-19-closes-all-five-facilities/.

Smith, S., Golden, A., Ramenzoni, V., et al. (2020). Adaptation and resilience of commercial fishers in the Northeast United States during the early stages of the COVID-19 pandemic. *PLoS ONE*, 15(12):e0243886.

Smolak, A. (2014). A meta-analysis and systematic review of HIV risk behavior among fishermen. *AIDS Care*, 26(3):282-291.

Sparks, J., Boyd, D., Jackson, B., et al. (2021). Growing evidence of the interconnections between modern slavery, environmental degradation, and climate change. *One Earth*, 4(2):181-191.

Standal, D., Grimaldo, E., & Larsen, R. (2020). Governance implications for the implementation of biodegradable gillnets in Norway. *Marine Policy*, 122:104238.

Stoll, E., Püschel, K., Harth, V., et al. (2020). Prevalence of alcohol consumption among seafarers and fishermen. *International Maritime Health*, 71(4):265-274.

Stringer, C., Whittaker, D., & Simmons, G. (2016). New Zealand's turbulent waters: the use of forced labour in the fishing industry. *Global Networks*, 16(3-24).

Stringer, C. & Harré, T. (2019). Human trafficking as a fisheries crime? An application of the concept to the New Zealand context. *Marine Policy*, 105:169-176.

Sweetman, A., Norling, K., Gunderstad, C., et al. (2014). Benthic ecosystem functioning beneath fish farms in different hydrodynamic environments. *Limnology and Oceanography*, 59(4), 1139-1151.

Sylvester, E., Wringe, B., Duffy, S., et al. (2018). Migration effort and wild population size influence the prevalence of hybridization between escaped farmed and wild Atlantic salmon. *Aquaculture Environment Interactions*, 10:401-411.

Szymańska, K., Jaremin, B, & Rosik, E. (2006). Suicides among Polish seamen and fishermen during work at sea. *International Maritime Health*, 57(1-4):36-45.

The Scottish Salmon Company (2021). The Western Highlands & Islands. Salmon's Natural Home, www.scottishsalmon.com/aboutus/whoweare.

Tickler, D., Meeuwig, J., Bryant, K., et al. (2018). Modern slavery and the race to fish. *Nature Communications*, 9:4643.

Toni, M., Manciocco, A., Angiulli, E., et al. (2019). Review: Assessing fish welfare in research and aquaculture, with a focus on European directives. *Animal*, 13(1):161-170.

UNEP – United Nations Environment Programme (2019). Global mercury assessment 2018. UN Environment Programme, Chemicals and Health Branch.

Valenzuela, R., Barrera, C., Cynthia, González-Astorga, M., et al. (2014). Alpha linolenic acid (ALA) from *Rosa canina*, sacha inchi and chia oils may increase ALA accretion and its conversion into n-3 LCPUFA in diverse tissues of the rat. *Food & Function*, 5(7):1564-1572.

Van Cauwenberghe, L. & Janssen, C. (2014). Microplastics in bivalves cultured for human consumption. *Environmental Pollution*, 193:65-70.

Van Hoytema, N., Bullimore, R., Al Adhoobi, A., et al. (2020). Fishing gear dominates marine litter in the Wetlands Reserve in Al Wusta Governorate, Oman. *Marine Pollution Bulletin*, 159, 111503.

Vázquez-Rowe, I., Moreira, M., & Feijoo, G. (2010). Life cycle assessment of horse mackerel fisheries in Galicia (NW Spain): Comparative analysis of two major fishing methods. *Fisheries Research*, 106(3), 517-527.

Vindas, M., Johansen I., Folkedal, O., et al. (2016). Brain serotonergic activation in growth-stunted farmed salmon: adaption versus pathology. *Royal Society Open Science*, 3(5):160030.

Volkoff, H. & Peter, R. (2006). Feeding behavior of fish and its control. *Zebrafish*, 3(2):131-140.

Watson, W. (2015). Molecular mechanisms in arsenic toxicity. In: Fishbein, J. & Heilman, J. (Eds.), *Advances in Molecular Toxicology*, 9:35-75 (Amsterdam, Netherlands: Elsevier).

Weitzman, J., Steeves, L., Bradford, J., et al. (2019). Chapter 11 – Far-field and near-field effects of marine aquaculture. In Sheppard, C. (Ed.), *World Seas: An Environmental Evaluation*, pp. 197-220 (Cambridge, US Academic Press).

West, B., Choo, M., El-Bassel, N., et al. (2014). Safe havens and rough waters: Networks, place, and the navigation of risk among injection drug-using Malaysian fishermen. *International Journal of Drug Policy*, 25(3), 575-582.

White, C. (2021). COVID-19 outbreaks force seafood processing shutdowns in Alaska, Chile. *SeafoodSource*, www.seafoodsource.com/news/supply-trade/covid-19-outbreaks-force-seafood-processing-shutdowns-in-alaska-chile.

WHO – World Health Organization (2010). Ten chemicals of major public health concern, www.who.int/ipcs/assessment/public_health/chemicals_phc/en/.

WHO – World Health Organization (2016a). Mercury and health, e-lactancia.org/media/papers/Mercury-WHO2016.pdf.

WHO – World Health Organization (2016b). Dioxins and their effects on human health, www.who.int/news-room/fact-sheets/detail/dioxins-and-their-effects-on-human-health.

WHO – World Health Organization & IARC – International Agency for Research on Cancer (2020). IARC Monographs on the Identification of Carcinogenic Hazards to Humans. Agents classified by the IARC Monographs, Volumes 1-128.

Wiech, M., Frantzen, S., Duinker, A., et al. (2020). Cadmium in brown crab *Cancer pagurus*. Effects of location, season, cooking and multiple physiological factors and consequences for food safety. *Science of the Total Environment*, 703:134922.

Wilhelm, M., Kadfak, A., Bhakoo, V., et al. (2020). Private governance of human and labor rights in seafood supply chains – The case of the modern slavery crisis in Thailand. *Marine Policy*, 115:103833.

Williams, A., Audsley, E., & Sandars, D. (2006). Determining the environmental burdens and resource use in the production of agricultural and horticultural commodities. Main Report. *Defra Research Project ISO205* (Bedford, UK: Cranfield University and Defra). Model available at: www.silsoe.cranfield.ac.uk and www.defra.gov.uk.

Worm, B., Barbier, E., Beaumont, N., et al. (2006). Impacts of biodiversity loss on ocean ecosystem services. *Science*, 314(5800), 787-790.

Wringe, B., Jeffery, N., Stanley, R., et al. (2018). Extensive hybridization following a large escape of domesticated Atlantic salmon in the Northwest Atlantic. *Communications Biology*, 1(108).

Wu, J., Lai, M., Zhang, Y., et al. (2020). Microplastics in the digestive tracts of commercial fish from the marine ranching in east China sea. *Case Studies in Chemical and Environmental Engineering*, 2:100066.

WWF – World Wildlife Fund (2021). Bluefin tuna, www.worldwildlife.org/species/bluefin-tuna.

Yue, S., Moccia, R., & Duncan, I. (2004). Investigating fear in domestic rainbow trout, *Oncorhynchus mykiss*, using an avoidance learning task. *Applied Animal Behaviour Science* 87(3-4):343-354.

Yue, S., Duncan, I., & Moccia, R. (2008). Investigating fear in rainbow trout (*Oncorhynchus mykiss*). Using the conditioned-suppression paradigm. *Journal of Applied Animal Welfare Science*, 11(1):14-27.

Zimta, A., Schitcu, V., Gurzau, E., et al. (2019). Biological and molecular modifications induced by cadmium and arsenic during breast and prostate cancer development. *Environmental Research*, 178:108700.

Chapter 5 references

Aghasi, M., Golzarand, M., Shab-Bidar, S., et al. (2019). Dairy intake and acne development: A meta-analysis of observational studies. *Clinical Nutrition*, 38(3):1067-1075.

Allen, M. (2019). World's third largest meat producer Cargill invests in Aleph Farms' cell-based steak. Good Food Institute, gfi.org/blog/cargill-invests-aleph-farms/.

Alwarith, J., Kahleova, H., Rembert, E., et al. (2019). Nutrition interventions in rheumatoid arthritis: the potential use of plant-based diets. A review. *Frontiers in Nutrition*, 10(6):141.

Alwarith, J., Kahleova, H., Crosby, L., et al. (2020). The role of nutrition in asthma prevention and treatment. *Nutrition Reviews*, 78(11):928-938.

Alzheimer's Association (2014). 2014 Alzheimer's & brain awareness moth international survey, www.alz.org/global/InternationalSurvey.pdf.

Aljuraiban, G., Chan, Q., Gibson, R., et al with INTERMAP Research Group (2020). Association between plant-based diets and blood pressure in the INTERMAP study. *BMJ Nutrition, Prevention & Health*, 3(2):133-142.

Barnard, N., Katcher, H., Jenkins, D., et al. (2009). Vegetarian and vegan diets in type 2 diabetes management. *Nutrition Reviews*, 67(5):255-263.

Barnard, N. (2020). *Our Body in Balance: The New Science of Food, Hormones, and Health* (New York: Hachette Book Group).

Beezhold, B., Radnitz, C., Rinne, A., et al. (2015). Vegans report less stress and anxiety than omnivores. *Nutritional Neuroscience*, 18:7:289-296.

Borg, C. (2019). Mental health is higher in the vegetarian/vegan demographic. What are we as practitioners missing? Is it simply nutritional? *Advances in Integrative Medicine*, 6(1):S8.

Craddock, J., Probst, Y., & Peoples, G. (2016). Vegetarian and omnivorous nutrition - comparing physical performance. *International Journal of Sport Nutrition and Exercise Metabolism*, 26(3): 212-220.

David, L., Maurice, C., Carmody, R., et al. (2014). Diet rapidly and reproducibly alters the human gut microbiome. *Nature*, 505(7484):559-563.

Dickerson, F., Stallings, C., Origoni, A., et al. (2019). Nitrated meat products are associated with suicide behavior in psychiatric patients. *Psychiatry Research*, 275:283-286.

Dobersek, U., Wy, G., Adkins, J., et al. (2021). Meat and mental health: a systematic review of meat abstention and depression, anxiety, and related phenomena. *Critical Reviews in Food Science and Nutrition*, 61(4):622-635.

Gaskins, A., Pereira, A., Quintiliano, D., et al. (2017). Dairy intake in relation to breast and pubertal development in Chilean girls. *The American Journal of Clinical Nutrition*, 105(5):1166-1175.

Glick-Bauer, M. & Yeh, M. (2014). The health advantage of a vegan diet: Exploring the gut microbiota connection. *Nutrients*, 6:4822-4838.

Good Food Institute (2021). State of the industry report: Cultivated meat, gfi.org/resource/ cultivated-meat-eggs-and-dairy-state-of-the-industry-report/.

Guan, X., Lei, Q., Yan, Q., et al. (2021). Trends and ideas in technology, regulation and public acceptance of cultured meat. *Future Foods*, 3:100032.

Haghighatdoost, F., Feizi, A., Esmaillzadeh, A., et al. (2019). Association between the dietary inflammatory index and common mental health disorders profile scores. *Clinical Nutrition*, 38(4):1643-1650.

Harrison, S., Lennon, R., Holly, J., et al. (2017). Does milk intake promote prostate cancer initiation or progression via effects on insulin-like growth factors (IGFs)? A systematic review and meta-analysis. *Cancer Causes Control*, 28(6):497-528.

IMARC (2021). Top companies in the plant-based meat industry, www.imarcgroup.com/ plant-based-meat-companies.

Juhl, C., Bergholdt, H., Miller, I., et al. (2018). Dairy intake and acne vulgaris: A systematic review and meta-analysis of 78,529 children, adolescents, and young adults. *Nutrients,* 10:1049.

Kahleova, H., Levin, S., & Barnard, N. (2017). Cardio-metabolic benefits of plant-based diets. *Nutrients*, 9(8):848.

Kahleova, H., Levin, S., & Barnard, N. (2018). Vegetarian dietary patterns and cardiovascular disease. *Progress in Cardiovascular Diseases*, 61(1):54-61.

Khoury, R., Chaaya, M., Waldemar, G., et al. (2014). The association between fish and meat consumption and dementia prevalence in Lebanon. *Alzheimer's & Dementia,* 10(4):P913-P914.

Kim, H., Rebholz, C., Hegde, S., et al. (2021). Plant-based diets, pescatarian diets and COVID-19 severity: a population-based case–control study in six countries. *BMJ Nutrition, Prevention & Health*, 4.

Król, W., Price, S., Śliż, D., et al. (2020). Vegan athlete's heart—Is it different? Morphology and function in echocardiography. *Diagnostics*, 10(7):477.

Lally, P., van Jaarsveld, C., Potts, H., et al. (2009). How are habits formed: Modelling habit formation in the real world. *European Journal of Social Psychology*, 40(6), 998-1009.

Le, L. & Sabaté, J. (2014). Beyond meatless, the health effects of vegan diets: Findings from the Adventist cohorts. *Nutrients*, 6:2131-2147.

Lynch, H., Johnston, C., & Wharton, C. (2018). Plant-based diets: Considerations for environmental impact, protein quality, and exercise performance. *Nutrients*, 10(12):1841.

MacDermott, R. (2007). Treatment of irritable bowel syndrome in outpatients with inflammatory bowel disease using a food and beverage intolerance, food and beverage avoidance diet. *Inflammatory Bowel Diseases*, 13(1):91-96.

Marrone, G., Guerriero, C., Palazzetti, D., et al. (2021). Vegan diet health benefits in metabolic syndrome. *Nutrients*, 13(3):817.

McCarty, M. (1999). Vegan proteins may reduce risk of cancer, obesity, and cardiovascular disease by promoting increased glucagon activity. *Medical Hypotheses*, 53(6):459-485.

McCarty, M. (2014). GCN2 and FGF21 are likely mediators of the protection from cancer, autoimmunity, obesity, and diabetes afforded by vegan diets. *Medical Hypotheses*, 83(3):365-371.

McCarty, M., Assanga, S. I., Lujan, L. L. (2021). Age-adjusted mortality from pancreatic cancer increased NINE-FOLD in Japan from 1950 to 1995 – Was a low-protein quasi-vegan diet a key factor in their former low risk? *Medical Hypotheses*, 149:110518.

Melina, V., Craig, W., & Levin, S. (2016). Position of the Academy of Nutrition and Dietetics: Vegetarian diets. *Journal of the Academy of Nutrition and Dietetics*, 116(12):1970-1980.

Mofrad, M., Mozaffari, H., Sheikhi, A., et al. (2021). The association of red meat consumption and mental health in women: A cross-sectional study. *Complementary Therapies in Medicine,* 56:102588.

Nagpal, R., Neth, B., Wang, S., et al. (2020). Gut mycobiome and its interaction with diet, gut bacteria and Alzheimer's disease markers in subjects with mild cognitive impairment: A pilot study. *EBioMedicine*, 59: 102950.

Nath, P. & Singh, S. (2017). Chapter 26 – Defecation and stools in vegetarians: Implications in health and disease. In Mariotti, F. (Ed.), *Vegetarian and Plant-based Diets in Health and Disease Prevention*, pp. 473-481 (Cambridge, US: Academic Press).

Qian, X., Song, X., Liu, X., et al. (2021). Inflammatory pathways in Alzheimer's disease mediated by gut microbiota. *Ageing Research Reviews*, 68:101317.

Rich-Edwards, J., Ganmaa, D., Pollak, M., et al. (2007). Milk consumption and the prepubertal somatotropic axis. *Nutrition Journal*, 6(28).

Sakkas, H., Bozidis, P., Touzios, C., et al. (2020). Nutritional status and the influence of the vegan diet on the gut microbiota and human health. *Medicina (Kaunas)*, 56(2):88.

Springmann, M, Wiebe, K., Mason-D'Croz, D., et al. (2018). Health and nutritional aspects of sustainable diet strategies and their association with environmental impacts: a global modelling analysis with country-level detail. *Lancet Planet Health*, 2:e451-61.

Sherzai, D. & Sherzai, A. (2017). *The Alzheimer's Solution: A Breakthrough Program to Prevent and Reverse the Symptoms of Cognitive Decline at Every Age* (San Francisco, US: HarperOne).

Viguiliouk, E., Kendall, C., Kahleová, H., et al. (2019). Effect of vegetarian dietary patterns on cardiometabolic risk factors in diabetes: A systematic review and meta-analysis of randomised controlled trials. *Clinical Nutrition*, 38(3):1133-1145.

Vitale, K. & Hueglin, S. (2021). Update on vegetarian and vegan athletes: a review. *The Journal of Physical Fitness and Sports Medicine*, 10(1):1-11.

Volpp, K. & Loewenstein, G. (2020). What is a habit? Diverse mechanisms that can produce sustained behavior change. *Organisational Behavior and Human Decision Processes*, 161(Supplement):36-38.

Willett, W. & Ludwig, D. (2020). Milk and health. *The New England Journal of Medicine*, 382:644-654.

Yang, J. & Yu, J. (2021). Chapter 3.43 - Relationship between microbiome and colorectal cancer. In Cifuentes, A. (Eds.), *Comprehensive Foodomics*, pp. 568-578 (Amsterdam, Netherlands: Elsevier).

Zhang, H., Greenwood, D., Risch, H., et al. (2021). Meat consumption and risk of incident dementia: cohort study of 493,888 UK Biobank participants. *The American Journal of Clinical Nutrition.*

ABOUT THE AUTHOR

Photo credit: Emmylie Cruz Photo & Film

Camila Perussello, Ph.D., is an extensively published Food Engineer with many years of postdoctoral experience. Her involvement with the animal liberation movement started decades ago when she first became vegetarian and later went vegan for ethical reasons. Dr. Perussello has worked as a food scientist in different parts of the globe, gathering evidence on animal food production's ethical and sustainability issues. She is the recipient of research grants by Brazilian and European governments as well as funding for vegan outreach from California-based *The Pollination Project.* On the verge of such pressing matters as animal rights, the climate and ecological crises, animal-origin pandemics, diet-related illnesses, and food insecurity, Dr. Perussello defends a shift towards plant-based living. She also collaborates with non-profit organisations to help food industries transition away from animal use. Dr. Perussello recently quit her job as a scientist in a renowned Irish university (linked to animal agriculture funding) to do independent research and animal rights activism. Her YouTube channel, *Gaia Vida Consciente,* can be found at www.youtube.com\channel\UC4lL0kiSrd94uAHFqXJEgIg.

ABOUT THE PUBLISHER

Lantern Publishing & Media was founded in 2020 to follow and expand on the legacy of Lantern Books—a publishing company started in 1999 on the principles of living with a greater depth and commitment to the preservation of the natural world. Like its predecessor, Lantern Publishing & Media produces books on animal advocacy, veganism, religion, social justice, humane education, psychology, family therapy, and recovery. Lantern is dedicated to printing in the United States on recycled paper and saving resources in our day-to-day operations. Our titles are also available as eBooks and audiobooks.

To catch up on Lantern's publishing program, visit us at www.lanternpm.org.

facebook.com/lanternpm
twitter.com/lanternpm
instagram.com/lanternpm